Mineral Resource Reviews

Mineral Resource Reviews is the official book series of the Society for Geology Applied to Mineral Deposits (SGA). It publishes selected conference and symposium proceedings, monographs, Short Course Volumes, In-depth Reports and Field Guides by researchers and professionals in the fields of mineral resources; their study, development and environmental impacts. The series will focus on providing in-depth descriptions, analyses and reviews of deposit types, metallogenic regions, ore forming processes as well as the analytical/investigative techniques needed to study them. High quality scientific contributions from experts in mineral deposit geology are invited for publication in this series.

Jamsrandorj Galsan

Uranium Deposits of Mongolia

SOCIETY for GEOLOGY APPLIED to MINERAL DEPOSITS

Springer

Jamsrandorj Galsan
Irkutsk Polytechnical Institute
Ulaanbaatar, Mongolia

ISSN 2365-0559 ISSN 2365-0567 (electronic)
Mineral Resource Reviews
ISBN 978-3-031-86480-3 ISBN 978-3-031-86481-0 (eBook)
https://doi.org/10.1007/978-3-031-86481-0

This Springer imprint is published by the registered company Springer Nature Switzerland AG
The registered company address is: Gewerbestrasse 11, 6330 Cham, Switzerland

Dedicated to the development of uranium geology and the uranium mining industry in Mongolia

Foreword

According to the IAEA, historically Mongolia has mined 550 tons of uranium and currently has resources of about 192,000 tons of uranium. Research on uranium in Mongolia began in the 1970s, and from the very beginning, it was classified and inaccessible to the public. Only recently, namely since 1993 after democratic reforms, information about uranium in the country became public.

At the moment, the volume of geological research for uranium in Mongolia is carried out on a limited scale. With the end of the Cold War and the decline in uranium prices on the world market, uranium issues concerned a very limited circle of people in Mongolia.

Nevertheless, interest in uranium has recently been restored, especially after nuclear energy was recognized as environmentally friendly, and without it, it's simply impossible to solve global energy problems.

An important role in determining the prospects for uranium mineralization in Mongolia was played by the airborne geophysical survey conducted in Mongolia by Soviet specialists, which subsequently laid the basis for assessing uranium resources and identifying promising areas for further geological exploration.

An assessment of Mongolia's prospects for uranium predetermined the need to create the national state company "Uran" in 1994, the joint Mongolian-Russian-American enterprise "Gurvansaikhan", and in 1996 the company "Central Asian Uranium".

The Mongolian government pays special attention to the hazards associated with uranium mining. In 2009, the People's Great Khural (Parliament) approved important documents such as the "State Policy on Radioactive Minerals and Nuclear Energy" and the "Nuclear Energy Law". At the end of last year 2024, Parliament made some fundamental changes to this law, so that foreign investment in this sector will undoubtedly increase.

In recent years, the French state-owned company ORANO LLC has made significant efforts to explore for uranium in Mongolia, leading to the discovery of two new uranium deposits that are currently being prepared for mining.

Over the last 30 years or so, Mongolia's uranium sector has undergone significant changes and achieved significant success.

Many deposits have been discovered and prepared for development. Our country has introduced new methods used for uranium exploration and also used in the uranium sector of such developed Western countries as the USA,

Canada, France, Japan, and Australia. Many young Mongolian geologists and scientists born in this new era for Mongolia and have become national collaborators. With the help of Western investors, information about almost all uranium deposits discovered and explored by Soviet geologists was largely revalued in accordance with the methods, technologies, and standards required in those countries. As part of this work, additional geological exploration was conducted at the Gurvanbulag uranium deposit and in ore bodies 2B, 2V, and 7 of the Dornod deposit. ORANO Group, a French state-owned company, has discovered and explored two new uranium deposits in Mongolia-Dulaan Uul and Zoovch Ovoo and is preparing to start mining operations. The Gurvansaikhan, Ulziit, Khairkhan, and Kharaat deposits have been explored and meet the requirements as ore bodies by international investors. The Mongolian company NEM carried out exploration of the Dalt uranium deposit in the Chuluut region, increased the reserves, and prepared it for production in accordance with international standards.

Much of the work on the geology of uranium in Mongolia was carried out by Yu. B. Mironov. Under his editorship, a monograph was published in Russian in 2003. It is worth noting that this book represents an invaluable summary of the work accomplished by Russian researchers in the field of uranium geology in Mongolia until the 1990s.

This book, authored by Jamsrandorj Galsan, is a new summary of the results of work performed by Soviet, and then Russian, Mongolian, and foreign specialist geologists who have worked over the past 30 years on the uranium deposits of Mongolia. This publication highlights the successes achieved in locating and defining the deposits of uranium, provides the most recent and scientifically sound basis for the metallogenic zoning of Mongolia for uranium, and provides acceptable recommendations for further geological exploration.

D. Dalaijargal
Executive Director
Mongolian State Company "Monatom"
Ulaanbaatar, Mongolia

Preface

This book contains geological investigation conditions for uranium and the geodynamic positioning of uranium mineralization for the country of Mongolia. The primary regularities in the distribution of uranium mineralization based on the geodynamic position of tectonics are discussed. Uranium-bearing provinces and zones and districts are characterized, uranium ore epochs are noted, the geological structures of uranium deposits are given and the types of uranium deposits, according to the IAEA classification are identified for the uranium reserves in Mongolia.

A quantitative and cost estimate of the uranium potential has been conducted. This book is written for geologists, specialists, and scientists in the field of uranium geology and for metallogeny.

A lot is known about the uranium geology of Mongolia, from the work of the Russian scientist Yu. B. Mironov, published in Russian in 2003 and 2012. It is worth noting that this book represents a summary of the work of Russian researchers in the field of uranium geology in Mongolia until the 1990s.

In this book, we tried to show the patterns of distribution of uranium deposits in Mongolia based on geodynamic considerations of the tectonic development of Mongolia. In a new way, the territories of Mongolia were classified into uranium provinces and regions and were given a new classification of uranium deposits. We have come up with a completely new estimate for the uranium potential of Mongolia in the amount of 1004 thousand tons.

Ulaanbaatar, Mongolia
January 2025

Jamsrandorj Galsan

Acknowledgments

We are very grateful to Springer for the opportunity to publish this book. This book is a summary of the results of work performed by Soviet, and then Russian, Mongolian, and foreign specialist geologists who studied and explored the uranium deposits of Mongolia. This publication highlights the progress achieved in the field of uranium and provides scientifically based recommendations for further geological exploration work for uranium in Mongolia.

We would like to especially thank Yuri Borisovich Mironov, Doctor of Geological and Mineralogical Sciences, employee of the All-Russian Scientific Research Geological Institute named after A. P. Karpinsky, for permission to use the series of books *Uranium Deposits of Mongolia* and *Uranium of Mongolia* in Russian. The author is grateful to Dr. J. Byamba for permission to use the book *Geodynamics and Uranium Metallogeny of Mongolia*. With gratitude, I thank F. Tungalag, A. A Budunov, M. N. Grechukhin, S. Enkh-Amgalan, and A. Champan for permission to use drawings from articles and books published by them in the Mongolian and Russian.

We are grateful to the American geologist Donn Pillmore for his editing and generous, constructive corrections that improved the text of the book.

Contents

Abbreviations

AGS	Airborne gamma spectrometric survey
AGSM	Airborne gamma spectrometry
DVTS	Dornod volcano-tectonic structure
EVTS	Engershand volcano-tectonic structure
IAEA	International Atomic Energy Agency
IGEM	(Instut geologii rudniikh mestorojdenii, petrographii, mineralogii i geokhimii) Institute of Geology of Ore Deposits, Petrography, Mineralogy and Geochemistry
MGSE	(Mongolskaya geologosemochnaya ekspeditsiya) Mongolian Geological Survey Expedition
MORB	Mid-ocean ridge basalt
NE	Northeast
NNE	North-north-east
NRE	Natural radioactive elements
NW	Northwest
OIB	Ocean Island basalt
"Priargunsky GOK"	(Gorno-obagatitelni kombinat) Priargunsky enterprise Mining and processing plant of Russia
PUD	Potential uranium district
TVTS	Turgen volcano-tectonic structure
USSR	Union of Soviet Socialist Republics
UTVS	Upper Turgen volcano-tectonic structure
UVTS	Ugtam volcano-tectonic structure
VIMS	(Vserossiiskii naucho-issledovatelskii institute mineralnogo sirya) All-Union Institute of Geology and Mineral Resources
VIRG	(Vserossiiskii institute razvedochnoi geofiziki) All-Russian Research Institute of Exploration Geophysics named after A. A. Logacheva
VIRG currently FGU NPP	Federal state unitary scientific and production institution named after "Geological Exploration"

VSEGEI	(Vserossiiskii naucho-issledovatelskii geologicheskii institut) All-Russian Research Geological Institute named after A. P. Karpinsky
VTS	Volcano-tectonic structure

List of Figures

List of Tables

1 Story of Studying the Uranium Potential in Mongolia

1.1 The Current State of Investigation of Uranium Mineralization in Mongolia

Geological exploration of Mongolia has been extremely inconsistent. From the first regional exploration by Russian explorers of Asia to the present, little more than 41% of the country's territory has been covered by areal geological surveys in combination with prospecting at a scale of 1:50,000. Regional geological surveys at a scale of 1:200,000, 1:500,000, and 1:1,000,000 cover 100% of the country. An aeromagnetic survey at a scale of 1:200,000 was carried out on 70% of the country and a gravimetric survey at a scale of 1:200,000–1:100,000 covers only large depressions in the south of the country. A medium-scale hydrogeological survey was carried out on 35% of the country. Eastern Mongolia has experienced the most recent and comprehensive exploration for uranium resources.

The results of exploration and research conducted since the beginning of the 1970s, in Mongolia, have been summarized by N. A. Marinov and R. A. Khasin, co-authors, in the three-volume monograph "Geological Structure of the Mongolian People's Republic". Under the editorship of the same authors, a geological map of the Mongolian People's Republic (1972) and a map of minerals (1973) at a scale of 1:1,500,000 were published. By 1975, geological and structural-formational maps of Eastern and Central Mongolia were compiled at a scale of 1:500,000 (Blagonravov, Borzakovsky); based on these maps, forecast metallogenic maps for gold, tungsten, tin, and fluorite were compiled (N. A. Marinov, R. A. Khasin, 1976–1980s). In 1980, metallogenic maps of the Mongolian People's Republic were published at a scale of 1:500,000 for copper, lead (D. Dorjgotov), and gold (G. Dejidmaa).

In recent years, Mongolian scientists have summarized the results of geological exploration and published a series of geological and hydrogeological maps of Mongolia on a scale of 1:100,000, as well as published an 8-volume monograph "Geology and Minerals of Mongolia" in the Mongolian language (Byamba 2011).

Against the background of relatively poor geological knowledge and fragmentary prospecting of Mongolia for certain types of minerals, information on the uranium content of Mongolia is relatively good. There are three stages in the study of uranium resources in Mongolia.

First Stage. This stage began in 1948 by the Eastern Geological Expedition of the USSR Ministry of Geology (1946–1970). Studies were carried out along with planned exploration work by the Eastern Expedition.

J. Galsan, *Uranium Deposits of Mongolia*, Mineral Resource Reviews,
https://doi.org/10.1007/978-3-031-86481-0_1

Based on the results of this exploration no large deposits of uranium were discovered, which was due to the very modest budgets allocated for this work, insufficient depths of prospecting, and inadequate laboratory facilities. However, many occurrences of uranium mineralization were found in Mongolia. Genetically they belong to two types of deposits—exogenous and endogenous (hydrothermal). Elevated uranium grades were found in brown coals of the Early Cretaceous age at the Zuunbulag, Bayanbulag, Sukhbaatar, Ulziit, and other coal deposits, as well as at the Tumentsogt rare-metal deposit.

In 1955–1959, V. A. Bobrov, N. A. Volkov et al. found uranium occurrences associated with brown coal seams in the terrigenous deposits of the Choir depression. The prospecting and exploration carried out at that time led to the conclusion that there was no large-scale mineralization present. In 1959, a group of Soviet specialists who provided technical assistance in conducting prospecting and exploration for uranium was led by S. N. Kondakov. The field party under his leadership discovered a number of new occurrences of uranium in the coal-bearing deposits of the Nyalga, Ulziit, and Choir depressions. The formation of the Suljur-Khudag and Shand-Bulag occurrences in the northern part of the Ulziit depression were associated with oxidation zones at the contacts of Early Cretaceous coal seams according to geologists of the party. Uranium mineralization was observed only in coal seams and was absent in host rocks and rocks that make up the sides and foundation of intermountain depressions. The areas that have uranium mineralization, in plan, look like isolated jets, bands, and spots, confined mainly upperpart of one of the coal seams that completes the geological section. As the uranium-bearing layer subsides, uranium concentrations, as a rule, gradually decrease and practically disappear in the pinch-out zone. Uranium mineralization has been observed both above and below the modern water table, but almost always in the upper areas of the coal seams.

During its activity, the Eastern Geological Expedition discovered seven small deposits and ore occurrences of exogenous uranium. However, none of them, due to the insignificant scale of mineralization and low uranium grades, were considered to be economic. A more detailed description and classification of the ores of these deposits is given below.

N. I. Khomizuri studied the radioactivity of pegmatites widely distributed in Mongolia. He came to the conclusion that these formations contain low-grade concentrations of uranium and thorium but only a few of them form concentrations worthy of further study (the pegmatite bodies of Gorkhi and Tsagaan Ovoo). Structurally, the most favorable are zonal pegmatites, in which radioactive elements are separated into small zones.

The most prospective, according to his conclusion, for the discovery of new occurrences of uranium mineralization are differentiated granitic bodies, and the less favorable are interstratal and small intrusions of a hypabyssal nature. The uranium grade in the six pegmatite fields studied rarely exceeded 0.03%.

Yu. A. Bobrov, according to observations at the Tumentsogt occurrence, notes that the uranium mineralization is confined to fissure zones that cut quartz-tungsten-molybdenum veins. Uranium mineralization associated with fluorite veinlets and hydrothermally altered limestones was found in the southwestern Khubsgul region at the Argol occurrence. A more detailed review of the data determined that this type of uranium mineralization was insignificant.

D. I. Fedorovich, who observed rare-metal ore occurrences of the Saikhanuul zone (Yugozyr area), pointed out that radioactive mineralization is closely related to a large latitudinal tectonic disturbance. K. P. Stolbkov visited the area and drew attention to the increased radioactivity of fluorite, calcite, and their crusts on the surface of debris and tectonic clay. A study of exogenous occurrences in Eastern Mongolia suggests that the source of uranium in the brown coals of the Choibalsan deposit are the Jurassic granites of the Aduun-Chuluun-Uul hill, as well as quartz porphyries developed in the setting of a coal-bearing trough.

All of this research led to the conclusion that the formation of uranium mineralization in the region occurred due to uranium leaching from late Permian-early Mesozoic quartz porphyries and granites highly enriched in radioactive elements, followed by its precipitation in sedimentary basins. An example of this is the Choir granite massif, where about ten areas containing uranium mineralization were later discovered in the Lower Cretaceous sedimentary deposits along the flanks of the depression. These exogenous occurrences of uranium were formed at the stage of primary sedimentation and its subsequent epigenetic redistribution by ground and surface waters. Factors contributing to this deposition are the presence of coals and carbonaceous rocks with a large reduction capacity, high permeability of sedimentary deposits, and their favorable grain size distribution.

Figure 1.1 shows a summary of prospecting and exploration conducted by the Eastern Geological Expedition for uranium deposits.

The beginning of the most important **Second Stage** was conducted by the Intergovernmental Agreement between the USSR and the Mongolian People's Republic (MPR) "On carrying out prospecting and exploration work for uranium in the Mongolian People's Republic" of June 9, 1970. The scientific justification for setting up specialized work on uranium was the memorandum of the employees of All-Russian Research Geological Institute named after A. P. Karpinsky (VSEGEI) and the Sosnovskaya expedition (I. S. Ozhinsky, M. I. Itsikson, P. A. Strona, O. N. Shanyushkin, G. A. Shatkov), M. I. Itsikson, P. A. Strona, O. N. Shanyushkin, G. A. Shatkov). In accordance with the order of the USSR Ministry of Geology, since 1970, prospecting, evaluation and scientific-thematic work in the MPR was carried out by the Mongolian Geological Survey Expedition (MGSE) of the

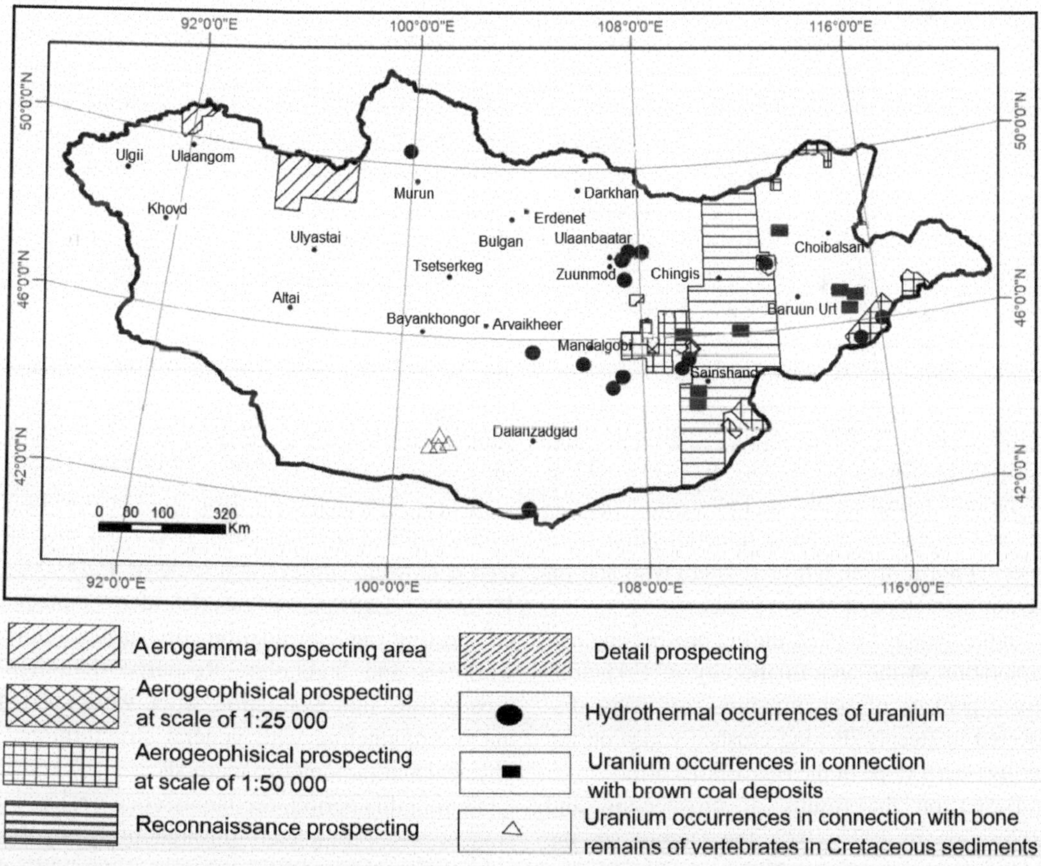

Fig. 1.1 Locations of uranium research activities up to 1970 (Mironov 2007)

Zarubezhgeologiya trust with the participation of the scientific-thematic groups of All-Russian Research Geological Institute named after A. P. Karpinsky (VSEGEI), All-Union Institute of Geology and Mineral Resources (VIMS), All-Russian Research Institute of Exploration Geophysics named after A. A. Logacheva (VIRG), Institute of Geology of Ore Deposits, Petrography, Mineralogy and Geochemistry (IGEM). Yu. M. Shuvalov (VSEGEI).

From 1970 to 1973, joint work by MGSE and VSEGEI in Eastern Mongolia revealed more than 15 ore occurrences and mineralization occurrences, as well as a large number of radiometric anomalies.

Detailed prospecting and exploration was carried out by MGSE in 1970 in the North-Kerulen region at two sites (the Ikh-Bulag ore occurrence and the Delger-Khaan site). This work established only minor occurrences of uranium mineralization. In the same year, occurrences of uranium-phosphorus mineralization Ikh-Khet, Baruun-Mandal, and Tsodoin-Tsagaan were recorded in the South Kerulen region.

As a result of the 1970–1971 prospecting and exploration on the largest of these occurrences (Ikh-Khet), a non-commercial type of mineralization was established. In 1971–1972, several ore occurrences of volcanogenic type were discovered in the Khar-Airag (Khongor, western flank of the Ulaan-Nuur site) and North Gobi (Khavir-Khudag, Ikh-Uul) regions. In addition, occurrences of uranium-phosphorus type were found in the Khar-Airag (Ulaan-Nuur, Khartolgoi) and North Gobi (Khashaat) regions. In 1971–1972 in the North Choibalsan region, radiometric anomalies and secondary uranium mineralization were detected north of the hill of Erkhet Ovoo and on the northern slope of Mount Ulaan.

In the autumn of 1972, during prospecting and exploration in the area of the city of Erkhetiin-Ovoo, pockets of rich uranium ores were discovered in steeply dipping crushed zones on the northwestern slope of the Bag-Erkhet hill.

Based on the results of prospecting and exploration in 1971–1972, the Dornot (in the Erkhet Ovoo area) and Ulan ore occurrences in the North Choibalsan district, as well as Ulaan-Nuur in the Khar-Ayrag district, were recognized as the most prospective.

The new data confirmed the prospects for uranium mineralization of the Lower Cretaceous basins of Eastern Mongolia associated with zones of in-situ oxidation. In 1971–1972, exploration specific for uranium was concentrated in the North Choibalsan district. Due to this shift in priorities, drilling operations in the Khar-Ayrag region were stopped.

In 1973–1974 during prospecting work by the MGSE at the Dornod ore occurrence (near the hill of Erkhet Ovoo), the economic parameters of uranium mineralization were established, and the occurrence was transferred to the category of a uranium deposit. At the same time, a polymetallic deposit was discovered in 1974, in the area to the west of the Ulaan ore occurrence, and the Gurvanbulag uranium deposit was discovered during prospecting and exploration work.

In 1988, the Mongolian Geological Expedition conducted airborne spectrometry searches and surface surveys in the river basin of the Turgen volcanic-tectonic structure. This work resulted in the discovery of the Baruunkhooloi, a uranium-sulfide-fluorite occurrence, and the Tsagaanchuluut, a uranium-phosphorus occurrence.

The primary objective of this work was to study uranium ore deposits associated with the processes of Mesozoic syn-, post-collision rifting. To meet this objective, the following exploration work was conducted, including airborne gamma spectrometric (AGS) and aeromagnetic surveys at scales of 1:25,000–1:200,000, ground gamma spectrometric profiles, areal prospecting, hydrochemical study of small rivers and specialized geological surveys at a scale of 1:200,000–1:10,000. Exploration work also included prospecting and evaluation of surface mine workings and boreholes. Resource definition, prospecting, and evaluation work were carried out in conjunction with general structural-geophysical and radiometric methods.

During this period, reconnaissance geological studies were completed throughout Mongolia, accompanied by AGS surveys, sampling of small rivers, and radiohydrogeological work.

The area available for aerial work is covered by AGS and aeromagnetic surveys at scales of 1:25,000 (171 thousand km^2), 1:200,000 (580 thousand km^2), and 1:1000,000 (300 thousand km^2). Numerous specialized geological surveys have been conducted at scales of 1:200,000, 1:50,000, and 1:10,000 (about 50 thousand km^2 in total), including ground-based geological and radiometric searches and detailed exploration. In the eastern part of the country, on an area of over 500,000 km^2, resource assessment and geological studies were carried out on a scale of 1:500,000 and a metallogenic map with potential resources was compiled, providing a general assessment of the prospects for the uranium content of this territory (Fig. 1.2).

Exploration conducted in Eastern Mongolia resulted in the discovery of four uranium deposits (Gurvanbulag, Dornod, Mardaingol, Nemer), which belong to the most prospective molybdenum-uranium ore formation. Together with other occurrences of uranium, they form the North Choibalsan uranium ore region, undoubtedly of economic importance. A number of regions, zones, and units characterized by similar features of the geological structure have been identified as potential targets for discovering additional uranium deposits.

A distinct emphasis was given to the assessment of prospects for exogenous uranium mineralization associated with the zones of oxidation in terrigenous deposits of the Late Mesozoic. Secondary mineralization was revealed in the pinch-out of oxidation zones in tectonically sheared and deformed horizons of sands, gravelstones, and conglomerates of Early and Late Cretaceous age, demarcating more than two dozen areas prospective for this type of uranium. Further work significantly expanded the prospects for the development of this type of uranium mineralization.

A large amount of factual material on the uranium mineralization of Mongolia, mainly in its eastern part, obtained during the forecast, prospecting, and exploration work of the MGSE with the participation of VSEGEI, was reflected in many reports.

During the specialized predictive studies for uranium of the MGSE and the May Party of VSEGEI in 1970–1985 occurrences

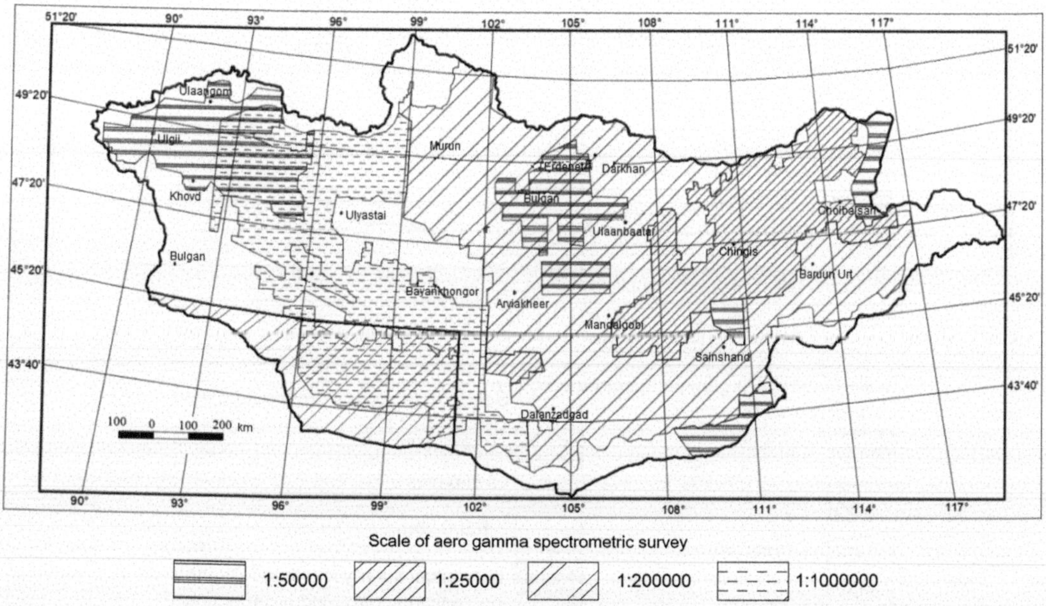

Fig. 1.2 General coverage of airborne gamma spectrometric (AGS) surveys (Mironov 2003; Jamsrandorj 2019)

of other minerals were discovered, including fluorite, lead, zinc, gold, coal, etc. Some of the identified occurrences (Maikhant I and II, Tsagaan-Takhilch, Ulaan, etc.) were explored by the MGSE and transferred to the category of deposits with quantified resources. Others (Tsav, Bayan-Uul, Byambar, Olongiin, etc.) were transferred to the Mongolian side for evaluation.

Of fundamental importance is the discovery and detailed exploration of the Ulaan polymetallic and uranium deposit in the Dornod ore district.

The complex nature of the work of the MGSE and VSEGEI made it possible to outline the general patterns of the distribution of uranium and other types of mineralization associated with it and use them in the course of geological exploration and resource estimation.

The study of the uranium mineralization in Mongolia was conducted in close cooperation with MGSE and a number of leading institutes in Russia. Groups of VSEGEI employees, led by G. M. Vladimirsky, G. A. Shatkov, G. B. Kochkin, N. S. Solovyov, focused on the regional study of the territory and the assessment of its uranium mineralization, including endogenous and exogenous occurrences (Fig. 1.3).

The team of "Rudgeofizika" (research and production organization) under the leadership of E. B. Vysokoostrovskaya, together with specialists from "Sosnovskaya" and "Zelenogorsk" (the production and geological organizations)

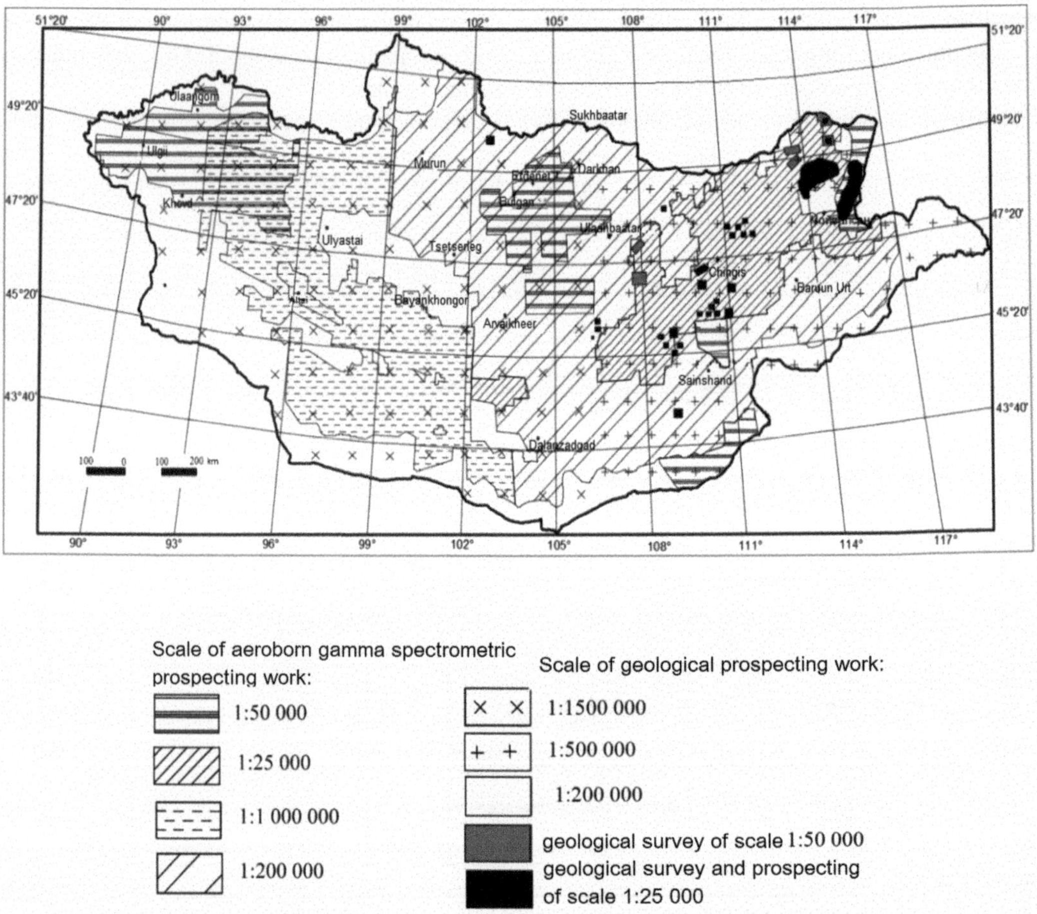

Fig. 1.3 General distribution of exploration for uranium MGSE for 1970–1991 (Mironov 2012)

compiled small-scale radiogeochemical maps of Mongolia at scales of 1:200,000 and 1:50,000 of its most prospective structures. Employees of the "Rudgeofizika" also participated in the aerospectrometric exploration work and the assessment of the identified anomalous targets.

A group of All-Union Institute of Geology and Mineral Resources employees led by G. A. Mashkovtsev and V. I. Shetochkina made important contributions to the study of exogenous uranium mineralization, and the group of I. S. Modnikova, I. V. Sycheva, and others, who worked together with specialists of Institute of Geology of Ore Deposits, Petrography, Mineralogy and Geochemistry (IGEM) including L. E. Tolkunov, Yu. D. Filonenko, and V. Yu. Petrov, contributed significantly to the interpretation of structural features controlling the localization of endogenous mineralization of the Dornod ore targets. The group of Melnikov V. A. and Golovin A. A. (IGEM) has made a significant contribution to the study of the mineral composition of ores, and the wall-to-wall alteration zonation.

Since 1992, the **Third Stage** of the study of uranium mineralization of Mongolia began, which is of an international nature. This stage is inextricably linked with the collapse of the socialist system and the end of the Cold War on a global scale, which made it possible for Mongolia to cooperate with the Western international community in the uranium field.

When this occurred, foreign companies from the USA, France, Canada, China, and other countries began participating in the prospecting and exploration of uranium occurrences in Mongolia. In 1994, the Mongolian State Company "Uranus" was established.

In 1996, the Mongolian-Russian-American joint venture "Gurvansaikhan" was created by the American company "Energy Fuels", the Russian state consortium "Geologorazvedka" and the Mongolian state company "Mongol-Erdene" for the exploration and exploitation of exogenous uranium deposits in the Mongolian Gobi region. In 1996, the American company Energy Fuels, the state enterprise-Mining and processing plant of Russia "Priargunsky GOK" and the state company "Mongol-Erdene" of Mongolia established a joint venture "Central Asian Uranium" with equal shares for each participant to heap leach low-grade ore stock piles from the Dornod deposit.

In 1997, the French company "Cogema" created a joint venture "KojeGobi" and now "Badrakh Energy" with the Mongolian joint-stock company "Govgeo" and began searching for exogenous uranium deposits in southern and eastern Mongolia. By 2003, the interest of foreign investors in the Mongolian uranium sector increased, and about ten companies received licenses for prospecting and exploration of uranium in Mongolia.

These include Canadian Stock Exchange-traded uranium companies such as Western Prospector and Khan Resources. Western Prospector established Eleelt Mines Co in Mongolia and obtained a license for the Gurvanbulag uranium deposit, while Khan Resources established Khan Resources Mongolia and obtained a license to explore some of the Dornod uranium ore bodies. These companies have carried out the work of translating Russian exploration data in accordance with Western standards.

Since 2007, Rosatom has repeatedly expressed interest in cooperation in the uranium sector of Mongolia, and in 2009, the Mongolian-Russian joint venture Dornod-Uran was established.

As a result of prospecting and exploration work carried out by these companies, uranium reserves in Mongolia have increased due to the discovery of new deposits such as Dulaan-Uul, Zoovch Ovoo, Ulziit, and Gurvansaikhan. As of 2014, Mongolia ranks 12th in the world in terms of uranium resources, which aroused the interest of foreign investors.

The Government of Mongolia pays special attention to the health and social issues of developing uranium deposits. In 2009, the Great People's Khural (Parliament) approved the following legislation: "State Policy on Radioactive Minerals and Nuclear Energy" and the "Nuclear Energy Law". The Nuclear Energy Law states the following: *"The Law gives the government the right to take ownership without payment of not less than 51% of the shares of a project or joint venture if the uranium mineralization was discovered by state funded exploration, and not less than 34% if state funding was not used to find the mineralization. It gives the State Administrative Authority the responsibility to implement and enforce state policy on the exploration and development of deposits of radioactive minerals and establishment of nuclear energy, including the power to grant, suspend or revoke any licenses granted pursuant to the Nuclear Energy Law. Licenses to conduct uranium exploration and production of any radioactive minerals must be obtained under this law.*

Parliament also passed legislation regarding the re-registration of existing exploration and mining licenses. Existing license holders had to apply to the State Administrative Authority by 15 November 2009 and comply with all of the conditions and requirements set out in the Nuclear Energy Law, including acceptance of the state's 34% or 51% share participation in the license holder, at no cost to the government."

In 2008, the state-owned company "Monatom" was established, to represent the interests of Mongolia in relations with foreign and domestic companies operating in the uranium sector, in accordance with the "Nuclear Energy Law". Up to this point, foreign and national companies have invested $XXUSD into the uranium sector.

Due to the 2009 legislation, the government's focus on the uranium sector has not brought the expected results. Major investors in the Gurvanbulag uranium project sold their shares to Czech investors before the Mongolian Nuclear Energy Law came into force. The Canadian company Khan Resources abandoned their uranium interests in Mongolia due to this legislation.

The first public information about uranium deposits in Mongolia was provided in 1986 by Yu. B. Mironov and R. M. Oleinikov on the Dornod volcanogenic-tectonic structure. V. A. Shelder, Yu. B. Mironov, N. S. Solovyov provided new information about uranium deposits in Mongolia in 1990, and Yu. B. Mironov, G. Jamsrandorj, and O. Chuluun in 1994 at the IAEA on the prospects for uranium mineralization in Mongolia. Since then, several papers have been published on uranium mineralization of the country. The most detailed information about the occurrence of uranium mineralization in Mongolia was published in the form of a book by Yu. B. Mironov in 2006, 2009, and 2012 in Russian and in 2005 in English. This book is a significant scientific contribution to the study of uranium mineralization in Mongolia and provides a direction for future exploration.

References

Byamba J (ed) (2011) Mongolian geology and mineral resources. In: Book 4. Plit tectonic of lithosphere, p 487 (in Mongolian)

Jamsrandorj G (2019) Uranium deposits of Mongolia, p 244 (in Mongolian)

Mironov YuB (2003) Uran Mongolii. Izd-vo Anatoliya, Spb, p 326

Mironov YuB (2007) Uranium deposits in Mongolia: emplasement patterns and potential // Proc. of the Nineth Biennial Meeting of the Society for Geologi Applies to Mineral Deposits Dublin. pp 67–74

Mironov YuB, Afanasьev AM, Bulosev AV, Buzovкin SV, Vysoкoostrovsкaya EB, Gerel O, Karpunin AM, Korshunov AM, Tamiraa A, Shor GM, Shuvalov YuM (2012) Uranovie mestrojdeniya Mongolii. Sanкt-Peterburg, p 304 (in Russian)

2 Geodynamics and the Relationship of Uranium Metallogeny in Mongolia

Geodynamics is a science which studies all the deep-seated processes of the earth including ravel of material and involved forces, as well as a reflection on the lithospheric plates and earth's crust. As a result of these forces, the deep materials in the earth shift, both horizontally and vertically. The morphological processes differ and can be effusive and intrusive, as well as forming sedimentary rocks during different geodynamical conditions. Origin of these geodynamic environments formed one after another successively replacing each other. For example, first formed was a spreading situation in a destructive tension oceanic crust and an oceanic plate was created. Then that plate plunged into azone of a deep-water trench and created a subduction environment and here a convergent process began Then, weak subductants, in this lithospheric plate in the zone of Bienof, partially melts into andesite magma, part of which rises to the surface of the Earth and forms an island arc in the boundaries of the ocean (transitional crust) of marginal-continental arcs in limits of the continent. It creates an accretionary environment. As a result, crust grows both vertically and horizontally. After the accretion process, felsic and acidic-alkaline products of a collision situation are generated. Thus, within the limits of this transitional crust, a continental crust forms on the edge of the ocean. Genetically, related geodynamic environment replacements are based on the theory of tectonics lithospheric plates.

The newly formed continental crust is going through another orogenic stage, developed under the influence of the surrounding oceanic plate and the development of subcortical mantle processes. This stage is divided into post-collision elevation, orogenic (late orogenic), anorogenic, post-genic (tafrogenic), and a plateau stage of development of folded structures.

In this work, we consider the geodynamic conditions of the divergent decay of the supernatural Rodinia, the formation of ancient oceanic basins, and then during their convergence, establishment of uneven-aged continental structures in Mongolia.

It should be noted that, with the circulation of the lithospheric substance (geodynamic settings), there is a cycle of all chemical elements at the same time, sometimes more chemical elements, sometimes less depending on their composition and depth. In the upper levels of the lithosphere, or even the continental cords itself, there are more mobile lithospheric elements, including uranium. This is one of the purposes of research layered on genetically linked geodynamic conditions of the development of the lithosphere evolution of Mongolia.

J. Galsan, *Uranium Deposits of Mongolia*, Mineral Resource Reviews,
https://doi.org/10.1007/978-3-031-86481-0_2

2.1 Accretionary-Collisional Tectonics and Formation of the Continental Crust on Mongolia

Let us first consider the divergent and convergent structures formed after the breakup of the supercontinent Rodinia, starting from the Baikal time. As for the pre-Baikalian ancient structures, they will be considered in the course of the characterization of folded areas and younger structures. In the process of the disintegration of the supercontinent Rodinia within Central Asia, including Mongolia, separate small-sized massifs and crystalline blocks were formed, such as the Sangilen, Buteel, Gargan, Baidrag, and Central Mongolian (Fig. 2.1).

In the course of accretion-collision processes and tectonic amalgamations, these ancient blocks and uplifts were combined into a single structure. For example, within the Paleo-Asian Ocean of the Bayannuur-Zavkhanmandal Early Baikal and Kerulen Early Caledonian accretionary-collisional sutures, separate ancient structures seemed to be "soldered" molded into a single massif. Thus, most of the links in this chain of ancient blocks form the Late Baikalian continental structure similar to the island-arc system. This indicates that the formation of the Central Mongolian massif as a single structure was finally completed in the Early Caledonian cycle of the development of the earth's crust.

Later, during the clockwise rotation of the Siberian craton, the newly formed arc acquired a southwestern orientation of the convex side as it developed. Starting from the Late Baikal time, subduction processes occurred around this arc from the outer side and the Ikh Jargalant, Urgamal, and Khantayshir active continental margins developed. At that time, the Paleoasian Ocean extended from the Ulutav and Kokchetav ancient massifs to the western shores of the Sea of Okhotsk, located between the Siberian craton in the north and the North China and Tarim cratonic blocks in the south.

Oka Khug Folded Area. The Oka Khug fold area is located between the Sangilen-Salig uplift in the west and the Gargano-East-Khubsugul block in the east. The paleotectonic history of the Baikal and Early Caledonian stages of development of the Tuva-Mongolian massif are considered in detail in the works of Kuzmichev (2004) and other researchers (Ilyin 1973, 1982; Gibsher and Terleev 1992; Fedotova and Khain 2002; Byamba 1990, 2012). There are two stages in the development of the region. In the early stage (1000–800 Ma), within the Oka basin, the ophiolites of the Dunzhugur island arc obduct onto the Gargan block, and the Early Baikal folded area is formed.

In the south, the second Late Baikal stage, characteristic of northern Mongolia (800–600 Ma years), continues to develop. The following morphological structures form in the Khug basin: The Shishkhid island arc in the west of the basin with the Kharberin back-arc trough. In the east, above the Khug accretionary prism, the setting of the Darkhad active continental margin is forming (Fig. 2.1). Thus, the closure of the Khug Basin corresponds to the end of the Ediacaran, although no accretionary-collisional granitoids were emplaced within the Oka-Khug fold zone by this boundary. However, the Khug late Baikalian folded structure is overlain by the Late Vendian-Early Cambrian terrigenous-carbonate cover through local erosion. This terrigenous-carbonate cover is associated not only with the accumulation of huge reserves of Khuvsugul bedded phosphorites, but also with the formation of uranium mineralization in bituminous and vanadium-bearing shales.

Starting from the Late Vendian-Early Paleozoic, along the periphery of the Tuva-Mongolian massif, both from the southwest, south, and from the west, processes of subduction of the oceanic plate of the Paleoasian Ocean occurred, active continental margins and back-arc riftings formed in the Middle Paleozoic. At that time, in the south, between Central Mongolian and Tuva-Mongolian massifs, the Ider small oceanic basin was formed, connecting the Lake zone with the Jid zone (Zonenshain et al. 1973; Blagonravov and Zaitsev 1974). In the Middle Paleozoic, in the central section of the Tuva-Mongolian massif, within marginal-continental

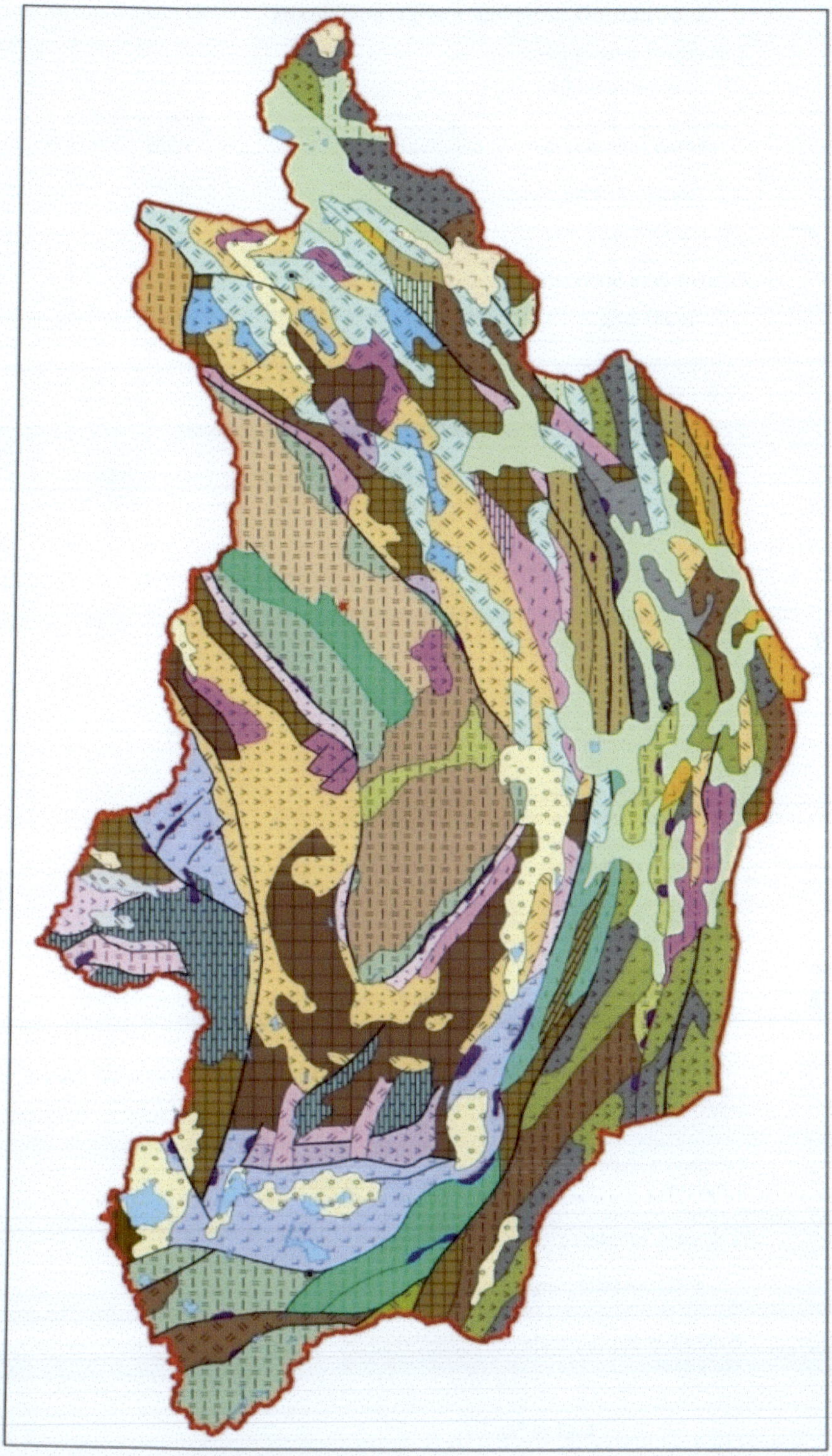

Fig. 2.1 Geodynamic map of Mongolia (Byamba 2018)

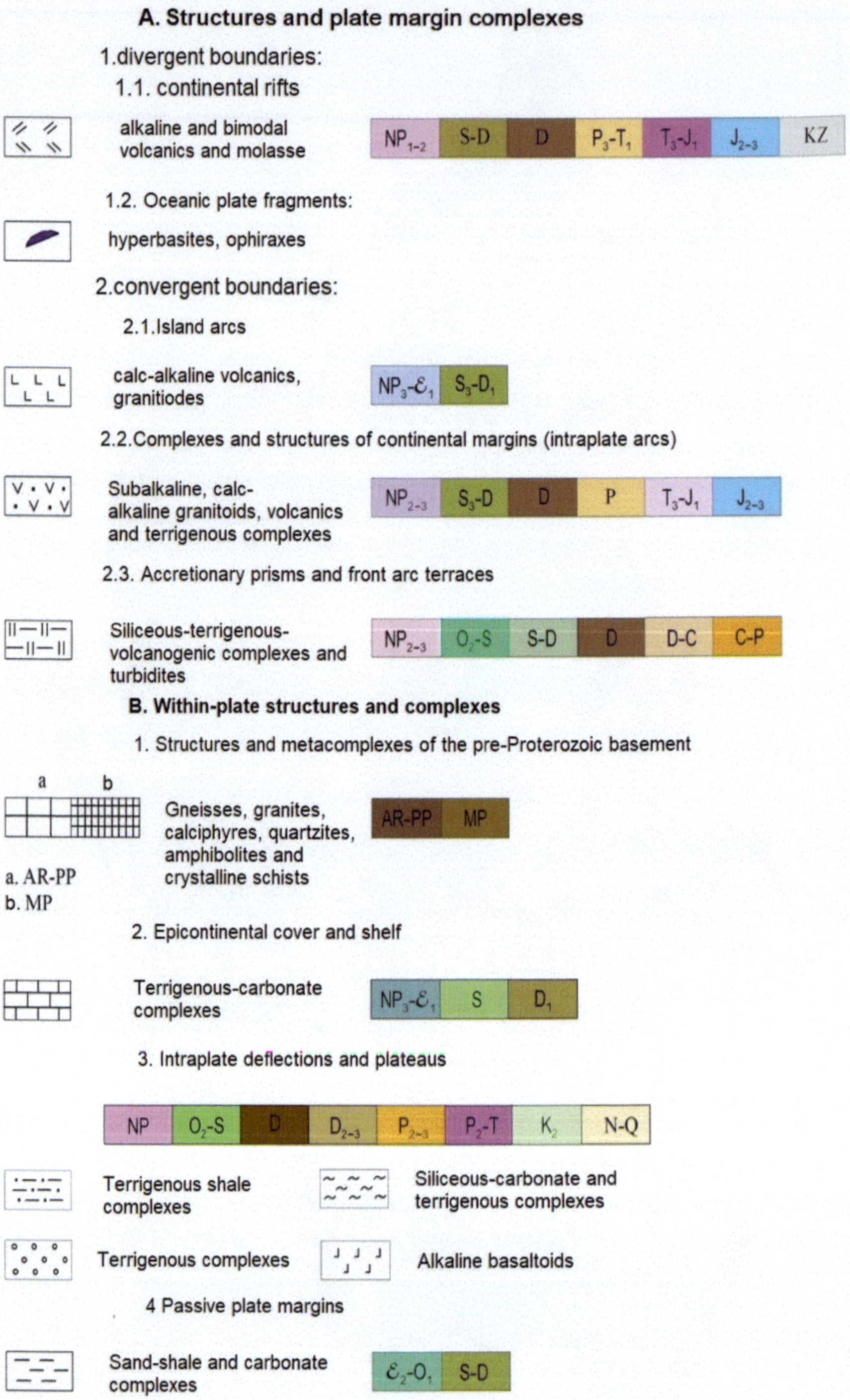

Legend Fig. 2.1.

(continued)

arcs and back-arc troughs, alkaline and subalkaline intrusions with uranium and rare-earth mineralization were emplaced (Mironov and Shuvalov 2009; Yarmolyuk et al. 2016).

The lake, Altai, and Jid (Egiingol) folded zones of the Caledonides, which developed on an oceanic-type crust, extended in the west from the Mongolian-Altai Mountains to the eastern border of Mongolia, located between the Tuva-Mongolian in the north and the South Gobi massif in the south. At the beginning of the opening of the Paleo-Asian Ocean, the Central Mongolian massif was located within its boundaries. A significant part of this massif was covered by a Vendian-Early Cambrian cover with remains of stromatolites, oncolites, and archaeocyaths. The Lake, Altai, and Egiingol zones were located in the west and northeast of the Central Mongolian massif and skirted it almost from all sides.

Starting from the end of the Late Proterozoic, the Ikh Jargalant, Urgamal, Khantaishir active continental margins of the tonicryogen, as well as the Ikhbogd-Undurshil suture zone between Central Mongolian and South Mongolian ancient massifs, formed along the periphery of this massif. In the Lake zone, at the end of the Vendian and in the early Cambrian, subduction of the oceanic crust took place with an age of 570–568 Ma (Gibsher 2001) with the formation of the Ulaanshand island arc, which extended to the southern and southeastern edge of the Shargyn Gobi basin. The age of accretionary granitoids of this island arc is 495–435.440 Ma (Kovalenko et al. 1996).

Collision of the Late Cambrian-Early Ordovician Ulaanshand island arc with the Late Baikalian folded framing of the Central Mongolian massif in the northern part of the Lake zone occurred at the end of the Cambrian—the beginning of the Ordovician. It is recorded by syncollisional granitoids of the last phase of the Togtokhinshil complex (488–484 Ma, Kovalenko et al. 1996; 494–465 Ma). In the south of the Lake Zone, the collision was marked by the thrusting of ophiolites from the Khantaishir marginal basin onto the Gobialtai block of the Central Mongolian massif.

To the west of the Lake zone in the Ediacaran-Early and Middle Paleozoic, the Altai zone develops, which differs in its monotonous composition of sandy-shale strata (7000–5000 m), sedimentary-volcanogenic formations and the manifestation of magmatism, as well as in its geodynamic settings. This thick sequence of terrigenous-shale rock is considered a transitional crust formed in the process of lateral accretion. In this regard, the Tsagaan Shuvuut, Tsengel Khairkhan (Mongolian-Altai), and Khovd active continental margins developed on the transitional crust.

The final formation of the continental crust in the Altai zone corresponds to the Middle Silurian and is marked by palingenic potassium granitoids of the "S" type of the Khovd complex (428–372 Ma, Gavrilova et al. 1975; Dergunov et al. 1971).

At the end of the history of the development of the Lake and Altai fold zones, it should be noted that in Devonian time, during movement of the lithospheric plate, a hot spot of the mantle was blocked. This caused the formation of the Altai-Sayan three-beam rift structure, which is associated with Ta, Nb, Zr, and rare-earth elements, as well as, possibly, uranium mineralization with Khaldzanburgedtei and other alkaline granitoids (Kovalenko et al. 2001).

The Jid (Egiingol) zone is located on the northeastern continuation of the Lake zone between the Tuva-Mongolian massif in the north and the Central Mongolian massif in the south. Within this zone, the following morphological elements have been identified: the Eg-Uuriin island arc, the Argyingol back-arc basin, and the Zunaigol forearc trough, as well as the Jid active continental margin in the northeast, which formed at the boundary with the Khamardavaa block.

Thus, at the end of the Late Caledonian, north of the main Mongolian lineament, with the exception of the Khangai, Khentei, and Duchgol troughs, the entire territory of the Baikal and Caledonian crust was attached to the Siberian craton. To the south of the Siberian continent, at the beginning of the Cambrian (520 ± 5–519 ± 4 Ma), possibly at the end of the Ediacaran time (Mingshuai Zhu et al. 2014; Jian et al. 2014), the Neo-Asian Ocean was formed during the extension of the lithospheric plate of the South Gobi massif. It should be noted here that this ocean was considered by former researchers (Dergunov et al. 1971) as the Paleotethys Ocean. However, in recent years, in the zone of subduction accretion of the Gurvansaikhan Range and along the Manlai Fault, the age of ophiolites has been determined as Early Cambrian. If we take into account that the divergence of the Paleo-Asian Ocean began somewhat earlier, in the middle of the Cryogen (745 Ma), than in the Neo-Asian (Southern Biysk) Ocean, then spreading and subduction processes began here only at the end of the Silurian-Early Devonian, while convergent processes began in the Paleo-Asian Ocean.

The South Mongolian Folded Area of the early Hercynides, formed at the end of the Carboniferous, has the following morphological elements. In the central section, the South Mongolian folded zone stands out, in the northern frame edge, the Gobialtai-Sukhbaatar uplift, is the northern shelf of this ocean. The southern shelf is composed of the Atasbogd, Khashaat, and Tsohiot suprasubduction zones, which form the active continental margins of the South Gobi massif. In the South Mongolian folded zone, subduction zones dipping on both sides were established, both in the south in Silurian-Devonian time along the Gobi Tyanshan, Zoolon, Urgen, and Modon-ovoo faults along the northern framing edge of the South Gobi and Nukhetdavaa massifs. And in the north in the Silurian-Lower-Middle Devonian along the North Mongolian superblock and along the Bulgan, Transaltai, Saikhandulaan, and Matad faults. Accretionary prisms were formed along these subduction zones, and above them the Butul-Khudag active continental margin of the same age and with magmatic products of the andesitic series (Nud-Khudag complex) with a high content of SiO_2 and K_2O were studied.

It should be emphasized that, probably in Early Carboniferous, coeval Middle-Late Devonian chains of Zoolon, Nemegt, Ekhiingol, and Tsagaanshavart occurrences of mafic and intermediate island-arc volcanic rocks, which gradually subside from west to east, were accreted along the Gobi Tyanshan subduction zone. The last one, the Zoolon arc, is completely unsubducted. Almost everywhere these volcanics are underlain by hypermafic mélanges. Also in the South Mongolian zone, early Carboniferous granitoids and volcanic rocks are calc-alkaline ($329 \pm 1.331 \pm 2$ Ma). Their introduction ends the transitional period of development. In the South Mongolian zone, the intrusion of collisional phases of granitoids occurred only at the end of the Carboniferous (305 ± 1, 310 ± 1, 310 Ma). The Zambelgekh, Mandakh, Matad, Shuteen, and other regions are marked by large multiphase granitoid massifs.

The closing of the South Mongolian Ocean at the end of the Carboniferous coincides with the opening of the small Sulinkheer (Solonker) ocean. Judging by the age of the oceanic crust, it can be assumed that the spreading of the Sulinkheer zone occurred in the period $279 \pm$ Ma (Miao et al. 2007), 275 Ma (Byamba et al. 2012). It can be assumed that it was at this time the melanocratic structure of the Paleotethys Ocean was revealed. At the same time, the formation of the subduction zones began in the south: Undursum under the North China craton and Khangi-Khar-Erdene, to the north, under the South Gobi ancient massif. This small ocean existed for about 30–10 Ma, forming the Sulinkheer suture zone at the end of the Permian. This stage completed the formation of the Late Hercynian fold zone at the end of the Permian—beginning of the Triassic on the border of Southern Mongolia and Northern China (Wenja Xiao, Brain F. Wendeley, Jiehao, Mingguo Zhai).

Central Mongolian Folded Area. This folded area occupies almost the entire central part of Mongolia. It is bordered on the north by the Tuva-Mongolian massif and the Jid zone, and along the Khangai, Zelter in the west. It is bordered on the southwest by the Lake zone along the Zavkhan and ShargynGobi deep faults, and also on the south by the Gobialtai-Sukhbaatar uplift along the Main Mongolian lineament. Within the folded area, the Precambrian ancient framing is prominent, and in the inner part, the Late Caledonian, Late Hercynian, Indosinian, and Cimmerian folded structures are prominent.

The Central Mongolian massif, bending around in an arc, frames the Central part of Mongolia from the Buteel massif in the southwest to the basin of the Khungui and Zavkhan rivers and to the North Kherlen uplift in the northeast.

The Buteel, Tarvagatai, Zavkhan, and Baidrag ancient blocks are composed of highly metamorphosed gneisses, granulites, amphibolites, charnockites, marbles, quartzite, and crystalline schists of various compositions. The Late Archai and Paleoproterozoic absolute age of these metamorphites are 3070–2650 Ma (Kozakov et al. 2011; Mitrofanov et al. 1987). The North Gobi, South and North Kherlen uplifts are primarily represented by crystalline schists of various compositions, phyllites, marbles, granite-gneisses, and metaterrigenous rocks. The age of these Mesoproterozoic rocks ranges from 1200 to 900 Ma (Zonenshain et al. 1973; Blagonravov 1978).

In late Baikal time, the Central Mongolian massif was surrounded by active continental margins almost on all sides. Starting from Early Baikal time, at the end of Late Proterozoic, the massif practically represented a single chain of islands parallel to the southern and southeastern margins of the Siberian continent, with the exception of the South and North Kerulen uplifts, which were "soldered" by the accretionary-collisional suture of the Early Kaleidian stage of development.

The inner section of the Central Mongolian massif, starting from the Late Caledonian orogeny, separated the Khangai-Khentei (Mongol-Okhotsk) ocean from the Paleoasian Ocean (Sengor et al. 1993).

Late Caledonian time somewhat complicated the relief of the ocean floor of the Mongolian-Okhotsk Ocean. Probably, by this time, the Kharkhorin transverse uplift had formed between the Khangai and Khentei troughs. Late Caledonian weakly metamorphosed accretionary-collisional complexes are widely developed along the marginal parts of the Khangai-Khentei trough and within the Kharkhorin uplift. They are represented by silicic-sedimentary-volcanic and terrigenous-schist formations of the Late Ordovician–Silurian, and also intruded by calc-alkaline granitoids of the accretionary-collisional type of the Boroogol and Borkhairkhan complexes of the Middle Silurian (480–450, 410 Ma).

The Late Hertzian accretionary-collisional history of the Central Mongolian fold system is represented primarily by Devonian siliceous-terrigenous-volcanogenic formations. In the Carboniferous and Permian in the Khangai and Khentei troughs, the rocks somewhat differ in composition and in the age and composition of intrusive formations. In the Khangai trough, there are more layers of siliceous rocks in the Carboniferous sediments than in the Khentei trough. In addition, riftogenic structures with alkaline-bimodal volcanics, with flora remains, appeared in the Permian in the Khangai trough, while marine shelf terrigenous sediments with marine fauna developed in the Khentei trough. As for the intrusive rocks in the Khangai trough, in the Permian and Early Triassic (249–200 Ma), mesocratic plume granitoids of plagiogranite composition of the Khangai complex were emplaced, and plume granitoids of the Baga-Khentei batholith complex (T3-J1) of calc-alkaline were emplaced.

On the periphery of the Khentei trough, Triassic-Early Jurassic volcanic-terrigenous orogenic troughs are known. The 87Sr/86Sr isotope ratios of both the Khangai and Baga-Khentei intrusive complexes are less than 0.7060, and therefore, it is believed that they are of plume origin (Kovalenko et al. 2001; Yarmolyuk et al. 2013, 2016).

Thus, with the intrusion of these plume granitoids and other tectonic factors in the Khangai trough, the continental crust was formed in the Permian-Early Triassic (Late Hercynides), and in the Khentei Trough, at the end of the Triassic-Early Jurassic (Indosinides).

It should be noted that both formed above the Late Caledonian and Late Hercynian subduction slabs within the Central Mongolian massif. Additionally, marginal-continental volcanic arcs and extensional structures formed, filled with molasse and calc-alkaline volcanic rocks. As for the rocks of carbonaceous accretionary prisms that were formed on both sides of the Khangai and Khentei troughs, their exit from the Benioff zone turned out to not be during the Carboniferous, but in Early Permian within the structures framing the ancient massif at different distances from the subduction site, up to a maximum of 500 km (Orkhon-Selenge zone).

The Mongol-Okhotsk Ocean continued to develop through the Jargalantiingol graben-synclinal trough to the Duchgol basin. Here, the Late Permian and Triassic-Jurassic full sections have been studied. Studies found in the Jargalantiingol trough that the axis consistently shifted south, and that it is possible the subduction zone dipped south under the North Kerulen uplift.

The Triassic-Jurassic rocks are represented by strongly deformed, mainly marine terrigenous formations with interlayers or thicker members of gray siliceous rocks and red jasper, and less often, primarily along the northern border of the trough, rhyolites, dacites, basalts, andesitic lavas and their tuffs. In addition, to the south under the North Kerulen block, according to the observations of P. Khosbayar and T. Binderyaa, strongly deformed terrigenous shales and interlayers of siliceous rocks are subducted, less often with thin interlayers of hyperbasite shales. Thus, it is assumed that in the Late Mesozoic in the Duchgol trough, a small oceanic basin is present in the north with an island arc, and in the south with an active continental margin. In this trough, the Yamlakh subduction-accretion intrusive complex was studied, which consists of Middle Jurassic calc-alkaline granodiorites (185–132 Ma) and collisional leucogranites of the Late Jurassic Chuluunkhoroot Complex (162–147 Ma). These events complete the Cimmerian accretion-collision processes within the Duchgol trough.

Thus, at the end of the Jurassic, the Mongolian-Okhotsk Ocean gradually leaves the central territory of Mongolia to the northeast, leaving behind successively formed continental crusts of different ages of Late Caledonian-Precambrian age.

2.2 The Orogenic Stage of the Development of Mongolia and the Continental Collision of the North and South Asian Massifs

The orogenic stage is understood as the process of formation of mountain-fold belts at the site of the closure of the oceanic basin during the final stage of its development.

In this work, for the structures of the continent, the principles of lithospheric plate tectonics were used, which emphasizes the fact that the oceanic plates are in constant motion and subduct, increasing the size of the continental plates. In addition, continental plates also unite with each other or break up into separate blocks. After each collision, any territory begins to develop in the regime of the continental (orogenic) stage of development. However, the orogenic stage has its own stages, which are "geodynamic settings" for specific folded areas.

To characterize the stage of orogenic development, the following scheme of the collisional developmental stage was adopted: syncollisional, post-collisional uplift, orogenic, anorogenic, postorogenic (taphrogenic), and plate (epiplatform alignment).

It should be noted that after the collision stage of both the Baikal and other collision cycles, after the completion of their development, successively developing intermountain troughs and destructive rift structures are formed. For a more substantiated identification of continental structures identified and studied on Mongolia, a classification of all riftogenic structures that are most important for the purposes of metallogeny and genetically related to geodynamic settings is proposed. The most uranium-bearing structures turned out to be granitoids of marginal-continental arcs and back-arc rift zones marked by alkaline-bimodal volcanics of basic, intermediate, and felsic composition.

The formation of such back-arc destructions in Eastern Mongolia occurred in the Late Permian, at the end of the Triassic-Early Jurassic, and at the end of the Mesozoic, including:

- in the late Paleozoic at the end of the Permian, within the Middle Gobi-Kerulen marginal-continental arc, riftogenic zones were established from the Zartkhudag region of alkaline volcanic rocks in the southwest to the Avdarkharaat and Tsav massifs of alkaline granitoids in the northeast;
- the next zone of rifting is the Late Triassic-Early Jurassic Bayan-Uul-Gurvanzagal destructive structure with a northeast direction;
- the third rift zone is the Middle-Late Jurassic marginal-arc structures-the North Choibalsan, Berkh, Onon, Tsagaandelger, composed of alkaline volcanic rocks, which are also promising for uranium. Other arc structures are known in the Ulz, South-Kerulen, Galshar regions for the development of rift formations. The rift structures of Eastern Mongolia, especially in the northeastern section, are significantly different from other orogenic structures that developed during the formation of accretionary-collisional geodynamic settings of folded areas. Eastern Mongolia, starting from the Mesoproterozoic, did not go through the stage of oceanic development and has an ancient foundation. The last collisional stage here was manifested in the Mesoproterozoic. Later, with the exception of the Kerulen suture zone of the early Caledonides, peculiar marginal-arc and continental structures developed over the Asian hot spot (plume) in Eastern Mongolia. Within the stage, the products of the crust-mantle development are enriched in U, K, Na, Sn, W, Mo, Pb, Zn, and other lithophile elements. Features of riftogenic structures of Late Mesozoic (J_3–K_1) age differ from similar riftogenic marginal-arc structures not only in age (J_3–K_1) of volcanic rocks, but also in composition. Thus, the structures developed here belong to the OIB and MORB types, which emphasize the upper mantle source and the relatively high content of uranium. In addition, at this time, the lithospheric plate of Northeast Mongolia drifted to the southwest and appeared above the fading hot spot of Northeast Asia (plume).

Additionally, it should be noted that in this region of Eastern Mongolia, in the process of crustal stretching, its thickness was reduced to 42.5 km, and the thickness of the subcrustal lithospheric plate increased to 47.5 km with an increase in thermal permeability of, more than, 70 mW/m^2.

On the basis of these and other geological data, it is considered justified, that the process of uplift of the mantle diapirus (plume) is the cause of the formation of foci of crustal-mantle subalkaline and acidic magmatism with an affinity for uranium, as well as other rare elements (W, Sn, Ta, Nb, Li) and metals (Mo, Zn, Pb, Au).

As for the reason for the introduction of a diapir (plume), namely in Northeastern Mongolia, it can be assumed that the Permian Khangai plume, which migrated in the Triassic to the Baga-Khentei region, also ended up in this region at the end of the Jurassic-Early Cretaceous. Thus, at the end of the Jurassic-Early Cretaceous, the lithospheric plate of Northeastern Mongolia drifted over the "hot spot" expressed by the crust-mantle mixture. However, the magnitude of the rise of this plume in Eastern Mongolia turned out to be less due to the closure of the Mongolian-Okhotsk Ocean outside this region. In this regard, the Late Jurassic-Early Cretaceous rifts of northeastern Mongolia were formed in connection with the migration of the plume and the increase in the continental crust above it.

It should be emphasized that taphrogenic structures are promising for uranium mineralization. They are formed in the process of syncollisional uplift and coincide with the transition zones between the rift and plate stages of the development of continental block-fold structures. Taphrogenic structures developed in valley-continental conditions and are filled with terrigenous-sedimentary deposits formed in a redox environment.

After the termination of collisional processes between the North and South Asian continental plates in the south, within the folded structures of the South Mongolian zone, the Gobi plate with a subplatform regime was formed at the end of the Cretaceous.

In conclusion, it can be noted that metallogenic studies and prognostication of any types of minerals, including radioactive minerals, should be based on the principles of geodynamics, which determine the possibility of using a practical application of the theory of lithospheric plates in scientific and applied geology.

References

Blagonravov VA, Zaitsev NS (1974) Dzhida zone of the early Caledonides of Mongolia. Geol Mongla 8:173–174

Byamba J (1990) Main stages of tectoic development territory of Mongolia Izd-vo VUZ. Geol Geophys 29–38

Byamba J (ed) (2012) Mongolian geology and mineral resources. In: Book 4. Plit tectonic of lithosphere, p 487 (in Mongolian)

Byamba J (2012) The continental crust of the Ediacaran period (Baikalid) in the Tuva-Mongolian massif. In: Geology and minerals of mongolia, Volume IV, plate tectonics of lithospheric structures. Ulaanbaatar, Soyombo Printing, pp 80–95 (in Mongolian)

Byamba J, Mironoy YB, Jamsrandorj G (2018) Geodynamics and uranium metallogeny of Mongolia, Ulaanbaatar, p 302 (in Mongolian)

Dergunov AB, Zaitsev NS, Mossakovsky AA et al (1971) Hercynides of Mongolia and the paleotethys problem. In: Problems of theoretical and regional tectonics. Moscow, Science, pp 87–103

Fedotova AA, Khain EV (2002) Tectonics of the southeastern sayan and its position in the Ural-Mongolian belt. Moscow, Scientific world, p 175

Gavrilov SP, Zaitsev NS, Pavlov VA et al (1975) Alkaline granitiod formations in the structures of Western and Northern Mongolia. Moscow, Science. p 285

Gibsher AS (2001) Vendian island-arc intrusive magmatism of the lake zone of western Mongolia (geological, geochronological, and petrochemical data). Russ Geol Geophys 62(6):619–632

Gibsher AS, Terleev AA (1992) The stratigraphy of the upper precambrian and lower Cambrian in southeastern Tuva and northern Mongolia. Russ Geol Geophys 33:19–25

Ilyin AV (1973) Khuvsugul phosphorite-bearing basin. Moscow, Science, p 166

Ilyin AV (1982) Geological development of Southern Siberia and Mongolia in the Late Precambrian-Cambrian. Moscow, Science, p 113

Kovalenko VI, Yarmolyuk VV, Kozlovsky AM et al (2001) Sources of magmas of alkaline granitiods and related rocks of intraplate igneous associations of Central Asia. II Dokl. RAS. Vol 377, No.5, pp 672–676

Kovalenko VM, Yarmolyuk VV, Pukhtel IS et al (1996) Igneous rocks and sources of ophiolitic magmas of the Lake Zone of Mongolia. Petrology 4(5):453–495

Kozakov IK, Didenko AN, Azimov PYa et al (2011) Geodynamic settings and conditions for the formation of crystalline complexes of the South Altai and South Gobi metamorphic belts. Geotectonics, No. 3. pp 7–30

Kuzmichev AB (2004) Tectonic history of the Tuva-Mongolian massif: Early Baikal, Late Baikal and Early Caledonian stages. Moscow, p 191

Miao L, Zhang F, Fan WM, Liu D (2007) Phanerozoic evolution of the Inner Mongolia-Daxinganling orogenic belt in North China: Constraints from geochronology of ophiolites and associated formations. Geol Soc London Spec Pub 280:223–237

Mingshuai Zhu, Munkhtsengel Baatar, Laicheng Miao et al (2014) Zircon ages and geochemical compositions of the Manlay ophiolite and coeval island arc: Implications for the tectonic evolution of South Mongolia. Article in Journal of Asian Earth Sciences

Mironov YuB, Grushevoi (2008) Rol noveishikh geodinamicheskikh protsessov v formirovanii girdotermalnogo uranovogo oredeneniya na yuge Evroasiiskogo kontinenta. Kazatomprom, Uran Kazastana, Alma Ata (in Russian)

Mironov YuB, Shuvalov YuM (ed) (2009) Uranium deposits of Mongolia. VSEGEI, p 304 (in Russian)

Mitrofanov FP, Kozakov IK, Paley IP (1987) Precambrian of Western Mongolia and Southern Tuva. Leningrad, Science, p 152

Jian P, Kröner A, Jahn BM, Windley BF, Shi Y, Zhang W, Zhang F, Miao L, Tomurhuu D, Liu D (2014) Zircon dating of Neoproterozoic and Cambrian ophiolites in West Mongolia and implications for the timing of orogenic processes in the central part of the Central Asian Orogenic Belt. Earth-Sci Rev 133:62–93

Sengor AMC, Natal'in BA, Burtman VS (1993) Evolution of the Altaid tectonic collage and Paleaozoic crustal growth in Euroasia. vol 364, pp 299–307

Smislov AA et al (1981) Geolocheskoe stroenie I rudonosnost Vostochnoi Mongolii. Izd-vo VSEGEI, p 156 (in Russian)

Soloev KhS, Shatkov GA, Yakobson LN (1977) Priargunsko-Mongolskii vulkanicheskii poyas. Geol Geoph 3:20–31 (in Russian)

Yarmolyuk VV, Kozlovsky AM, Salnikova EV et al (2013) Participation of alkali-granitoid magmatism in the formation of the Khangai batholith: Data of geological and geochronological studies. Reports of the Academy of Sciences. Geology Vol 452(5):543–547

Yarmolyuk VV, Lykhin DA, Kozlovsky AM, Nikiforov AV, Travin AV (2016) Composition, sources and mechanisms of formation of rare-metal granitoids of the Late Paleozoic East Sayan zone of alkaline magmatism (on the example of the Ulan-Tolgoi massif). Petrology 24(5):518–536

Zonenshain LP, Amantov VA, Borzakovsky et al (1973) Central Mongolian fold system. Geology of the MPR. T.II. Moscow, Science, pp 609–647

3 Distribution and Location of Uranium Deposits in Mongolia

3.1 Main Patterns of Distribution of Radioactive Elements in the Geological Complexes of Mongolia

The study of the distribution of natural radioactive elements (NREs) in the geological structures of Mongolia in the twentieth century was conducted by teams of the Mongolian Geological Survey Expedition (MGSE), the All-Union Scientific Research Geological Institute (VSEGEI), the State Enterprise Sosnovgeologiya, the All-Union Institute of Exploration Geophysics (VIRG-Rudgeofizika, currently FGU NPP Geological Exploration), the All-Union Institute of Mineral Resources (VIMS), and other organizations. The purpose of the studies was to assess the uranium content of Mongolia. Information about the distribution of radioactive elements in various geological settings throughout Mongolia is given in the works of Yu. M. Vladimirsky, G. A. Shatkova, Yu. M. Shuvalov, and other researchers Visokoostrovskaya et al. 2007).

The characteristics of the radiogeochemical features of geological complexes of Mongolia are given sequentially for geodynamic settings of various types in the North and South Mongolian superblocks, Middle-Late Paleozoic, riftogenic intracontinental volcanic belts, subplatform sedimentary basins, and depressions of Late Mesozoic-Cenozoic.

The basis of radiogeochemical framework is presented, based primarily on the results of detailed airborne gamma spectrometric surveys (AGS) carried out in 1970–1991 by Soviet specialists using the latest (for that time) equipment and methods. The surveys in east and west Mongolia were generally at 1:10,000, 1:25,000, 1:50,000 scale, whereas in central Mongolia surveys were conducted on a broader scale at 1:200,000 and 1:1,000,000. In 1995, the data from airborne gamma spectrometric surveys were compiled and equalized, and integrated with maps of adjacent territories from Kazakhstan, Russia, and China, and on this basis, maps of NRE content were produced at different scales. The consolidated maps showing the distribution of uranium, thorium, and potassium facilitated the estimation of exposure dose rates of gamma radiation throughout Mongolia. The derivative maps were created at 1:1,500,000 scale with more detailed insets at 1:200,000 and 1:500,000 scale (responsible editor Visokoostrovskaya et al. 1999, 2006).

When analyzing the NRE and identifying formations enriched in uranium and defining radiogeochemical zones, precise NRE measurements, in the laboratories of the MGSE and VSEGEI, were also taken into account. Statistical calculations of the NRE distribution parameters were carried out for individual tectonic structures;

The original version of the chapter has been revised. A correction to this chapter can be found at https://doi.org/10.1007/978-3-031-86481-0_5

J. Galsan, *Uranium Deposits of Mongolia*, Mineral Resource Reviews,
https://doi.org/10.1007/978-3-031-86481-0_3

certain NRE analytical QA/QC procedures, standardized throughout the Soviet block were utilized to assure the accuracy of the results.

Radiogeochemical Characterization of Precambrian and Paleozoic Structures

The radiogeochemical characterization of pre-accretionary complexes composing structures of different ages and different types is given separately for the North and South Mongolian superblocks demarcated by the Main Mongolian lineament.

It should be emphasized that blocks dominated by highly metamorphosed Precambrian formations are considered, in plate tectonic terms, as cratonic or metamorphic massifs-relics of the base of the Siberian and North China plates.

North Mongolian Superblock. The most ancient rocks are distributed in cratonic and metamorphic blocks (see Fig. 2.1) including Gargan, Zavkhan, Tarbagatai, Baidrag, Ereendavaa, Tsagaan-Shiveet, Sangilen, Khamar-Davaa, Butel, and Dariv. The main feature of such cratons is the uneven and limited development of uranium-bearing formations of different ages. The Neoarchean-Proterozoic granulite-gneiss-amphibolite complex and the phosphate-bearing terrigenous-carbonate formations correspond to the earliest, distinctly uranium-mineralization epoch. The internal heterogeneity of the AGS-mapped fields of terrains is primarily due to post-accretionary granite formations and redistribution of NRE due to postmagmatic processes, as well as the redistribution of the NRE during Riphean and Paleozoic metamorphism.

The most ancient granulite-gneiss-amphibolite complexes of northern Mongolia, sometimes containing graphite, are distinguished by elongated zones of U (up to 4 ppm), Th (up to 20 ppm) and K (up to 3.5%) contents. The carbonate-quartz (with amphibolites) siliceous-carbonate formations of the Sangilen and Khamar-Davaa metamorphic cratons of the Khubsugul region are characterized by low value NRE (U 1–2 ppm, Th up to 2–6 ppm, K 0.5–2.0%). The NRE zone increases slightly in the overlying carbonate-terrigenous and volcanogenic Vendian-Cambrian complexes located to the east-southeast of the Khubsugul region in the Khamar-Davaa craton (U 4–8 ppm, Th 4–8 to 10 ppm, K 1–2%).

The eastern part of the superblock is occupied by the Ereendavaa craton, which is part of the Kerulen megablock, the western fragment of the Argun composite massif. The NRE contents here are low (U from 1.8 to 3.0 ppm, Th 8–10 ppm, K 2.2–3.0%). The Proterozoic and Ediacaran-Cambrian complexes in the eastern part are distinguished by low and relatively uniform radioactivity. In the central part of the craton, the NRE contents are slightly elevated (U from 2 to 5 ppm, Th from 4 to 12 ppm, K 2–3%, and as high as 4%); the most radioactive formations are developed in the western part of the Ereendavaa craton. Younger terrigenous deposits of the Ordovician and Silurian, terrigenous carbonate complexes of the Devonian, Carboniferous, and Devonian molasse are characterized by 2–3 ppm uranium background values, thorium at 8–12 ppm, and potassium around 2–3%. The molasse complexes of the Upper Paleozoic in the Sangilen craton are also characterized by background NRE contents of U 2–3 ppm, Th 8–16 ppm, and K 2–4%.

Paleozoic, moderately acidic, igneous rocks, gabbroids, and ultramafic rocks of the Ereendavaa, Khamar Davaa, Gargan, and Ilchir cratons are distinguished by relatively low contents of U, Th, K, respectively, 1–2 ppm, 4–8 ppm, and 0.5–1.5%. Slight increases in NRE contents are observed in massifs of the Ordovician granite-migmatite formation (U 2.0–3.5 ppm, Th 8–12 ppm, K 2.3–3.0%), Silurian granodiorite-granite formations (U 2–3 ppm, Th 6–10 ppm, K 1.8–2.8%) and in small post-accretion Middle-Late Paleozoic and Mesozoic massifs of leucogranites and granosyenites (U 3–4 ppm, Th 16–20 ppm, and K 3–4, rarely 5%). Devonian acid volcanics are less radioactive (U 2–3 ppm, Th 8–12 ppm, K 2–3%). The rhyolitic and dacite-rhyolitic complexes of the Altanbulag, Sant, and Khangol VTS of Permian and Triassic Jurassic age are moderately radioactive (U from 2.0 to 4 ppm, Th 8–12 ppm, K 2–3%).

The western part of the superblock contains the Altai and Agardag forearc structures, the Khovd accretionary prism, and the Lake and Turgen island-arc structures. The first three structures (up to the Tsagaan-Shiveet fault zone) are composed of Proterozoic-Early and Middle Paleozoic formations and are characterized by complex geochemistry with average contents of uranium 2–3 ppm, thorium 8–12 ppm, and potassium 1–3%. Areas of elevated values of NRE (U up to 6–12 ppm, Th 12–70 ppm, K 3–8%) map into isometric shapes, and mimic the overall shape of post-accretion Devonian granitoid massifs.

The island-arc Lake and Turgen structures are distinguished by relatively low and uniform, weakly differentiated NRE zones in outcrops of basic Cambrian volcanic rocks marked by ophiolites and pillow lavas. These structures are characterized by low and fairly consistent NRE levels of U (0.5–2 ppm), Th (2–6 ppm), K (1–2%). Local areas of slightly elevated NRE values are associated with small outcrops of post-accretionary subalkaline granitoids and Late Paleozoic syenite porphyries where NRE ranges are: U 3–5 ppm, Th 12–20 ppm, and K 3–8%. Terrigenous marine deposits of Cambrian-Ordovician age differ in the background values of radioactive elements (U 2–3 ppm, Th 10–12 ppm, and K 13%). In the variegated sedimentary-volcanogenic sequence of the Devonian, the uranium content sometimes increases to 4–6 ppm. Carboniferous and Jurassic-Cretaceous terrigenous coal-bearing deposits of small depressions (Umnu-Gobi, Achitnuur, etc.) superimposed on older structures are generally low in radioactivity (U 1.5–2.5 ppm, Th 12–16 ppm, and K from 1 to 3%).

Abnormally high NRE contents are characteristic of Late Paleozoic granite massifs such as Sagsai, Turgengol, Bogot, Bayan-Zurkh, Durgun, Kulagash, Kharkhiraa, Achitnuur, Orgui, Bayannuur, which are often associated with quartz-wolframite and scheelite greisen-vein mineralization (Fig. 3.1).

Large massifs (Kharkhiraa, Sagsay, etc.) are characterized by an uneven distribution of NRE with an increase in the content of radioactive elements from the center to the periphery. Small granite massifs in the north (Boguut, Turgengol) are characterized by high uranium content (5–8 ppm). The maximum concentrations of radioactive elements were noted in the granites of the Turgengol massif, near the Askhat silver deposit (U 7.2 ± 1.4 ppm, Th 56 ± 12 ppm). Volcanic and subvolcanic formations of the Kyzyltau, Achitnuur, Narinnuur, Dulaan-Kharuul VTS are characterized by a NRE field containing U 4–6 ppm, Th 16–20 ppm, and K 4–5%. The formations with the highest enrichment in uranium are the Kyzyltau (6.0 ± 0.7 ppm), Achitnuur (4.8 ± 0.6 ppm) and Dulaan-Kharuul VTS (4.6 ± 0.9 ppm).

Terrigenous formations of the Upper Paleozoic are weakly radioactive and contain, as a rule, Clarke-level background concentrations of radioactive elements. In the granitoids of the Altai, Khovd, and Lake zones, the redistribution of NRE displays multistage characteristics. High-temperature metasomatic transformations redistributed individual NRE elements within ore zones, which led to the predominance of local anomalies of uranium associated with tungsten and tin deposits at Kobdogol and Kyzyltau. The thorium values are typical for blocks hosting complex thorium-uranium-rare-earth mineralization at, for example, Khundelengol, Shar-Tolgoi, Ulaan-Tolgoi, Tsakhir, and others.

In the turbidite basins of Central Mongolian massif (Khangai-Khentei and others), Riphean-Paleozoic sedimentary rocks with elevated uranium are absent.

From the west, the Tsag block of the passive continental margin adjoins the Khangai-Khentei turbidite basin, composed of weakly radioactive Paleozoic terrigenous rocks of greenschist facies metamorphic grade (U 1.8–2.3 ppm, Th 6–8 ppm, K 0.2–2.4%). The Bayakhongor block is characterized by the lowest NRE values of typical ophiolite complexes (U 0.5–1 ppm, Th 2–3 ppm, K 0.5–1.2%). Further west the Baidrag cratonic metamorphic block, consisting of ancient metamorphic complexes and migmatized gneisses, contains near-Clarke contents of NRE (U 3.0–4.2 ppm, Th 8–12 ppm, K up to 2.8%).

The Idermeg passive continental margin in the southern part of the Northern Superblock, near the Main Mongolian lineament, has a composition like the Tsag block of weakly radioactive Riphean-Lower Paleozoic carbonate-quartzite complexes and deep-sea sediments. It is overlain

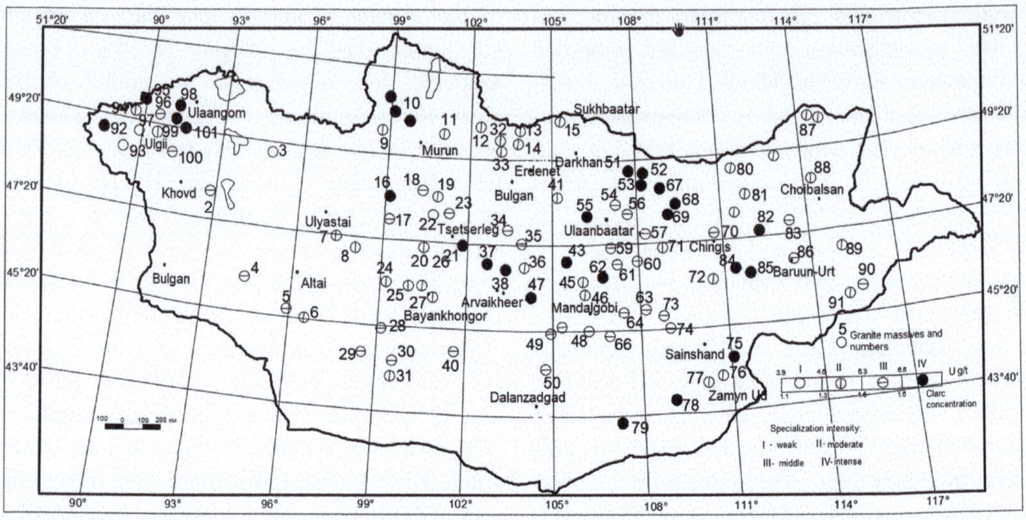

Fig. 3.1 Map of main occurrences of uranium in Mesozoic granitoid massifs (Vysokoostrovskaya et al. 2007). Granite massifs (numbers on the scheme): 1—Bayannuur, 2—Chandman, 3—Underkhangai, 4—Bujiingol, 5—Burun, 6—Gashuungol, 7—Baruunard, 8—Otgon, 9—Delger-Muren, 10—Khubsugul group, 11—Egingol, 12—Erengol, 13—Adragin, 14—Tavtyngol, 15—Bulagtai, 16—Solongin, 17—South Terkhingol, 18—Chuluut, 19—Khailast, 20—Egindavaa, 21—North-Tamir, 22—Undurulaan, 23—Tamir, 24—Ulziitgol, 25—Bayanburd, 26—V. Ulziit, 27—Taatsingol, 28—Adgin-Tsagaannuur, 29—Nukh Khudag, 30—Shine Jinst, 31—Khudag, 32—Tashingol, 33—Khutag, 34—Tsetserleg, 35—East Khangai, 36—South-Bor Khudag, 37—Dulaan-Khairkhan, 38—Bor Khudag, 39—Tugreg, 40—Uldzit, 41—Naryntolgoi, 42—Erdenesant, 43—Ongonkhairkhan, 44—Ikh-Khairkhan, 45—Adaatsag, 46—Bagagazaryn, 47—Ogiingol, 48—Baruunujir, 49—Delgerkhangai, 50—Uizen-Khudag, 51—Buguntai, 52—Mezen, 53—Khukhuchulin, 54—Bayangol, 55—Tukhum, 56—East Zuunbayan, 57—Bayandelger, 58—Janchivlan, 59—Abdarbayan, 60—Bayanoovoo, 61—Tukhim, 62—Bayan-Baraat, 63—Kharaatuul, 64—Khairkhanuul, 65—Kharatsav, 66—Khashaat, 67—Shorootingol, 68—Jargalantiin, 69—Yudegin, 70—Muren, 71—Bayanuulin, 72—Noenuul, 73—Eastern Gobi, 74—Ikh-Nartyinkhiid, 75—Unegetin, 76—Bargyinovoo, 77—Ulaanuulin, 78—Ail, 79—Khanbogd, 80—Balj group, 81—Saikhanuul, 82—Tumentsogt, 83—Bulgan group, 84—Bambar, 85—Baruuntsogt, 86—Salkhit, 87—Agiin group, 88—Engershand, 89—Burkhanovoo, 90—Tsagaanuul, 91—Yugodzyr, 92—Dungurkhiid, 93—Kulagash, 94—Bayan-Zurkh, 95—Turgengol, 96—Boguut, 97—Narynnuur, 98—Achitnuur group, 99—Orgiu, 100—Namarjaa, 101—Kharkhiraa

by moderate to low NRE-bearing sedimentary sequences of Devonian, Carboniferous, and Permian volcanics.

The South Mongolian Superblock, located south of the Main Lineament of Mongolia, includes nearly east–west oriented, cratonic, ophiolite, island-arc, forearc, and back-arc structures. The vast majority of the complexes that make up the structures of the superblock are characterized by low radioactivity. Slightly increased radioactivity is only noted in the Tsagaanuul (39) and Khutag-Uul (42) blocks, where it is associated with Precambrian gneisses, migmatites, pegmatites (U 3.5–5.6 ppm, Th 10–22 ppm, K 5.8–3.5%) and post-accretionary Permian granitoid massifs (U up to 4 ppm, Th 22–25 ppm, K 3.5%). In the west, "cross-linking" granitoid complexes of Silurian and locally Carboniferous monzonites, are highly radioactive. In granitoids, the uranium grade is 3.0–3.6 ppm, thorium up to 20 ppm, and potassium 3.5–4.0%. In separate small massifs, the uranium grade can increase up to 10–12 ppm, thorium up to 30–35 ppm, and potassium up to 5.5%. There are local occurrences enriched in NRE with uranium grades in gneisses up to 12 ppm and thorium up to 50 ppm, and in pegmatites up to 50 ppm uranium and up to 70 ppm thorium.

Early and Middle Paleozoic massifs of granodiorite and gabbro, in the axial part of the South Mongolian superblock, are characterized by background to low levels of radioactivity with uranium, thorium, and potassium approximately 2–3 ppm, 4–8 ppm, and 1–2, less often 3%, respectively. The Permian granite massifs are somewhat more radioactive: the associated areas of the anomalous zones, with an area of up to 50×70 km, are characterized by uranium grades of 2–4 ppm, thorium from 8–12 to 16–30 ppm, and potassium 3–4%. The content of radioactive elements in "crosslinking" granitoid intrusions are often at background levels: i.e., 2–3 ppm uranium, 12–16 ppm thorium, and 2–3% potassium.

The Late Paleozoic primary volcanic rocks of the Tsagaan-Suburga and other regions, containing mineralization of non-ferrous, ferrous, and precious metals, are weakly radioactive: 1–2 ppm uranium, 2–4 ppm thorium, 1–2% potassium. Permian subalkaline volcanics are more radioactive: 4–8.0 ppm uranium, 12–30 ppm thorium, and 2–3 to locally 5% potassium.

Terrigenous deposits of the Ordovician and Silurian, as well as terrigenous-carbonate and molasse complexes of the Devonian, Carboniferous, and Triassic are characterized by background values of uranium 2–3 ppm, thorium 8–12 ppm, and potassium 2–3%.

More contrasting NRE contents are observed in post-accretionary Permian nepheline syenites and alkaline granites of the Khan-Bogd massif in the southeastern part of the Gurvansaikhan island-arc block, where U ranges from 5–6 to locally 8–12 ppm, Th ranges from 16–25 and up to 40 ppm, and K ranges from 4 to 5%.

Thus, the pre-accretionary complexes of various structures of the North and South Mongolian superblocks are characterized by low to near-Clarke NRE contents. Anomalous zones of uranium and other metaloids are noted within cratonic and metamorphic structures, where they are associated mainly with the products of post-granitization hydrothermal-metasomatic alteration. Evidence for uranium mineralization has been established in the collision and post-collision massifs of the Middle and Late Paleozoic. The contrast of the AGS anomalies, observed in some cases, is associated with differences in the NRE content in smaller Mesozoic and Cenozoic formations.

Radiogeochemical Features of Late Mesozoic Intracontinental Volcanic Belts

The Mesozoic and Cenozoic of Mongolia are divided into two large regions. Fold-thrust structures of the Khangai-Gobi-Altai orogen are developed in the north and northwest of the country. In the center and east portions, along the Main Mongolian lineament, collisional rifting occurred accompanied by the formation of volcanic belts with broad development of bimodal (basalt-rhyolite) volcanism and associated hydrothermal uranium mineralization (Fig. 3.2). In the south and southeast, under conditions of a similar process, depression structures of the Gobi Plate developed which are very favorable for the formation of infiltration processes and hold the greatest interest for the exploration of exogenous uranium mineralization.

Acidic and subalkaline volcanics (rhyolite, trachyrhyolite, vitrophyre) of the late Jurassic-early Cretaceous basalt-rhyolitic formation are highly enriched in uranium. The NRE content varies significantly: U 4–10 ppm, Th 16–50 ppm, K 2.5–6%. The highest level of radiogeochemical mineralization is characteristic of felsic volcanic glasses and volcanics of the Kerulen megablock, a fragment of the Ereendavaa block (Tables 3.1, 3.2 and 3.3).

In basic and intermediate volcanic rocks of the basalt-rhyolite formations of eastern Mongolia, the content of uranium, thorium, and potassium are close to Clarke; slightly elevated uranium content is noted in basalts of the North Kerulen and Middle Gobi block volcano-tectonic structures (VTS).

The level of radiogeochemical mineralization of felsic volcanic rocks in different VTS varies widely. The average content of uranium in acidic volcanic rocks in VTS is divided into six types: (1) intensely mineralized in uranium

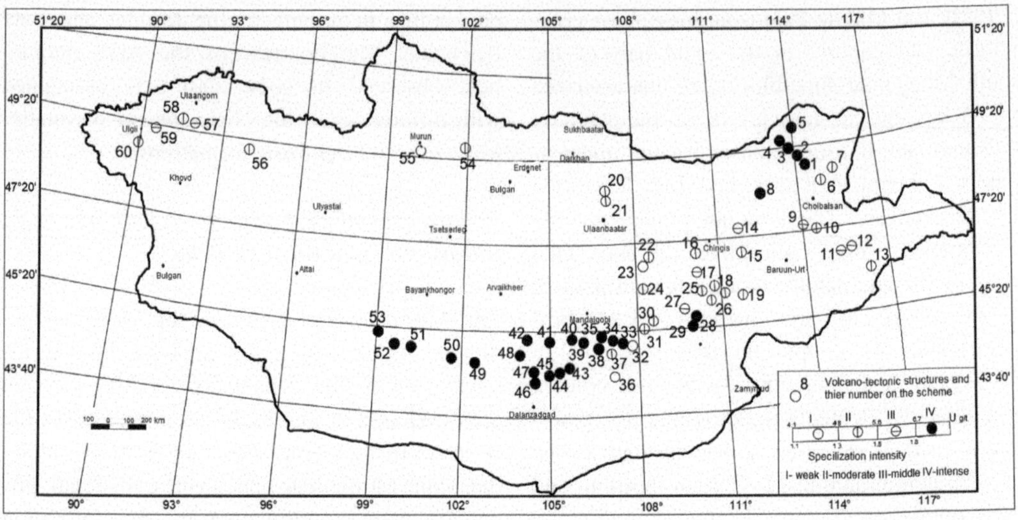

Fig. 3.2 Map of uranium anomalies in Late Mesozoic volcano-tectonic structures (after Vysokoostrovskaya 1985). Volcano-tectonic structures (numbers on the figure): 1—Dornot, 2—Ugtam, 3—Turgen, 4—Upper Turgen, 5—Duchgol, 6—South Engershand, 7—North-Engershand, 8—Ulzii-Saikhanuul, 9—Garshiinuulin, 10—Urtynayrag, 11—Bayan-Ovoo, 12—Znar, 13—Urtyn-Tsagaanuul, 14—Berkh, 15—Bayan-Khurai, 16—Tsenkhirgol, 17—Salhitovoo, 18—Zuun bargin, 19—Tuvshinshiree, 20—Kharin, 21—Zuunmod, 22—Tsagaanovoo, 23—Shivee, 24—Tsagaan-Delger, 25—Ikhk Khet, 26—Buyant, 27—Dalanjargalan, 28—Bayankhongor, 29—Ulaan-Nuur, 30—Bayanjargalan, 31—Egen Khudag, 32—Shuluetin, 33—East Khashaat, 34—Khashaat, 35—Centuul, 36—Khavtagaiuul, 37—Modot ovoo, 38—Zuun-Khar tsav, 39—East Khuldov, 40—Khuldov, 41—Delgerkhangai, 42—Saikhanovoo, 43—Tsogtovoo, 44—Durbendert, 45—Bayan-Khushy, 46—Khetsuu teeg, 47—Mushgai-Khudag, 48—Ulzii bulag, 49—Ulziit-Uul, 50—Khets-Uul, 51—Hever, 52—Bayanbulag, 53—Sangin dalai, 54—Ikh uul, 55—Delgermuren, 56—Khyargisnuur, 57—Bayrim, 58—Achitnuur, 59—Kyzyltau, 60—Khyargis

(> 1.8 Clarke); (2) mineralized (≥ 1.8 Clarke); (3) weakly mineralized (> 1.5 Clarke); (4) moderately anomalous (> 1.3 Clarke); (5) weakly anomalous (> 1.1 Clarke); and (6) background, with reduced uranium content (< 1 Clarke).

In the North Mongolian superblock, the following VTSs are classified as intensely mineralized in uranium (from west to east): Duchgol, Upper Turgen, Turgen, Ugtam, Dorno, Ulzii-Saikhanuul, Shuluet, Khashaat, Khiid-Ovoo, Zuun-Kharatsav, Tsagaanuul, Altash, Garshin-Uul, Berkh, and Salkhit-Ovoo.

In the South Mongolian superblock, intensely mineralized tracts include the Dalanjargalan (U 8–10 ppm, Th 34 ppm), Bayandjargalan (U 8 ppm, Th 30 ppm), Ulaannuur (U 12 ppm, Th 32 ppm), and Bayankhongor (U 6–8 ppm, Th 25–30 ppm). In VTS settings, moderately mineralized occurrences include the Burkhan-Ovoo and Zuun Urt (U 5.6–6.1 ppm, Th 22–33 ppm), the weakly mineralized Tsant and Khud occurrences (4.2–4.6 ppm uranium in rhyolites), and anomalous Tuvshin-Shiree and Bayan-Ovoo (3.4–3.7 ppm). The remaining VTS contains only background or reduced concentrations of uranium.

In the spatial distribution of Late Mesozoic volcanics, lateral zoning of radioactivity is apparent. In the eastern part of Mongolia, most of the felsic volcanic rocks are uranium-bearing and local AGS anomalies are dominated by uranium content (i.e., U dominates over Th and K). In the west, volcanic rocks with uranium–thorium and thorium-only mineralization predominate (*cf.* Zun-Khartsav, Khets-Uul, and other VTS).

In the main volcanic tracts of late Mesozoic and Neogene-Quaternary age, the concentration of radioactive elements is generally low, without higher spikes. On the Dariganga plateau, in the southeastern part of Mongolia, the average contents are in the range of 1.2–1.3 ppm uranium,

Table 3.1 Radiogeochemical characterization of rhyolites of Mongolia (Mironov et al. 2012)

Volcano-tectonic structure	U (ppm)	Th (ppm)	K (%)	Number of definitions
Engershand	5.6±0.6	20.3±1.0	3.0±0.6	33
Ikhbulag	5.7±0.5	18.6±2.4	3.1±0.4	34
Tsodoi-Tsagaan	5.4±0.9	19.7±2.2	2.8±0.6	29
Tsagaan-Delger eastern	5.8±0.8	21.4±1.3	3.4±0.8	29
Tsenkhergol	5.6±0.7	20.3±0.9	3.3±0.3	27
Tsagaan-Delger western	5.4±0.6	17.6±1.1	3.2±0.4	32
Burkhan-Ovoo	5.8±0.7	20.1±0.9	3.2±0.7	30
Yun-Urt	5.7±0.3	18.2±2.3	2.8±0.4	28
Dornod	6.9±0.7	28.0±1.9	3.2±0.5	38
Ugtam	5.9±0.9	22.0±2.1	3.2±0.3	40
Turgen	6.8±0.7	26.5±2.1	2.9±0.5	38
Upper Turgen	6.2±0.6	28.9±1.1	3.0±0.7	32
Ulziu-Saikhanuul	6.7±1.4	29.0±1.7	3.3±0.8	35
Bayankhongor	6.8±0.7	28.0±1.1	3.4±0.4	38
Ulaannuur	7.2±0.7	34.2±0.7	3.8±0.8	36
Salkhit Ovoo	6.5±0.8	32.0±1.0	3.9±0.6	42
Bayanjargalan	7.0±1.1	25.7±1.2	3.2±0.5	34
Elgen-Khudag	6.7±1.1	27.0±1.4	2.8±0.4	33
Shuluetn	7.2±0.9	28.0±2.0	3.0±0.8	41
Khashaat eastern	7.0±0.8	27.6±1.6	3.2±0.5	32
Khiid-Ovoo	6.9±0.7	24.0±1.8	2.9±0.7	29
Tsant-Uul-1	6.6±0.9	22.6±0.9	4.0±0.8	28
Tsant-Uul-2	6.7±0.8	24.4±1.6	3.9±0.6	27
Zun-Khartsav	7.0±1.0	26.3±1.8	3.8±0.5	26
Khashaat western	7.1±0.9	24.2±2.1	3.2±0.6	30

5–6 ppm thorium, and 0.8–1.2% potassium. Relatively elevated uranium grades are recorded only in sodium subalkaline basalts of the southeastern part of the plateau (1.7–3.5 ppm) and in potassium basalts of the Sumiyngol depression (up to 1.6–3.2 ppm) in northern Mongolia.

The most important feature of the late Mesozoic VTS, where uranium-bearing rocks are present, is a large number of anomalies of relatively small size. This pattern is most clearly observed, in the North Kerulen uplift of the Kerulen block of the Ereendavaa Craton, in northeastern Mongolia.

Radiogeochemical Features of Terrigenous Formations of Sedimentary Basins of Late Mesozoic-Cenozoic

At the end of the Mesozoic period in Mongolia, against a background of relative tectonic quiescence, broad fault-bounded depressions formed as a result of post-collision rifting. The interpreted paleoenvironmental conditions were of the transcontinental Sahara-Gobi belt of arid weathering, as identified by a number of researchers (Volsky et al. 2004; Byamba 2012). A characteristic feature of the sedimentary deposits within these basins is the occurrence of mobilization and transportation/redeposition of the NRE under oxidizing conditions. The heterogeneity of the resulting AGS zones is primarily due to the redistribution of uranium, being highly mobile; differences in the content of thorium and potassium, in depression structure sediments, are minimal due to their lower mobility under basin hydrogeochemical conditions.

The early limnic gray-colored (coal-bearing) formation in north and west Mongolia is characterized by background grades of radioactive elements: 1.5–3.2 ppm uranium, 7–8 ppm thorium, and 2.2–3.4% potassium. In the red-colored terrigenous-carbonate formations of the Upper Cretaceous sedimentary basins, relatively low grades of NRE are present: uranium 2.0–3.0 ppm, thorium 4.0–8.0 ppm, and potassium from 2.0 to 3.0%.

Sedimentary deposits of the Early Cretaceous in the north and extreme southeast of the Gobi Plate are also characterized by moderate NRE

Table 3.2 Average grades of radioactive elements in acidic volcanic glasses (Mironov et al. 2012)

Volcano-tectonic structure	Number of definitions	Uranium (ppm)	Torium (ppm)	Samples authors
Dornod	20	17.9	50	Yu. B. Mironov, G. A. Shatkov
Ugtam (base of the section)	27	12.8	47	G. A. Shatkov
Turgen	12	10.0	48	G. A. Shatkov
Upper Turgen	6	8.8	37	N. S. Nightingale
Ulzii-Saikhanuul	5	14.6	50	CM. Mironov
Ikh khet	13	7.8	33	N. S. Solovyov
Shivee	18	12.0	38	I. I. Tsaruk
Ulaannuur (green perlites)	24	17.8	42	E. B. Vysokoostrovskaya
Ulaannuur (black perlite)	11	19.8	46	E. B. Vysokoostrovskaya
Ail	8	12.0	35	E. B. Vysokoostrovskaya
Zunkhar tsav (green perlite)	18	22.4	57	E. B. Vysokoostrovskaya
Zunkhar tsav (black perlite)	8	32.3	96	E. B. Vysokoostrovskaya
Tsagaanuul	6	21.4	70	Yu. S. Shmelev
Khurion Airag	28	18.0	46	E. B. Vysokoostrovskaya
Dalanjargalan	5	7.0	37	N. S. Solovyov, G. A. Shitkon, Yu. A. Grishin
Salkhit	8	6.3	22	L. M. Alekseev
Tuvshin-Shiree	6	6.2	27	L. M. Alekseev
Shine Khudag	1	2.2	13	L. M. Alekseev
Matad	3	6.0	25	N. S. Solovyov
Delgertsogt	5	6.2	31	P. S. Nightingale
Mandal	3	7.0	27	G. A. Shatkov
Tsenhergol	5	7.0	17	Yu. A. Grishin
Akhar zeeg	8	17.4	42	Yu. S. Shmelev

grades. The average uranium grade in the structures of the central part of Mongolia (2.8–3.7 ppm) is somewhat higher than the regional Clarke. In the west, in depressions superimposed on the Altai and Agardag structures, the level of uranium accumulation is somewhat lower, 2.3–2.6 ppm, while the concentrations of thorium and potassium range from 2 to 8 ppm, 1.8–2.5%, respectively.

The lacustrine-alluvial clay-sand formation is geochemically heterogeneous: in the red-colored subformation, the NRE grades are close to Clarke, and the variegated subformation is uniformly enriched in uranium. The grade of NRE in coal-bearing deposits of most Mesozoic depressions varies from uranium 2–3 ppm, thorium 8–12 ppm, and potassium 3–4%. The most radioactive (due to the uranium component) are the coal-bearing deposits of the Choirinsky (U 8–12 ppm), Choibalsan, Sainshand, Ulziit, Undurshil, and Bayanjargalan sedimentary basins, where the average uranium grade in sandstone ranges up to 5.5–6.5 ppm.

Table 3.3 Radiogeochemical characterization of black clays and terrigenous coal-bearing formation of the taphrogenic basins in Mongolia (Mironov et al. 2012)

Depression, area	U (ppm)	Th (ppm)	Number of definitions
Section of the Narst deposit	6.2±0.8	20.2±2.2	84
Ulaannuur	4.4±0.9	19.6±2.1	37
Oroinnuur	4.2±0.6	21.5±1.4	42
Khoshuu-Khid	6.0±0.8	23.8±1.6	40
Buginnuur	4.7±0.4	18.7±1.4	38
Ongiingol	4.3±0.7	19.6±1.8	35
Oldoh khiid	4.8±0.6	21.0±2.0	31

A large number of local AGS anomalies have been identified in the deposits of the coal-bearing Aptian-Albian formation, in the lower and middle parts of the gray-colored platform formation, and in the variegated subformation in Cenomanian and Turonian strata. Local AGS anomalies caused by epigenetic (i.e., secondary, due to remobilization) accumulations of uranium are often associated with deposits of brown coals and sandstones of the Lower Cretaceous concentrated in the Sainshand, Choir, Nemegt, and Khairkhan basins where large sandstone type uranium deposits have been discovered.

A specific feature of the black clays of the Lower Cretaceous Zuunbain suite is their increased thorium grade (up to 20 ppm), which makes it possible to define and correlate sections of various depressions and sedimentary basins of Mongolia (Table 3.4).

Radiogeochemical Zoning of Mongolia

An analysis of the primary distribution and subsequent redistribution of natural radioactive elements in the geological formations of Mongolia made it possible to identify the most important elements of regional radiogeochemical zonality—i.e., anomalous radiogeochemical zones (Fig. 3.3). Having studied the distribution of radioactive elements in Mongolia, Vysokoostrovskaya et al. (2007) identified

Table 3.4 Content of radioactive elements in the rocks of the Dornod VTS (J_3–K_1) (Mironov et al. 2012)

Rocks	U (ppm)	Number of samples	Th (ppm)	Number of definitions
Andesite-basalts, andesites	1.5±0.1	47	7.2±0.5	30
Conglomerates	4.5±0.5	35	18.7±1.0	17
Quartz-sanidine rhyolites, trachyrhyolites	3.5±0.7	49	20.2±2.0	24
Spherulitic felsite	5.9±0.4	36	25.0±0.1	18
Felsite flows	7.7±0.5	37	26.0±1.6	30
Volcanic glass (vitrophyre)	17.9±2.7	34	50.0±4.8	22
Quartz containing felsite	6.9±1.8	33	28.0±3.4	14
Ignimbrite-like rocks	6.6±2.6	35	23.2±2.8	37
Sandstones, siltstones	5.8±0.8	40	24.6±0.9	42
Andesitic basalts	1.8±0.4	50	5.2±0.9	42
Conglomerates, sandstones	2.6±0.7	43	7.0±2.0	36
Trachytes	4.3±0.5	23	17.0±2.9	18
Andesitic basalts	1.3±0.1	45	21.0±2.5	27
Ignimbrites, felsic tuffs	4.7±0.9	45	21.0±2.5	27
Felsite-porphyries (dykes)	7.0±0.9	14	20.0±2.6	20
Dacites, trachydacites	4.4±0.6	50	15.0±0.9	22
Trachyandesites, andesitic dacites	2.5±0.6	22	10.0±1.8	39
Andesitic basalts	1.5±0.3	15	5.4±0.5	19
Conglomerates	4.0±0.7	53	13.2±0.3	51

(continued)

Table 3.4 (continued)

Rocks	U (ppm)	Number of samples	Th (ppm)	Number of definitions
Diorite, granodiorites $PZ_{2\text{-}3}$	1.7 ± 0.5	26	14.7 ± 1.3	19
Granosyenites, syenite porphyries $PZ_{2\text{-}3}$	43 ± 1.0	40	14.3 ± 5.5	21
Monzonites, quartz monzonites $PZ_{2\text{-}3}$	4.6 ± 0.5	26	14.7 ± 1.3	19
Aplites, granite porphyries (dykes) C_2	2.2 ± 1.3	42	12 ± 1.6	31
Diorites, granodiorites PR	1.6 ± 0.3	39	11.0 ± 1.6	36
Gneissic granites PR	1.8 ± 0.3	39	11.0 ± 1.6	36

the following 19 radiogeochemical zones:: Tsagaan-Shiveet (1), Khangai (2), Khubsugul (3), Egingol (4), Butelinnuru (5), Bayankhongor (6), East-Khangai (7), Ikhkhairkhan (8), Nalaikh (9), Iluur (10), Middle Gobi (Oshinnuur) (11), Kharairag (12), Berkh (13), Choibalsan-Onon (14), Madat (15), Mushgia-Khudag (16), Khanbog (17), Ail (18), Nuketdavaa (19), all of which occur in various geodynamic settings.

In 2019, G. Jamsrandorj identified the Uvur Altai and Atas Bogd radiogeochemical zones in southwestern Mongolia.

Anomalous increases in the grade of NRE in the peripheral parts of ancient cratonic and metamorphic blocks are associated with later local occurrences of silicic-potassium metasomatism (i.e., alteration). Formations enriched in uranium (carbon-bearing and phosphate-bearing terrigenous-carbonate units and graptolitic shales) have been identified in sections of weakly radioactive Ediacaran and Paleozoic complexes.

Uranium and thorium grades close to Clarke or slightly elevated are characteristic of the rocks of Paleozoic subduction-accretion volcanoplutonic belts. Uranium and thorium anomalies have been mapped in a number of massifs of subalkaline and alkaline granitoids in the northern and southern Mongolian provinces.

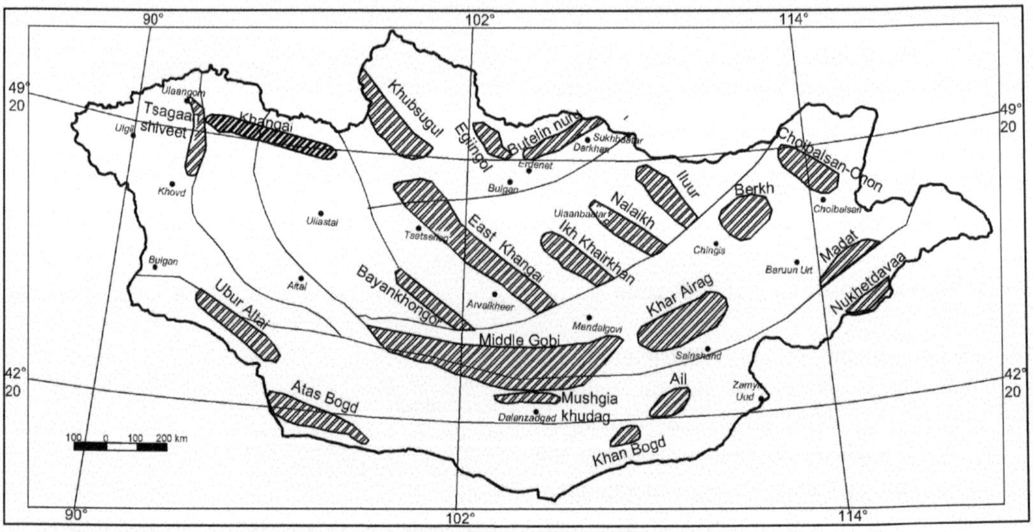

Fig. 3.3 Generalized map of radiogeochemical zones of Mongolia (after Vysokoostrovskaya et al. 2007; Jamsrandorj 2019)

Highly anomalous NRE and uranium-dominated formations and complexes are associated with Mesozoic back-arc and supraplume riftogenic (plume head doing) complexes, as well as with Late Mesozoic-Cenozoic taphrogenic (block faulting) structures. The enrichment of uranium in terrigenous deposits of the Late Mesozoic-Cenozoic sedimentary basins, superimposed on tectonic structures of various types, is due to its supply and redistribution, related to the processes of diagenesis and epigenesis of primary gray-colored (reduced) sediments. These processes are most clearly manifested in the riftogenic depression (back-arc graben) structures of southeastern Mongolia.

The processes that caused the formation of economic concentrations of uranium in the rift structures of Mongolia manifested themselves within a background of formations already anomalous in uranium. In carbonaceous shales, uranium mineralization was deposited during hydrothermal processes in areas of acidic and subalkaline magmatism. In the Late Mesozoic volcanic belts, economic uranium mineralization is associated with hydrothermal processes, manifested both in the VTS and in their basement. Ore-bearing VTS are characterized by the highest level of uranium mineralization in acidic volcanics, particularly in the volcanic glasses (vitrophyre). In late Mesozoic-Cenozoic sedimentary basins, large-scale hydrogenous (roll-front) uranium deposits are associated with redox processes in terrigenous sequences. There is reason to believe that the source of uranium in ore-bearing sedimentary basins is uranium-enriched granitoids of the substrate and rocks of the basin edges. In Mesozoic massifs of leucogranites, promising polygenic uranium mineralization is deposited during the hypergene transformation of highly radioactive granites in fault zones under the influence of descending meteoric waters.

Anomalous radiogeochemical zones are considered areas of the earth's crust, within which, the process of introduction and redistribution of radioactive elements have operated with maximum intensity. At the regional level, they act as favorable structures for the deposition of uranium mineralization. Calculations for Eastern Mongolia and Transbaikalia showed that 91% of uranium deposits and 84% of uranium occurrences are concentrated in such zones (Mironov 2006a, b).

The primary deposition of uranium mineralization leading to deposits of medium and large scale is the level of accumulation of uranium and the degree of differentiation and concentration of U–Th–K constituents in geological host formations. Understanding this basic principle necessitates the expediency of identifying anomalous radiogeochemical zones with similar formations within large prospective geological blocks. The anomalous radiogeochemical zones correspond to areas with a high (> 50%) coefficient of variation in the grades of individual elements and are characterized by dramatic local increases (and attendant decreases where elements are leached away) of U–Th–K. This is a result of hydrothermal metasomatism and hydrogenous mobilization/redeposition processes. A specific feature of anomalous radiogeochemical zones is an increased density of local AGS anomalies.

Regional features of the spatial distribution of radioactive elements and the position of anomalous radiogeochemical zones are largely determined by the influence of long-lived faults that delimit the main geotectonic structures of the region. Such faults (Main Mongolian, Mongol-Okhotsk, Choibalsan-Onon, Kharairag, etc.) not only control the location of Mesozoic volcano-tectonic structures and superimposed depressions, but often serve as natural boundaries for elements of radiogeochemical zoning. In a number of cases, zones of radioactive element inflow are confined to long-lived faults. The main parameters of the anomalous radiogeochemical zones are given in Table 3.5.

Within the Eastern Sayan-Mongol-Altai folded region, the Tsagaan Shiveet (1), Khangai (2), Khubsugul (3), Egingol (4), and Butelinnuru (5) radiogeochemical zones are distinguished. Tsagaan Shiveet and Egingol zones are localized within the Lake-Egingol folded area in the continental crust which was formed during the Early Ordovician. The remaining zones are

Table 3.5 Characteristics of anomalous radiogeochemical zones in Mongolia (based on airborne gamma spectrometric (AGS) surveys) (Mironov et al. 2012)

Radiogeochemical zones	Number of AGS anomalies			Status of uranium occurrence
	U	Th	K	
Tsagaan-Shibeet	32	25	12	Ore occurrences of uranium–thorium mineralization Buuralyn, Khagiin
Khangai	41	23	9	Uranium occurrences Khashaa-tyn-Khar, Ulaan-Tolgoi, etc.
Khubsugul	48	7	12	Ore occurrence Aragol, Ujiggol, Balbrig and other uranium–thorium-rare-earth formations, Erkhelnuur uranium occurrence
Egingol	22	14	31	Egingol uranium ore occurrence
Butelinnuru	13	17	7	Butulinnuru group of uranium and thorium ore occurrences
Bayankhongor	11	11	12	Anomalous uranium in quartz-gold formations
East Khangai	16	8	1	Uranium anomalies in salt halos
Ikhhairkhan	18	12	–	Uranium fault gouge/mylonite zones of leucocratic granites
Nalaikh	11	–	1	Uranium fault gouge/mylonite zones of leucocratic granites
Iluur	14	2	1	Uranium fault gouge/mylonite zones of leucocratic granites
Middle Gobi	108	31	6	Ore occurrences of Uljiit-Uul and other fluorine–molybdenum–uranium formations related to late Mesozoic VTS, Oshiynuur group of uranium occurrences in terrigenous deposits of Meso-Cenozoic depressions
Kharairag	61	29	4	Ore occurrences of Boroundur, Khongor, Ulaan-Nuur, Bujiger, Kharaat and more than 10 occurrences of supergene uranium in Choir and Ulaannuur depressions
Berkh	38	3	5	The uranium ore region of the same name. Ore occurrences of Tumen-Khan, Mizernoe, Ikh-Bulag, and Tanai of the fluorine–molybdenum–uranium formation related to Late Mesozoic VTS
Choibalsan-Onon	44	4	3	North-Choibalsan uranium ore district, including deposits of Dornot and other volcanics related to late Mesozoic VTS
Matad	18	–	–	Ore occurrences of Burkhan-Ovoo and other fluorine–molybdenum–uranium formations related to late Mesozoic VTS

(continued)

Table 3.5 (continued)

Radiogeochemical zones	Number of AGS anomalies			Status of uranium occurrence
	U	Th	K	
Mushgai-Khudag	16	12	2	Ore occurrences of Bayan-Khushuu, Khetsuu-Teeg and other formations of uranium-bearing (with Th, TK, and rare metals) metasomatites related to alkaline and subalkaline rocks; Teeg-Uul ore occurrence of fluorine–molybdeum–uranium formation related to late Mesozoic VTS
Khanbog	2	3	–	Ore occurrence of Khanbogd rare-metal-rare-earth system
Ail	7	3	2	Ail group of ore occurrences of the fluorine–molybdenum–uranium type related to late Mesozoic VTS
Khanbogd	2	–	–	Uranium occurrences Saikhanuul, Tsagaanuul in the ore field of the Yugodzyr tungsten deposit
Ubur Altai	12	6	–	Exploration was not completed
Atas Bogd	9	7	3	Exploration was not completed

confined to the Tuva-Mongolian folded area, in the older continental crust, formed at the end of the Ediacaran.

The Tsagaan Shibeeti zone is controlled by the submeridional (nearly north–south trending) Tsagaan Shibeeti deep fault system on the continental margin defining a structural zone more than 250 km long and up to 50 km wide. The zone has a variable radiogeochemical signature, but with a predominance of thorium. A significant role in the formation of the geochemical background is played by dikes and small intrusive bodies. Devonian syenites, syenite-porphyries, alkaline granites, felsic volcanics, and importantly, the alteration products of these rocks—riebeckite-microcline-albite metasomatites and albitites—are all enriched in radioactive elements.

According to the level of the NRE background, the density of local AGS anomalies of thorium-only and U–Th–K signal, and the nature of interpreted superimposed processes, Tsagaan Shibeet zone is considered to be a promising target for complex rare-metal-rare-earth (typically associated with thorium and uranium) mineralization. Uranium mineralization in the zones of crushing of leucocratic granitoids is most clearly manifested in the Dungurkhi massif; whereas uranium mineralization due to volcanism occurs within the Khyargisnuur and Achitnuur VTS.

The Khangai Zone is controlled by the Khangai deep fault. It extends in the sublatitudinal (east–west trending) direction for more than 300 km and is 20–25 km wide. The zone is distinguished by numerous and various AGS anomalies associated with metasomatites containing uranium–thorium mineralization. The redistribution of uranium and potassium is noted in biotite-amphibole, K-feldspar-sericite, and epidote-chlorite mylonites developed in Paleozoic volcanics.

The Khubsugul Zone (60–70 km wide, over 150 km long) in northern Mongolia corresponds to the southern part of the Khug accretionary-collisional prism and the southwestern section of the Ediacaran-Cambrian Khuvsgul phosphorite basin. The sources of radiogeochemical elements are Paleozoic alkaline intrusions that were emplaced during the destruction of the island arc. In the Ediacaran-Early Cambrian shelf basin, carbonaceous-graphitic carbonate units, with faulted, silicified, and sulfide-bearing shales, host the most intense anomalies. Radioactive mineralization (Erkhelnur, Zaanar, Nachalnoye, etc.) is mainly located in two settings: (1) carbonaceous-siliceous shales and (2) uranium-bearing (with thorium, rare metals, and rare earths) metasomatites (alteration zones) related to alkaline and subalkaline intrusions.

Uranium-bearing layers of siliceous- and clayey-carbonaceous, locally graphite-bearing shales can be traced for hundreds of meters and up to 2 km, their thickness is a few to tens of meters. Mineralization is represented by fine-grained pitchblende, lead, copper, zinc, and molybdenum sulfides; however, the uranium content is low (< 0.001%).

Uranium-bearing (with thorium, rare metals, and rare earths) metasomatites (alteration zones) related to Paleozoic alkali granite-syenite massifs of the Orkhon complex spatially correlate with siliceous-carbonaceous shale horizons. Apparently, under the influence of alkaline, essentially sodium metasomatism, mineralization was mobilized from black shale strata, redistributed, and concentrated in the most intense metasomatized (sodic-altered) zones. All known ore occurrences (Ugjiingol, Argol, Alag-Erdene, Yarisgol, Egiingol, Ustgol, etc.) are located in exocontact (outlying, wall rock) zones of intrusive massifs or in large xenoliths and roof pendants.

Nasturan-sulfide mineralization is host in metasomatites (alteration halo/zone) or along weakly developed fault zones in the host metasediments near intrusive massifs. Uranium mineralization is contained within zones of the most intense alteration. Mineralization is determined by AGS surveys as intense thorium, uranium–thorium, rarely pure uranium anomalies, creating a mosaic radiochemical field. A specific feature of local anomalies is a noted low potassium content, probably a result of K depletion in the alteration process.

The Egiingol Zone extends along the major fault separating the Egiingol island arc from the Ariggol sea margin. The large zone is 40–60 km wide and over 150 km long and has a mixed uranium–thorium–potassium signature. Local AGS anomalies of thorium and potassium are associated with post-granitization alkaline metasomatism (alteration), locally, also associated with gold mineralization.

The Butelinnuru Zone spatially coincides with the eastern section of the Buteel ancient metamorphic ridge ledge. The zone includes massifs of Proterozoic uranium-bearing granitoids (uranium up to 8–12 ppm, thorium up to 35–40 ppm, and potassium up to 7.5%), a result of late silicic-alkaline metasomatism, that consists of various xenoliths of granitized gneisses and numerous bodies of pegmatites and quartz-albite metasomatite (replacement) veins.

Local anomalies of uranium and uranium–thorium tend to favor long-lived faults of northwest and northeast strike along the margins of the ridge.

The Bayankhongor (6), East Khangai (7), Ikhhairkhan (8), Nalaikh (9), Iluur (10), Berkh (13) and Choibalsan-Onon (14) radiogeochemical zones occur within the Central Mongolian fold area.

The ***Bayankhongor Radiogeochemical Zone*** is developed along the Proterozoic-Paleozoic paleo-margin of Zag and Jirem, as well as in the Bayankhongor volcanic arc, over a distance of nearly 200 km in a narrow strip trending southeasterly. AGS anomalies spatially coincide with fields of gold-quartz mineralization and associated alteration. Within the zone, 11 uranium, 1 thorium, and 7 potassium anomalies are present.

The East Khangai Zone, stretching in a northwest direction for more than 400 km and approximately 20–25 km wide, strikes from northwest to southeast between the East and West Khentei regional faults. Starting from the middle of the Tarbagatai ancient ledge and extending through the Northern Khangai Lower-Middle Paleozoic continental arc, the Undurulaan Proterozoic-Early Paleozoic continental structure, the Arkhangai-Bayangol Late Proterozoic continental arc, the South-Tamir

Late Paleozoic volcanic arc, and the Khangai Phanerozoic accretionary structure generally trending across their strike.

The radiogeochemical zone was formed by metasomatic transformations (alteration) related to orogenic granitoids in these structures. The zone includes the Chuluut, Undurulaan, Tamir, Dulaan-Khairkhan, Boro-Khudag, and Ogiingol granite massifs with average uranium content (ppm) of 3.7 ± 1.2, respectively, 3.6 ± 0.8; 3.8 ± 1.1; 5.3 ± 0.6; 4.8 ± 0.7; 4.2 ± 2.1. The content of thorium ranges from 18 to 49 ppm, potassium 3–4%. Local AGS uranium and uranium–thorium anomalies are grouped into linear zones of northwest strike. Uranium mineralization is hosted in sheared zones in the Boro-Khudag, Dulaan-Khairkhan, and Solonga massifs.

In the northern part of the zone, in the Sumiyngol Cenozoic basin, superimposed on highly radioactive granites, ground-based verification of local AGS anomalies under the cover of basalts revealed the Dald deposit. Dald is localized in unconsolidated deposits of the Cenozoic paleovalleys and forms a part of the 450 km^2 Chuluut occurrence district.

Ikhkhairkhan, Nalaikh, and Iluur zones occur within large transverse (northwestern) fault zones of different ages that were intruded by uranium-bearing Mesozoic leucocratic granitoids. The intrusive massifs have anomalous concentrations of uranium (8–12 ppm), thorium (35–40 ppm), and potassium (3.5–4%). The enclosing Paleozoic wall rocks are weakly radioactive: 1.5–2.5 ppm uranium, 5–10 ppm thorium, 1.5–2% potassium. These zones are considered promising for the discovery of uranium mineralization in fault (mylonite) zones where subsequent granite-related alteration by meteoric waters may have concentrated the overall NRE content.

The Berkh Zone in the west of the South Khentei forearc trough extends northeastward for more than 200 km and is tens of kilometers wide. High AGS fields are associated here with the structures of the Mongol-Priargunsky volcanic belt with Late Mesozoic VTS. In particular, the rhyolites of the Ilzii-Saikhanuul VTS (5–8 ppm) values are very similar to the Ikhbulag dome (4–6 ppm) and the Nimgeni VTS (4–6 ppm). Numerous local AGS anomalies are composed of uranium-only and uranium–thorium. The zone corresponds to the uranium exploration target of the same name. The heterogeneity of the AGS anomalies is most pronounced in the eastern part of the zone, where elevated uranium contents are associated with intensely fault-sheared, argillized rhyolites. Within the zone, fluorite–molybdenite–uraninite ore occurrences are present at Tumenkhaan, Mizernoe, Ikh Bulag, and Tanai and a number of sandstone type uranium occurrences were identified in the Tanai and Ulziit depressions.

The Choibalsan Onon Zone is characterized by very high uranium content in the Late Mesozoic VTS including the Upper Turgen, Turgen, Ugtam, Dornod, and Engershand occurrences. The level of radiogeochemical mineralization in felsic volcanics is also high: up to 9–12 ppm of uranium and 40–45 ppm of thorium in rhyolites, up to 32 ppm of uranium, and up to 96 ppm of thorium in volcanic glasses (vitrophyre). The highest NRE mineralization is noted in the Dornod and Ugtam VTS. The increased density of local AGS uranium anomalies indicates a wide areal manifestation from the process of redistribution and deposition of radioactive elements in these preferred host rocks. Precambrian and Lower Paleozoic metamorphic rocks and metasediments of the VTS basement contain background or even depleted levels of radioactivity (1–2 ppm uranium). In the southern part of the zone, in the Tukhemiin, Sumiynuur, and Shinebulag Cretaceous depressions, broad and local AGS anomalies are associated with the concentration of uranium in lithified sandstones. The zone corresponds to the North Choibalsan uranium district, which includes commercial uranium deposits of the Dornod district.

Six anomalous radiogeochemical zones are distinguished within the South Mongolian fold region: Middle Gobi-Oshinuur (11), Kharairag (12), Madat (15), Mushgaikhudag

(16), Khanbogd (17), Ail (9), Nukhetdavaa (19). These zones extend along Late Paleozoic–Mesozoic Middle Gobi-Kerulen and Bayannuur-Gurvanzagal active continental margin suture zones and the North Choibalsan continental arc and its structurally dismembered fragments. Alkaline-carbonate intrusions are dominant.

The Middle Gobi Zone is located near the Main Lineament of Mongolia, along the border of the North and South Mongolian superblocks. The zone is elongated in the sublatitudinal (east–west) direction for more than 500 km with a width of 40–60 km. It covers the western part of the Late Mesozoic Mongolian-Priargunsky volcanic belt and a number of Late Mesozoic-Cenozoic basins superimposed on the structures of the forearc Gobialtai and passive-continental-marginal Idermeg terrane. The main uranium (up to 16 ppm) and thorium (up to 80 ppm) anomalies correlate strongly to outcrops of subalkaline volcanic rocks of the late Mesozoic basalt-rhyolite association. Uranium, thorium, and potassium mineralization, in the rocks of acidic composition (felsic rocks), are consistently high: in the Uldzitula VTS, respectively, 6–8 ppm, 25–40 ppm, 3–4%; in trachyrhyolites 6–12 ppm, 90–120 ppm, 4–6%; and in Tsagaanuul VTS 4–8 ppm, 45–50 ppm, 3–4%. It is worth noting the high content of thorium, sometimes reaching 8–10 times Clarke value.

In the Oshinnuur, Ongingol, Taatsingol, Ulaan-Toyrim, Tsagaannuur, Usug-Khudag, Khalzan-Uul, Undurshil, and other Cretaceous depressions where siltstones and calcareous sandstones dominate, brown coals and lignitized (organic carbon-rich) clays enriched in uranium are locally developed. Numerous local AGS uranium and nature-Th-K anomalies identify areas of epigenetically altered and enriched sediments. The greatest differentiation in NRE content is observed within the Oshinnuur and Khashukhild depressions, where uranium varies from 4 to 17 ppm, thorium from 8 to 60 ppm, and potassium from 1.5 to 6%. In the western part of the Middle Gobi zone, 16 uranium occurrences and 25 AGS anomalies of three types have been identified:

- endogenous (primary) uranium occurrences and anomalies in volcano-tectonic structures composed of subalkaline basalt-rhyolite formations of Late Jurassic-Early Cretaceous age (cf. Ulziit-Uul and others);
- uranium occurrences in Lower Cretaceous terrigenous deposits related to high uranium content in phosphates (Khuren-Tsav, Shiir, Zoolon, Bogd, etc.);
- sandstone type uranium occurrences and anomalies in Cretaceous terrigenous sediments. These uranium occurrences are related to surface, ground, and reservoir oxidation including occurrences at: Oshinnuur, Ikh-Khar, Dombiin, Khairkhan, Shavariin, Untsger, Talyn-Us, etc., and in the rocks of the Zuunbayan (K_2), Sainshand, and Bainshiree (K_2) formations along the contact of gray-colored and red-colored sediments (i.e., at the redox boundary).

The Kharairag Zone is located along the east-northeast continuation of the Middle Gobi-Oshiinnur radiogeochemical zone and is largely controlled by regional-scale and local geologic structures. It is characterized by reduced radioactivity of Precambrian carbonate deposits (uranium 0.5–1 ppm, thorium 1.5–3 ppm, potassium 0.5–1%) adjacent to uranium-mineralized horizons in graphitic shales, with average uranium content from 8 to 12 and reaching up to 120 ppm.

Within this zone, dome-shaped uplifts are recognized which mimic radiogeochemical halos such as Khalzan, Ulaan-Nuur, and Ulziit. On the periphery of the dome, there are anomalies of uranium-only and less often uranium–thorium, confined to pegmatites and metasomatites (alteration zones) of micaceous-feldspar, K-feldspar-epidote, and pyroxene-plagioclase.

Less radioactive rocks are developed in the center of the domes. Numerous dikes of syenite porphyry and quartz porphyry with uranium grades of 30–50 ppm and thorium grades of 40–80 ppm (up to 240–300 ppm) form a northeast trending belt. Massifs of Mesozoic granites are enriched in uranium and thorium: Kharaatuul–uranium 5–6 ppm, thorium 30–36 ppm; Ikhnartiin khiid–uranium 5–8 ppm, thorium 22–28 ppm.

The highest grades of uranium and thorium were found in the felsic volcanic rocks of the late Mesozoic VTS: Dalanjargalan, 8–10 and 34 ppm; in Bayanjargalan up to 8 and 30 ppm; in Ulaannuur 12 and 32 ppm; in Bayankhongor 6–8 and 25–30 ppm, respectively.

In the Choir and Ulaannuur depressions, highly uranium-enriched gray-colored (reduced) sediments are widely developed and correlate to numerous AGS uranium anomalies, confined mainly to permeable sediments and brown coals with zones of surface oxidation.

Within the zone, high-intensity uranium anomalies were identified (Choir, Idermeg, Kharaatuul, Boroundur, Tsagaanuul, etc.). Also included are the Kharaat hydrogenous (roll front type) uranium deposit and other uranium occurrences (Ulaannuur, Khongor, Bujigar, Khar-Tolgoi, Borundur, etc.) found in permeable sediments, along with more than a dozen occurrences of uranium in terrigenous coal-bearing strata.

The Mushugaikhudag Zone in the Late Mesozoic rift structure (Mushugai graben) can be traced in the sublatitudinal (north–south) direction for more than 100 km with a width of 25–30 km. It includes a number of small Late Mesozoic volcanic-tectonic structures (Mushgai-Khudag, Tag-Uul, Bayan Khoshuu, Noen-Ovoo) composed of sub-alkaline volcanic rocks, trachyrhyolites, trachytes, trachyandesites, and small exposures of syenites. The grade of uranium locally reaches 8–16 ppm, thorium 40–90 ppm, and potassium from 4 to 8%. Numerous local AGS anomalies of a mixed nature and thorium-only have been identified, and field sampling discovered rare-earth (with uranium) mineral occurrences of Teeg-Uul and Bayan-Khushuu and areas of fluorite mineralization.

The Teeg-Uul uranium occurrence, at the Hetsuu-Teeg site, is hosted in late Mesozoic volcanic rocks of flow dome and vent facies and it is controlled by a broad northeast trending zone of fluorite mineralization. Uranium mineralization (in the form of tyuyamunite, a member of the carnotite group) is localized in structural shear zones within carbonatized and fluoritized rhyolites, their autobreccias, trachyrhyolites, and weathering crusts (oxidized soil horizons). Uranium grades range from 0.01 to 0.042%. Elevated concentrations of Be, Li, Sn, Pd, As, Ag are present throughout the NRE-bearing zones.

The Bayan-Khushuu ore occurrence is confined to a small (6×3 km) VTS composed of sedimentary-volcanogenic formations. At the site of the ore occurrence, two parallel sublatitudinal (east–west trending) zones of crushing, intense fracturing and argillization were identified and traced for 1.3–1.5 km and spanning a width of 200–300 m. Numerous radiometric anomalies and uranium halos were found within the zones, generally confined to trachyte, trachyte porphyry, tuff of trachyrhyolite, siltstone, as well as veins and veinlets of hematite, magnetite, and barite-celestine composition. Exploratory trenches uncovered mineralized zones from 0.4 to 6.8 m thick with uranium contents from 0.019 to 0.040%.

The Khanbogd Zone in the central part of the Tsokhiot continental rift is structurally related to the emplacement of the Khanbogd massif, a post-accretional granite to alkaline-granite intrusive complex. In the western part of the zone, near the contact of the massif, local AGS anomalies with uranium contents up to 15 ppm and thorium up to 70 ppm form a well-defined ring structure. Elevated NRE contents and local AGS anomalies are generally confined to altered alkaline granites cut by pegmatite and quartz-carbonate veins with rare-metal and rare-earth mineralization. As a broad generalization, there is a certain asymmetry to the radiogeochemical halo whereby the southwestern sector is dominated by uranium, while the eastern sector is thorium-dominant.

The Ail Zone is located at the terminus of the Tsokhiot continental arc and is controlled by the Tavan-Khar-Ovoo fault. Based on distribution of local uranium anomalies, the radiogeochemical zone can be traced for more than

150 km with a width of 15–20 km. The enrichment of rocks with uranium is associated with the occurrences of hydrothermal-metasomatic (alteration) processes. Crystalline schist, gneiss, granite-gneiss contain numerous quartz-feldspar veins and pegmatite bodies with a uranium content up to 30 ppm and thorium content up to 50 ppm. Subalkaline granite and granosyenite of the Devonian and Permian complexes consist of dikes and small bodies with uranium, thorium, and potassium contents averaging 8–13 ppm, 30–45 ppm, and 3–4%, respectively. Local anomalies with uranium, thorium, and potassium contents up to 30–80 ppm, 400–600 ppm, and 6–8%, respectively, are confined to quartz-fluorite veins with rare-earth mineralization.

Uranium enrichment is most intense in the felsic volcanic rocks of the Ail VTS. In flows and flow-dome complexes of sanidine-euhedral quartz rhyolite, the content of uranium is up to 15 ppm, thorium 12–18 ppm, and potassium 4–5%. The carbonized sediments of Cretaceous age in the southeastern part of the zone are also enriched in uranium up to 30 ppm in lignitized (organic, carbon-rich siltstones). A specific feature of the zone is the wide distribution of uranium-bearing chalcedony (from 30 to 80 and up to 150 ppm). Chalcedony develops along a network of small subvertical fissures in weakly lithified rocks. In some places, during the chemical alteration of host rocks, uranium-bearing chalcedony forms thin surface rinds (up to 1 cm thick) on highly radioactive hard pan soils (tiles), which from air-borne surveys are marked by intense local AGS anomalies of a pure uranium.

The redistribution and concentration of uranium are most intense within a 400 km^2 block in the central part of the zone. Ground-based verification of local AGS anomalies leads to the discovery of the Khadiin, Ail, and Elgyn ore occurrences and many small uranium prospects in sandstone.

The Nukutdavaa Zone along the paleo-continental margin of the same name includes two highly radioactive Late Paleozoic tungsten-bearing leucogranites massifs-Yugodzyr and Tsagaanuul. The massifs are characterized by intense, complexly differentiated NRE deposits and high uranium (4–6 ppm), thorium (25–40 ppm), and potassium (3–5%). AGS anomalies, predominantly uranium-only, correlate with areas of greisenization and lower-temperature alteration. The terrigenous-volcanogenic rocks of the Nukut-Daban terrane are generally characterized by low, near-Clarke, NRE contents.

The Madat Radiogeochemical Zone stands out within the southern part of the Sukhbaatar continental shelf terrane. Mesozoic VTS zones are characterized by gamma spectrometric anomalies, and 28 significant anomalies have been identified, including the Burkhan-Ovoo uranium ore occurrence.

The Uvur Altai Zone covers the southwestern extremity of the Khovda continental margin and merges into the Gobialtai uplift, where Ordovician and Devonian granitoids are developed. Detailed exploration has not been conducted for uranium in the zone.

The Atas Bogd Zone in the active continental margin terrane of the same name includes areas of highly radioactive massifs of Early Carboniferous granitoids. But again, no detailed exploration has been conducted. These areas require exploration and study in order to understand their significance.

This section describes radiogeochemical features and shows the position of uranium ore regions and clusters in radiogeochemical zones of various types, which may be of practical importance in forecasting and exploring for uranium in the future.

Conclusions

1. The natural distribution of radioactive elements in Mongolia is close to background levels or slightly increased and can be attributed to variations in natural sources of

radioactive elements within the generally complex and rich geologic framework.

2. Rocks with close or slightly elevated levels of radioactive elements are present in all geodynamic settings, but uranium deposits and occurrences tend to favor structures of the active continental margin and their back-arc destruction (accretionary) zone. In the folded area of the northern superblock of Mongolia, Ediacaran-Cambrian carbonaceous shales and carbonate-sedimentary rocks have low natural radioactivity. The content of radioactive elements in blocks of metamorphic rocks is slightly increased due to later silicic-potassium metasomatism (alteration). Uranium and thorium levels tend to increase in volcano-tectonic structures of Devonian and Carboniferous-Permian systems. Certain massifs of alkaline and subalkaline intrusions are characterized by relatively high contents of uranium and thorium.
3. Rocks with highly anomalous uranium concentrations are often associated with continental (orogenic) rocks formed in the Mesozoic and later periods, mainly in the folded structures of Central and Southern Mongolia. These areas are poorly understood, and therefore, represent excellent targets for further research and, if warranted, exploration for new uranium resources.

3.2 Regional Distribution of Uranium Deposits

The resource estimation and metallogenic study of uranium mineralization in Mongolia is primarily the work of Soviet organizations such as VSEGEI, Sosnovgeologiya, and the Mongolian Geological Survey Expedition, in conjunction with the latest data collected by foreign geological companies from France, Canada, etc. The first information about uranium mineralization in Mongolia was presented in 1972 by Russian scientists Yu. M. Vladimirsky, G. A. Shatkov, and Yu. M. Shuvalov. Data on the patterns and distribution of uranium mineralization in Mongolia were first covered by Mironov (2003) in the monograph "Uranium of Mongolia" (Mironov 2006a, b), considered by many to be the "bible" on Mongolian uranium science and resources.

A study of the metallogenic zoning of Mongolia, for uranium, was carried out by VSEGEI research staff in 1985, and later by Mironov (1993) and Jamsrandorj et al. (2016). Based on their data, Mongolia was divided into four uranium ore provinces: Mongolian-Argun; Gobi; Khentei-Daur; and North Mongolian. These ore provinces were further supported, in 2009, taking into account the new results of terrane analysis (Fig. 3.4). Also based on the terrane analysis, the Mongolian-Argun uranium ore province, was subdivided into four distinct districts: the North Choibalsan, Berkha, East Gobi, and Middle Gobi.

In the Gobi ore province, the Sainshand and Tamtsag regions represent key uranium mineralization. The Khentei-Daur ore province is characterized by the presence of regions, districts, and nodes: notably represented by the Khentei and Khangai regions, and the Central and Chuluut districts. The North Mongolian ore province is large and consists of the Mongolian-Altai, Tsagaanshibeet, Lake, Tuva-Mongolian, Aragol, Butelinnuru, and Khubsugul zones. The Bayankhongor zone of uranium potential stands out as an independent region outside the main, better-known provinces.

The most studied uranium districts are the North Choibalsan, Berkh, East Gobi, and Middle Gobi. They are characterized by mineralization associated with volcano-tectonic structures. In addition, the Sainshand and Tamsag regions contain significant uranium mineralization hosted in sandstone, of typical roll front deposition. Prospective areas of uranium mineralization include Khangai, Khentii, Mongol-Altai, Butelinuru, Argol, and Bayankhongor—these areas represent targets for future exploration and possible resource identification.

The uranium districts of Chuluut in Khangai and Central in Khentei were identified as uranium-bearing systems associated with plume-related radiogenic granitoids. In the eastern section, of the Central Metallogenic Province,

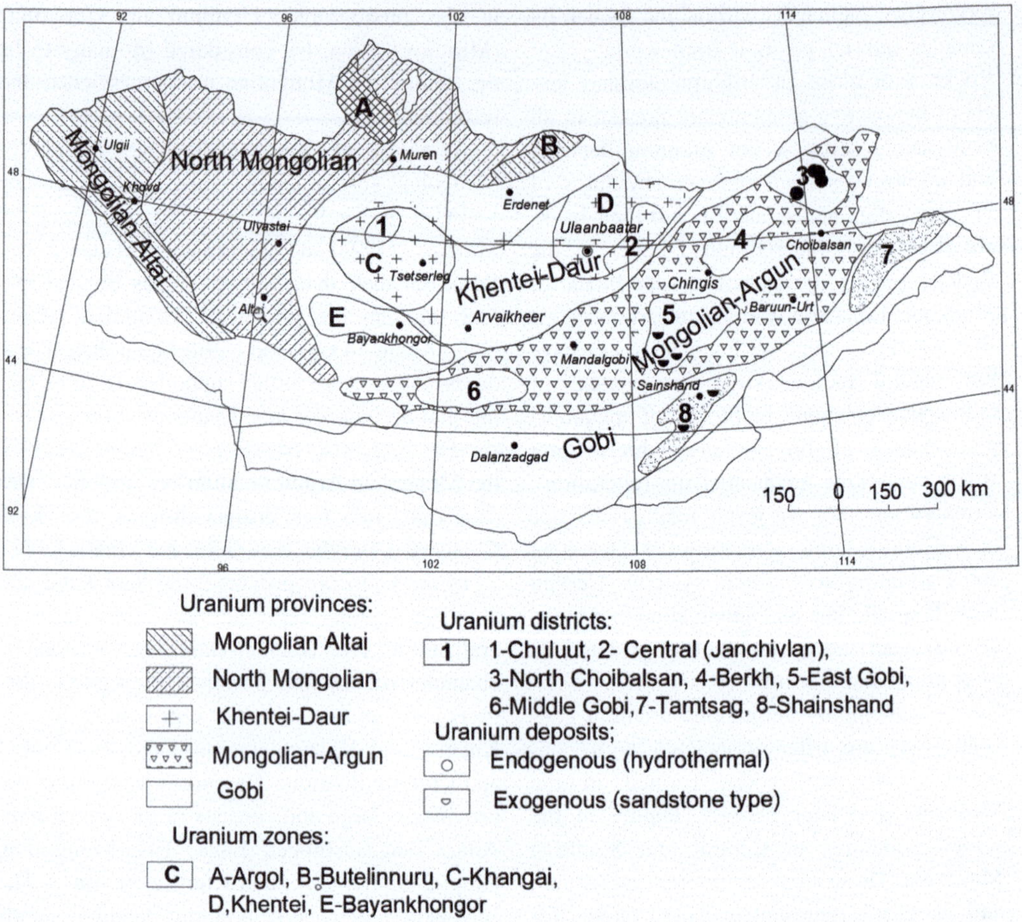

Fig. 3.4 Generalized map of uranium zones of Mongolia (after Mironov 2006a, b). 1–5—provinces: 1—Mongolian-Argun uranium ore, 2—Gobi uranium ore, 3—region of overlapping uranium-bearing structures of the Gobi and Mongolo-Argun provinces, 4—Khentei-Daur uranium ore, 5—North Mongolian potential uranium ore; 6—regions: 1—North Choibalsan, 2—Berkh, 3—East Gobi, 4—Middle Gobi; 7—districts: 1—Sainshand, 2—Tamtsag; 8—regions: Xr—Khangai, Xm—Khentei; 9—districts: 1—Central, 2—Chuluut; 10—zones, regions: MA—Mongolian-Altai, A—Argol, B—Butelinuru; 11—Bayankhongor zone (outside the province); 12—uranium deposits: a—endogenous (hydrothermal), b—exogenous (roll front type)

Yu Mironov distinguished a uranium ore zone with an associated assemblage of polymetals, gold, tin, tungsten, and fluorite. Within this province Gundsambuu (2007) singled out the Eastern Mongolian province or belt as the primary feature of interest.

In recent years, significant changes have been made to this zoning, by applying modern geotectonic concepts on the basis of recent exploration results. Several new sandstone type uranium deposits and a number of other targets were identified. The new work and new discoveries required modification to our understanding of the uranium resources and their setting. The first published revision of the metallogenic zoning was made by G. Jamsrandorj and Sh. Dandar in 2010 (Fig. 3.5).

In 2010, G. Jamsrandorj and Sh. Dandar made additional, more detailed revisions to supplement this zoning (Fig. 3.6).

The currently accepted metallogenic zoning of Mongolia is based on geodynamic concepts of the development of the continental crust (Fig. 3.7). To define metallogenic zones for uranium in Mongolia, geodynamic settings were determined based on the development

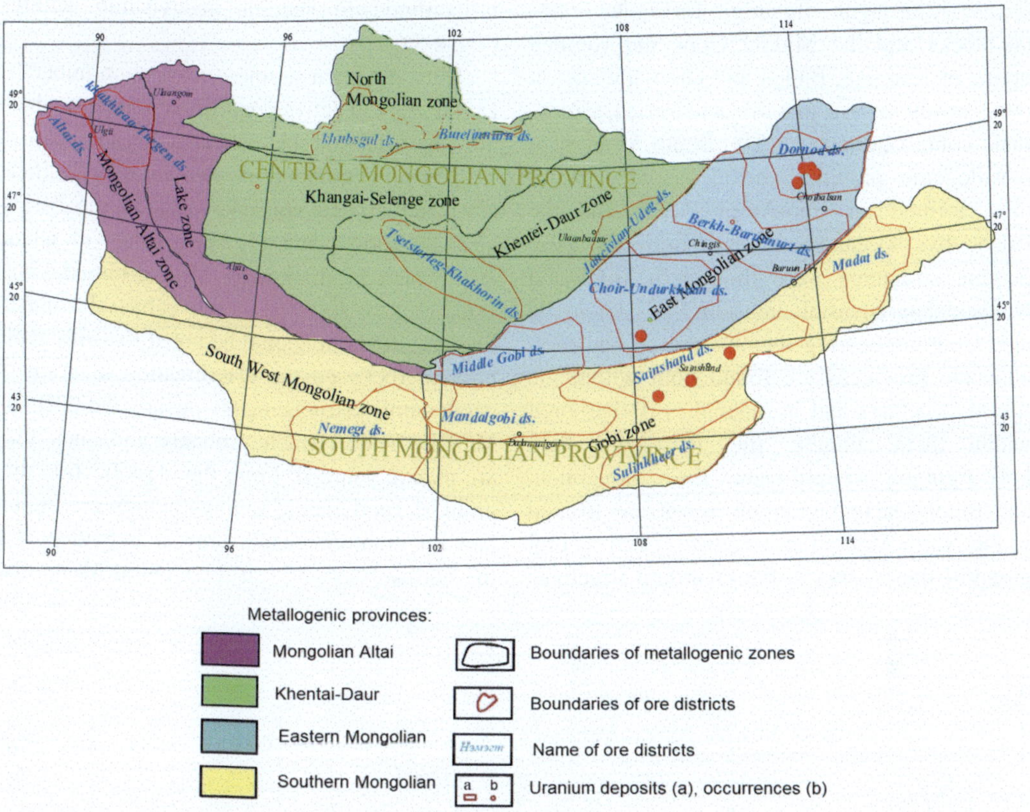

Fig. 3.5 Uranium-bearing zones of Mongolia (Jamsrandorj & Dandar 2010)

of geological complexes and formations with which uranium occurrences are related both spatially, and where possible, genetically. It should be noted that a large number of NRE radiogeochemical halos associated with various types of rocks, of different ages and genesis, were previously established. By integrating new results and geologic concepts with previous studies of the distribution and composition of these rocks showed that most uranium mineralization is associated with granitoids of active continental margins and their back-arc rift structures. Additionally, particularly in Eastern and Northern Mongolia, uranium mineralization is associated with alkaline-bimodal volcanic rocks of rift origin.

This pattern has been previously noted by various researchers. On the classification diagram, a significant part of the basaltic rocks of these regions is located within paleo-oceanic arcs (OIB) and mid-ocean ridges (MORB). This indicates that the back-arc and plume rift basalts are probably of mantle origin, and likely of tholeiitic affinity. Thus, on the basis of various materials collected by geologists from different countries who have studied uranium deposits in Mongolia in recent years, and, above all, Mironov (2006a, b, 2007a, b, 2009), a preliminary attempt to classify uranium deposits and manifestations by geodynamic settings based on paleotectonic reconstructions was presented by Byamba (2012).

At present, the two primary metallogenic provinces identified in Mongolia are the large North Mongolian and South Mongolian blocks. Within each of these provinces, metallogenic belts, zones, and regions, as well as numerous occurrences, are prominent.

It should be noted that the North Mongolian province includes folded structures formed on the location of the Paleoasian and Mongol-Okhotsk oceans, as well as the

Gargan-Khubsugul, Baidrag, Tarvagatai crustal blocks and the Middle Gobi and Kerulen uplifts of Karelian, Baikal and Grenville age; in other words, over a massive and varied geotectonic composite terrane. The South Mongolian metallogenic province includes folded structures formed on Neo-Asian Ocean and the Paleotethys Ocean basement. Between these oceanic structures were the South Gobi and Nukhetdaban-Ulyastai massifs of Grenville age. Uranium-bearing regions, zones, and metallogenic provinces are found both within the folded structures and in older massifs harbored within them. Finally, the Tuva-Mongolian, Lake-Egingol, Khentii-Daur, East Mongolian, and Butel-Khangai uranium zones are located in the North Mongolian province. Each natural group of these zones in the identified structures is confined to specific geodynamic settings (Table 3.6).

Thus, the main settings for uranium mineralization are: (1) active continental margins, which are characterized by calc-alkaline magmatism with a high content of potassium and silica; (2) differentiated volcanic rocks of the basalt-andesite series; and (3) back zones of island arcs, which are marked by alkaline-bimodal volcanic rocks associated with hypabyssal subalkaline granitoids. It should be noted that riftogenic basaltic rocks on the classification diagram of geodynamic settings mainly occupy the OIB and MORB fields, which probably indicates they are mantle sourced. From the classification diagram of geodynamic settings, which are important for uranium metallogeny, it is evident that the northwestern part of the Central Mongolian

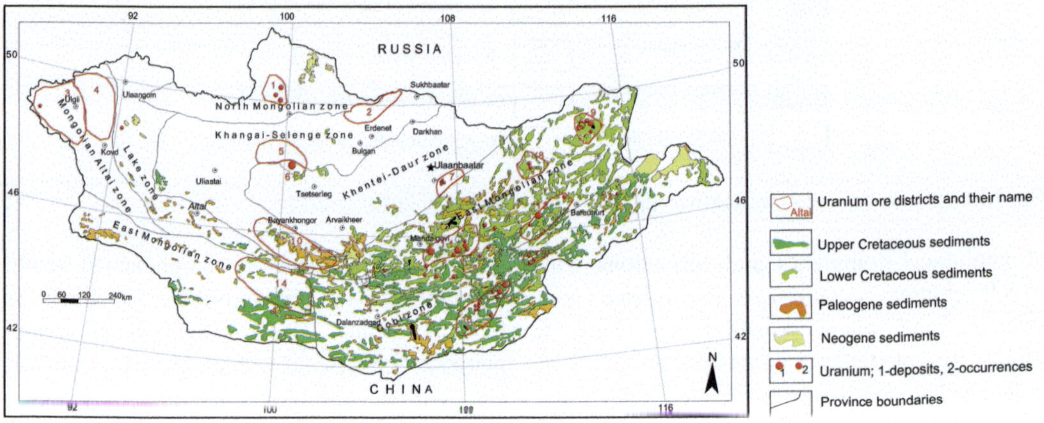

Fig. 3.6 Uranium zones of Mongolia revised (Jamsrandorj & Altankhuyag 2015)

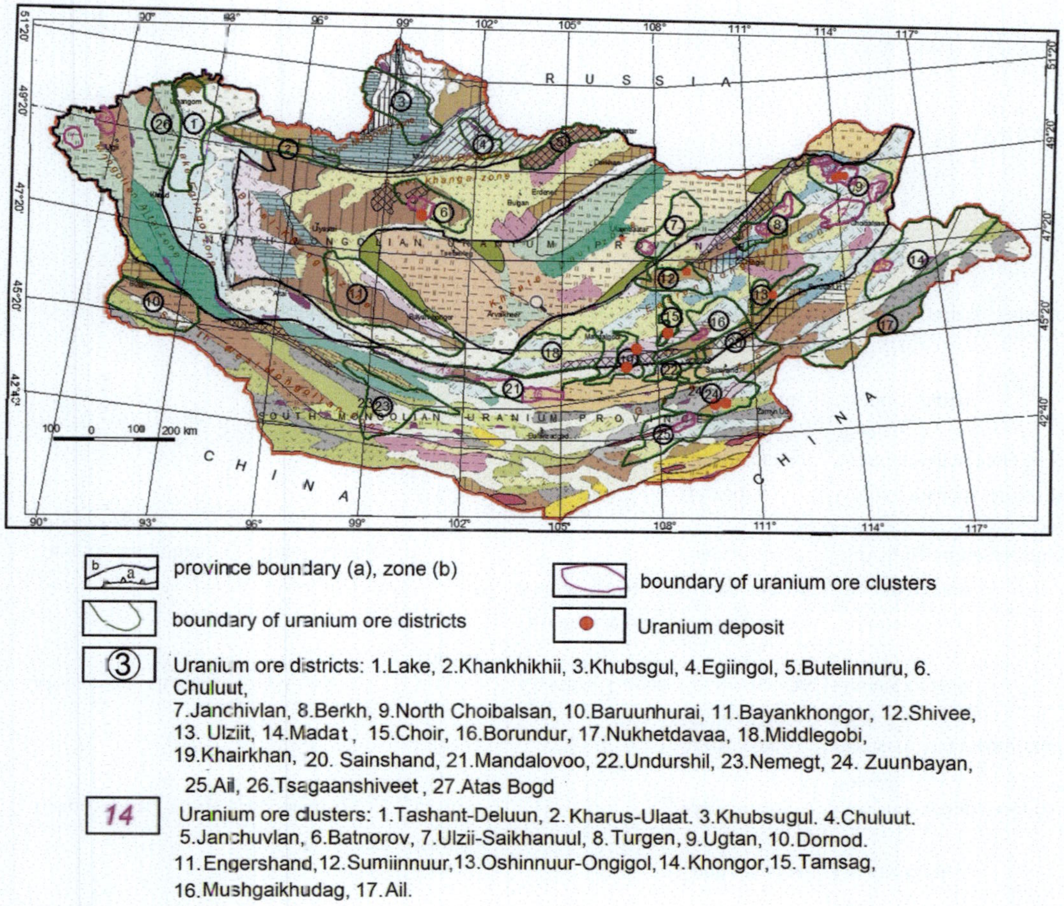

Fig. 3.7 Most recent scheme of uranium metallogenic zoning of Mongolia (Jamsrandorj & Mironov 2023). The geotectonic basis was developed by Byamba et al. (2018). The legend of this basis is shown in Fig. 2.1

Table 3.6 Metallogenic zoning of Mongolia for uranium (Jamsrandorj 2019)

According to geodynamic conditions provinces	Provinces (p), zones (z)		Districts	Clusters	Geodynamic setting	Predicted types of mineralization	Deposit examples
North Mongolian province	Tuva-Mongolian (late Baikalids)		Khubsugul		Continental shelf	Metasomatite related, phosphate, carbonate, black shall	Zaanar, Erkhel
			Khan Khukhii		Continental arc	Granite related	
	Lake-Egiingol (early Caledonides)		Egingol		Island arc	Intrusion and metasomatite related	
			Lake		Active continental margin	Intrusion and metasomatite related	
	Mongolian-Altai (late Caledonides)		Tsagaan Shiveet	Kharhira-Turgen	Continental arc	Granite related	
				Tashant-Delun	Continental arc	Granite related	
				Kharus-Ulaat	Continental arc	Granite related	
	Central Mongolian (multi-stage)	Butel-Khangai			Ancient ledge	Intrusion and metasomatite related	Ridge Butelinnuruu
			Chuluut	Chuluut	Accretion	Granite related, sandstone	Deposit Dald
					Continental arc	Granite related and black shall	
		Khentei-Daur	Janchivlan	Janchivlan	Environment of alkaline granitoids	Granite related	Occurrences Tamgat, Elstiin
		Eastern Mongolian (Mongol-Priargun)	Dornod, Ugtam, Turgen, Enger-Shand, Sumiinnuur	Dornod, Ugtam, Turgen, Enger-Shand	Late Mesozoic rift	Volcanite related, sandstone	Dornod, Gurvanbulag, Nemer, Mardaingol
				Sumiinnuur			

(continued)

Table 3.6 (continued)

According to geodynamic conditions provinces	Provinces (p), zones (z)		Districts	Clusters	Geodynamic setting	Predicted types of mineralization	Deposit examples
			Batnorov, Ulzii-Saikhanuul	Batnorov, Ulzii-Saikhanuul	Late Mesozoic rift	Volcanite related, sandstone	
					Mesozoic-Cenozoic taphrogen		Kharaat
			Nyalga, Shivee, Tavansuveet	Nyalga, Shivee, Tavansuveet	Mesozoic-Cenozoic taphrogen	Volcanite related, sandstone	Enger Ar
			Ulziit		Mesozoic-Cenozoic taphrogen		Ulziit
			Middle Gobi		Mesozoic-Cenozoic taphrogen	Volcanite related, sandstone	
			Khairkhan	Khairkhan, Gurvan-Saikhan	Mesozoic-Cenozoic taphrogen	Sandstone	Khairkhan, Gurvansaikhan
			Borundur, Ulaannuur, Khongor	Borundur, Ulaannuur, Khongor	Vocanite related, sandstone	Volcanite related, sandstone	
South Mongolian	Gobi (early Hercynides)		Mandal-Ovoo	Oshinnuur-Ongigol, Mushgaikhudag	Oshinnur-Ongigol, Mushgai	Volcanite related, sandstone	
			Nukhetdavaa		Active continental margin	Granite and volcanite related, sandstone	
			Sainshand	Narst	Mesozoic-Cenozoic taphrogen	Sandstone	Narst, Ingen, Myst, Durbulj

(continued)

Table 3.6 (continued)

According to geodynamic conditions provinces	Provinces (p), zones (z)		Districts	Clusters	Geodynamic setting	Predicted types of mineralization	Deposit examples
			Madat	Tamsag	Mesozoic-Cenozoic taphrogen	Granite and volcanite related, sandstone	Zunbulag
			Zuun-bayan	Unegt, Zuun-bayan	Mesozoic-Cenozoic taphrogen		Dulaan Uul, Zoovch Ovoo
			Iil	Iil	Mesozoic-Cenozoic taphrogen	Sandstone	Ail, Hud, Elgan
	Southwestern (early Hercynide)		Nemegt		Mesozoic-Cenozoic taphrogen	Sandstone	
			Atas Bogd		Mesozoic-Cenozoic taphrogen	Sandstone	
			Baruunk-hurai		Mesozoic-Cenozoic taphrogen	Sandstone	

massif—the North Mongolian uranium province, which is characterized by significant assimilation of crustal-mantle material from the late Paleozoic—Mesozoic, has a dominate role. The setting was initially a subduction-accretion convergence of the Mongolian-Okhotsk oceanic plate. Within Eastern Mongolia, in the late Paleozoic ($P_{2\text{-}3}$) and Mesozoic (T_3–J_1 and $J_{2\text{-}3}$), three stages of subalkaline magmatism occurred in an active continental margin setting. Initially, partial melting of the lithosphere and upper mantle is thought to have occurred, which led to the formation of calc-alkaline magmatism. During the intrusion of the rising melt, magma was enriched in potassium and silica, as well as in tungsten, tin, molybdenum, uranium, gold, fluorite, lead, zinc, and other lithophile elements. By the end of the Jurassic this stage of partial melting and magmatic processes gradually faded. As accretion, of the Mongolian-Okhotsk Ocean within northeastern Mongolia terminated in the Jura-Cretaceous, the final result was a decrease in the thickness of the earth's crust and the formation of an anomalously light mantle. These crust-destructive processes were accompanied by outpourings of rifto-related alkaline-bimodal basalts and related products over a large area of individual eruption sites.

Thus, the latter phenomenon is interpreted to be related to collision of the North and South Asian cratonic blocks at the beginning of the Triassic, resulting in a regionally significant suture zone and a large and sustained continental mobile belt. Further from the collision boundary, tectonic forces weakened and volcanism was diachronous in the north resulting in igneous rocks as young as 80 Ma.

As noted earlier, Mongolia is divided into the North Mongolian and South Mongolian metallogenic provinces.

The North Mongolian metallogenic province contains:

1. Tuva-Mongolian prospective mineral zone.
2. Lake-Egiingol prospective mineral zone.
3. Mongolian-Altai prospective mineral zone.
4. Buteel-Khangai prospective mineral zone.
5. Khentii-Daur prospective mineral zone.
6. East Mongolian prospective uranium mineral zone.

The South Mongolian metallogenic province includes:

7. Gobi prospective mineral zone.
8. Southwest prospective mineral zone.

The main deposits and occurrences of uranium are associated with Middle-Upper Jurassic carbonatite-alkaline basalt-andesite-rhyolitic volcanism in association with alkaline-granite magmatism. The intrusions of which formed above the subduction zone between the Central Mongolian uplift and the Mongolian-Okhotsk oceanic plate in the south. These regions include the Northern Choibalsan, Berkh, Shivee, and Middle Gobi regions. In these areas, the Dornod, Mardaingol, Nemer, and Gurvanbulag uranium deposits are associated with volcanic rocks, and the Tumenkhaan deposit is associated with granites; additionally, the hydrogenous (secondary) Kharaat, Khairkhan, and Gurvansaikhan deposits were formed.

In general, eastern Mongolia is thought to have been uplifting since the late Permian, with active continental calc-alkaline magmatism occurring three times: in the Early Permian; Late Triassic; and Early-Late Mesozoic. Each of these magmatic episodes was followed by periods of alkaline-bimodal volcanism. The process of volcanic eruptions, which covered a large area, reached its peak during the Jura-Cretaceous transition.

In Northeast Asia, against the general background of subduction of the Mesozoic Pacific Ocean plate under the Eurasian plates. Initially, the Mongolian-Okhotsk Sea was closed on one side in Eastern Mongolia and in two places along the Central Mongolian uplift; ultimately ongoing closure led to the destruction of the oceanic region.

Since the beginning of the Proterozoic, rift tectonics in Eastern Mongolia have been associated with the formation of uranium mineralization as a result of alkaline basalt-rhyolitic and alkaline-acid volcanism. Related to these processes, not only uranium, but also rare metals, rare-earth elements and zinc, fluorite, and molybdenum mineralization occurred, forming a single but complex metallogenic association.

Outside of our area of focus, in the Okhotsk-Chukotka region of Russia, hydrothermal formations of alkali-potassium volcanoes with scattered gold-uranium mineralization associated with the subduction of a mantle plume and its various roof pendants has been interpreted. In northeastern Mongolia, it has been suggested (Korobeinikov and Rikhvanov 2002) that within the ancient massifs distributed along rift depressions, an association of galena, sphalerite, molybdenite, fluorite, rutile, uranium, rare metals, and rare-earth elements is associated with acidic products of differentiated mantle and related magmatic products.

3.2.1 North Mongolian Metallogenic Province

The North Mongolian province includes folded structures formed on the site of the Paleoasian and Mongol-Okhotsk oceans, as well as the Gargan-Khubsugul, Baidrag, Tarvagatai ancient blocks and the Middle Gobi and Kerulen uplifts of Karelian, Baikal and Grenville age, located within them. In the North Mongolian province, the Tuva-Mongolian, Lake-Egingol, Khentii-Daur, East Mongolian, and Butel-Khangai uranium-bearing zones are prominent. Each group of these zones in the selected structures is confined to certain geodynamic settings.

3.2.1.1 Tuva-Mongolian Uranium Zone

Neoproterozoic and Cambrian metamorphic formations, Devonian sedimentary deposits, and Permian alkaline intrusive rocks are common in the Tuva-Mongolian uranium-bearing zone. The metallogeny of the zone is characterized by aluminum (as bauxite), phosphorites, rare-earth elements, rare metals, gold, and occurrences of iron.

The primary uranium occurrences, such as Zaanar and Kherbestei, are associated with vanadium-bearing carbonaceous limestone with bitumen; however, there are also significant uranium occurrences in Khuvsgul and the Khankhukhii associated with Paleozoic subduction and associated magmatism along the paleocontinental margin. Radiogeochemical zones are well-defined in these regions.

Khubsugul Potential Uranium District

In the Khubsugul uranium district, hydrothermal and metasomatic zones are noted at the endo- and exo-contacts of alkaline massifs, which intrude the Ediacaran-Early Cambrian carbonate rocks and host uranium–thorium mineralization. Stock-shaped, nested, lenticular, and vein mineralization with haloes of disseminated mineralization are localized in zones of intense albitization, K-feldspathization, silicification, chloritization, and hematitization and are accompanied by sulfide, fluorite, and less often beryllium mineralization.

The areas of metasomatically altered rocks vary from 0.2–0.3 to 1.5–2 km^2 and enclose vein-like bodies from several tens to a few hundreds of meters long with a thickness of 10–20 m. The veins are composed of disseminated orthite, thorite, orangite, zircon, monazite, less often uraninite and are typical of Ujiggol, Aragol, Balbrik, Alag-Erdene, Yarkhisgol occurrences and early-stage anomalies including Usutgol, Berkhim Uul, Berkh, Egingol, Aduungol, Khagin nuur, and others. The best zones contain uranium from 0.001 to 0.23%, thorium from 0.01 to 13%, zirconium 0.1–0.2% and locally up to 10%, niobium 0.2–0.5%, tantalum 0.03–0.2%, beryllium 0.01–1%, lithium 0.08–0.2%, lanthanum 0.05–2%, cerium 0.1–2%, yttrium 0.1–2%, gadolinium 0.5–0.06%, ytterbium 0.02% (Mironov 2006a, b).

Uranium mineralization in carbonaceous-siliceous shales has been identified in Proterozoic and Vendian-Cambrian black shale deposits of the Sub-Khubsugul and Sangilen regions, which are primarily enriched in uranium, vanadium, and molybdenum. Black shale strata are developed here over a large area, composed of individual litho-stratigraphic packages up to hundreds of meters thick and up to 1.5 km long.

Elevated uranium grades are confined to areas of siliceous, siliceous-carbonaceous, and vanadium-phosphorus-containing shales within the fault shear and hydrothermal-metasomatic alteration zones. As a rule, such zones have a submeridional (north–south) or northeast strike and are accompanied by silicification, hematitization, and carbonatization. The distribution of increased radioactivity (up to 100–400 μR/h) in tectonic zones is highly variable and may comprise just a point, or a larger cluster, or lenticular body; uranium and thorium concentrations in the zones vary within; uranium ($10–200 \times 10^{-4}$), and thorium ($2–5 \times 10^{-4}$%). In uranium-bearing formations, elevated metal concentrations are recorded (all in %): nickel—0.01–0.15, molybdenum—0.001–0.02, copper—0.01–0.3, zinc—0.01–0.15, and silver—$1.1–2.0 \times 10^{-4}$. The concentration of these elements (including uranium) does not correlate with the phosphorus content, as might be expected. The length of the ore bodies is a few hundred meters, the thickness is from 1 to 12.8 m, the uranium grade is 0.013–0.099% and vanadium pentoxide is 0.72–2.70%. In most cases, mineralization is represented by secondary uranium minerals: francswillite, paraskupit, carnotite, and uranophane; vanadium is concentrated in sulvanite (a copper-vanadium sulfide). The most interesting occurrence in the Zaanar cluster (Fig. 3.8) is confined to the apparent sag zone of the roof of the alkaline granitoid intrusion, largely hosted in the terrigenous-carbonate rocks of the Doodnuur suite. The Doodnuur suite consists of marbled and dolomitic limestones with thin horizons of dolomites, calcareous conglomerates, with subordinate carbonaceous-siliceous and argillaceous-siliceous shales, which are dissected by northeast, northwest faults with a submeridional (north–south) strike.

Fault zones are the primary factor controlling areas of increased radioactivity (up to 400 μR/h and more) and are generally accompanied by low to medium temperature silicification, hornfelsing, argillization, sericitization, and limonitization. Potential ore zones (i.e., the best grades) occur at fault intersections within the carbonaceous-siliceous shale host rocks. In two subparallel mineralized bodies 120 and 100 m long, 1 and 11 m thick, respectively, the uranium grades are 0.099 and 0.032%, vanadium pentoxide 2.7 and 2.5%, with increased concentrations

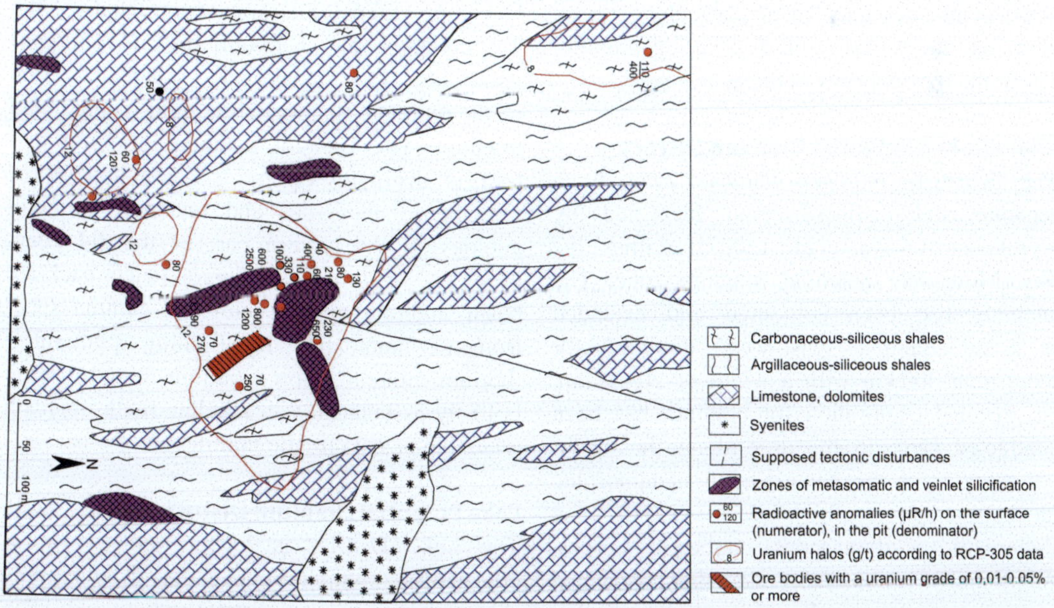

Fig. 3.8 Simplified geologic map of the Zaanar mineralized zone

of molybdenum—up to 0.15%, copper—0.2%, chromium—0.1%, nickel–15%, zinc—0.5%, and silver—> 10 ppm.

West of Lake Khuvsugul lies the Argol region with small occurrences of uranium (with Mo and V) mineralization in widely dispersed in blocks of carbonaceous-siliceous shale. In general, the Proterozoic black shale strata and Ediacaran-Cambrian phosphorus- and vanadium-bearing metamorphic rocks have a high uranium background that may indicate an early phase of regional enrichment. The Bayan-Khongor zone is in a similar geological setting and is characterized by similar mineralized features. Within the Aragol region, late Paleozoic-early Mesozoic subalkaline granite and syenite intrusions are present, accompanied by metasomatite (alteration) zones of disseminated rare-earth, uranium, and thorium minerals. These occurrences are grouped together into a uranium-bearing (with thorium, rare metals, and rare earths) metasomatite (alteration zone) type that is genetically related to alkaline and subalkaline rocks.

In the Khubsugul region, occurrences of uranium mineralization associated with metasomatites (altered zones), phosphorites, black shales of the Ediacaran-Cambrian strata and in small pegmatites may be of particular interest. However, the prospect of discovering uranium sandstone type deposits in the area is poor.

Khankhukhii Potential Uranium District

The Khangai radiogeochemical zone is a separate and distinct region identified by Vysokoostrovskaya et al. (2007), within the Tuva-Mongolian potential uranium zone. The anomaly had been contoured and extended in a sublatitudinal (east–west) direction for more than 300 km with a width of 20–25 km. No detailed exploration for uranium has been conducted here. In the area along the northern margins of the Khan-Khuhii metamorphic block, uranium locally increases to 0.0X% in Proterozoic carbonaceous carbonate rocks. Additionally, various AGS anomalies associated with metasomatites (alteration zones) of uranium–thorium character are probably associated with granosyenite, syenite, and subalkaline granite of diverse ages from early Paleozoic to the Permian.

The redistribution of uranium and potassium is notable in biotite-amphibole, K-feldspar-sericite, and epidote-chlorite blastomylonites developed in Paleozoic volcanic protolith. The radon sources along the Khangai deep fault remain unverified and require further study. The Khankhukhii region may be characterized by the dominance of thorium mineralization, but with lesser occurrences of thorium-uranium-rare-earth mineralization (occurrences Shar Tolgoi and Ulaan Tolgoi).

3.2.1.2 Lake-Egingol Potential Uranium-Bearing Zone

In the Lake-Egingol zone, the continental crust was formed at the beginning of the Ordovician, and two potential uranium districts, Egingol and Nuurin, are noted.

Egingol Potential Uranium District

The most important rock units in the Egingol area include ophiolites of the Neoproterozoic (Tonian), volcanics of the Ediacarian-Lower Cambrian Badargol formation, Middle-Late Cambrian gabbro-diorite-granodiorites, Devonian granosyenite-granite intrusions of the Tes complex, and Carboniferous (Miss-Penn) granitoids. The metallogenic characterization of the region is dominated by gold, copper, and uranium. The Egingol potential uranium district is distinguished by the combination of radiometric anomalies and favorable uranium geology in the eastern part of the Jid zone, developed within the Argingol paleo back-arc basin. Judging by the nature of gamma spectrometric anomalies highlighting potassium-thorium mineralization, uranium mineralization is likely associated with alkaline metasomatism related to postmagmatic granites.

Lake Potential Uranium District

The Lake District is located on the western natural continuation of the Khankhukhii potential uranium ore district, where radiogeochemical anomalies have been established. The geological structure of the area includes volcanics, siliceous-carbonate rocks of Ediacarian-Lower Cambrian age, sedimentary formations of Ordovician and Silurian age, which are intruded by ultrabasic, basic, and plagiogranite intrusions of Cambrian and Carboniferous age, respectively. The district is metallogenically rich and contains iron, copper, gold, and uranium.

Thick Mesozoic-Cenozoic formations of sedimentary cover are widespread in the district. Oxidation and reduction processes controlling the concentration of uranium in the form of sandstone deposits can be developed in such a setting. Particular attention is drawn to the anomalies of uranium 88, 90, and 91, revealed by an airborne gamma spectrometric (AGS) survey in the valley of Lake Khyargas nuur in the Upper Jurassic-Lower Cretaceous deposits. Further detailed exploration work is required to advance knowledge of these anomalies.

3.2.1.3 Mongolian-Altai Uranium-Bearing Zone

In the Mongolian-Altai uranium-bearing zone, terrigenous and volcano-terrigenous deposits of the lower and middle Paleozoic that are intruded by Triassic and Jurassic granitoids with high potassium and silicon content are developed. The zone is characterized by tungsten, rare-earth elements, rare metals, gold, silver, iron, and uranium. In radiogeochemical terms, the zone has marked high variability.

Higher NRE grades within continental rocks are observed in late Paleozoic post-accretionary granite massifs (Sagsai, Turgengol, Bogot, Bayanzurkh, Durgun, Kulagash, Kharkhiraa, Achitnuur, Orgui, and Namarjaa) which are often associated with quartz wolframite and scheelite greisen-vein mineralization (Fig. 3.4). Large massifs (Kharkhiraa, Sagsai, etc.) are characterized by a highly variable distribution of NRE, but typically display an increase in the content of radioactive elements from the center to the periphery, that possibly indicates a composite plutonic structure. Small granite massifs (Bogot, Turgengol) in the northern parts of the terrane are characterized by high uranium content (5–8 ppm) but may lack the concentric zonation of grades exhibited in the larger complexes. The maximum concentration of radioactive elements was noted in the granites of the Turgengol massif, near the Asgat silver deposit (U 7.2 ± 1.4 ppm, Th 56 ± 12 ppm).

The redistribution and concentration of uranium in the granite massifs of the Mongolian-Altai zone is complex and apparently multistage. High-temperature metasomatic transformations (alteration) include biotitization, greisenization, and albitization. There are apparently only weak correlations between uranium, thorium, and potassium and the alteration assemblages. In areas of some deposits of tin, tungsten and silver, signs of uranium remobilization with local re-concentrations (up to 0.X%) via late stage low-temperature processes were noted (Borburgas ore occurrence, Kyzyltau deposit, eastern flanks of Asgat).

In the Kyzyltau, Achitnuur, Narinnuur, and Dulaan-Kharuul VTS volcanogenic and subvolcanic formations are distinguished by a differentiated NRE field with the following average components: U 4–6 ppm; Th 16–20 ppm; and K 4–5%. The majority of uranium-enriched VTS are found in the Kyzyltau (6.0 ± 0.7 ppm), Achitnuur (4.8 ± 0.6 ppm), and Dulaankharuul (4.6 ± 0.9 ppm). Terrigenous formations of the upper Paleozoic are moderately radioactive and contain, as a rule, Clarke-level concentrations of radioactive elements. In the granitoids of the Altai and Khovd subzones, the NRE apparently experienced multistage remobilization and redistribution.

High-temperature metasomatic transformations (alteration) redistributed the NRE concentrations, which led to the predominance of local uranium anomalies with associated tungsten and tin deposits, such as at Kobdogol and Kyzyltau. Numerous (> 25) anomalies of uranium–thorium–potassium nature are also noted here. The Mongolian-Altai potential uranium zone can be considered a southeastern extension of the uranium-bearing structures of the Altai-Sayan region of Russia, where large-scale exogenous uranium deposits are known (Ust-Uyuk, Onkazhinskoye, etc.). Also endogenous occurrences with lesser grades such as the Karasug and Uliastai prospects are related to acidic subalkaline and alkaline magmatism. No deposits or significant ore occurrences of uranium have yet been identified in the Mongolian extension of this zone.

According to the results of an airborne gamma spectrometric (AGS) survey, potential uranium mineral occurrences, Tashant-Delun, Kharaus-Ulaat, and the Tsagaan Shibeet, are known within the zone. In the Tashant-Delun area VTS, above background concentrations of uranium are confined to tectonic faults within acidic (felsic) volcanic rocks. These signs indicate the possibility of the development of uranium mineralization in the fault shear zones of various rock types.

Tsagaan Shiveet Potential Uranium District

The Tsagaan Shiveet radiogeochemical zone is part of the Altai-Sayan uranium ore province and continues northward into the Russian Federation to comprise the large Altai-Sayan uranium province.

Geologically, the area is composed of Proterozoic and Paleozoic metasedimentary formations, which are overlain by Devonian potassium-alkaline and felsic volcanics, intruded by Late Devonian-Early Carboniferous syenites and granites formed in a three-axis rift (i.e., triple junction, and classic example of an aulocogen setting). The district is enriched in gold, copper, iron, manganese, polymetals, molybdenum, and uranium. Radiogeochemical anomalies are localized in the southern part of the region. They have not been studied by geological prospecting due largely to their remote setting and relatively poor infrastructure.

The Tsagaan Shiveet district geotectonically corresponds to the Tsagaan Shiveet continental arc and includes the northeastern part of the Khovd paleo continental margin. The foundation of the geological structure is created by Precambrian and Paleozoic metasediments. Overlying Devonian volcanics are predominantly of a rhyolite composition, and cross-cutting (and locally "stitching") leucocratic and subalkaline granites and syenites of the Late Devonian-Early Carboniferous are significant metallogenic features.

The anomalous radiogeochemical zone is comprised predominantly of thorium, less often mixed with uranium and potassium, and includes a significant number of anomalous uranium–thorium fields and local anomalies of a purely thorium nature. Above background concentrations of uranium are concentrated in six areas—Ulzii, Khalzanburged, Ugoomur, Khalzan, Kharausnuur, and Urgot; all related to and occurring around, areas of alkaline, essentially sodic metasomatites (altered zones), less often to tectonically disturbed leucocratic granite massifs.

The district is considered promising for rare-earth and rare-metal mineralization with uranium and thorium. In further studies of the geological structure and mineralization of the Tsagaan Shiveet zone, special attention should be paid to fault zones within leucocratic granites and enclosing Paleozoic terrigenous rocks, as these are likely fluid conduits that may have transported and deposited radioactive elements.

3.2.1.4 Central Mongolian Metallogenic Province

Butel-Khangai Uranium Zone

The Butel-Khangai zone differs from others in that it consists of many of the most ancient cratonic metamorphic blocks, such as the Zavkhan, Tarbagatai, Baidrag and Butelinnuru, which surround (having been accreted) the Central Mongolian massif on the northwest and north sides. The geological structure of the zone involves volcanogenic-sedimentary rocks of the Proterozoic and Paleozoic, intrusive formations of Paleozoic and Mesozoic age, with metallogenic signature of gold, copper, ferrous, and non-ferrous metals and uranium. The main feature of the rocks in this zone is the uneven and limited development of formations of different ages that display uranium mineralization. Radiogeochemical anomalies were only observed in the Butel, East-Khangai, and Bayankhongor regions. However, this work ultimately identified the Chuluut uranium deposit and Butel and Bayankhongor uranium prospects. The Chuluut uranium deposit is a significant feature at the western end of the East Khangai radiochemical zone, where the Dald uranium deposit has also been discovered in recent years.

Mironov (2009) identified the Khilok uranium occurrence and Chikoi uranium regions as strike extensions of the Butel-Khangai uranium ore zone across the border into the Russian Federation.

Chuluut Uranium Ore District

This area was identified by Soviet geologists (Mironov 2009) in the northwestern part of the East Khangai radiogeochemical zone. It includes the Dald uranium deposit (the Bayar and Sul occurrences are merged and renamed the Dald deposit) and a number of uranium occurrences. In the zone, eruption and outflowing of Cenozoic basalts, underlain by weakly lithified terrigenous deposits of the Neogene-Pleistocene age, occurrences of Vitim-type uranium (Russia) in gray-colored (reduced) Quaternary channel deposits were mapped. This was a breakthrough for effective exploration of the Dald deposit.

The most well-studied area in the district is the Chuluut uranium deposit located in the intermountain basin of the Khangai and Tarbagatai ranges at the confluence of the Sumiin-Gol and Chuluut rivers. During reconnaissance work in 1982, radioactive anomalies were discovered and further assessment and ground campaigns lead to discovery of numerous anomalies and ultimately to the promising uranium occurrences at Daagin-Gol, Bayar, and Sul. The footprint of the deposit covers three paleovalleys of northwestern strike (310–320°), the length of which is 10, 16, and 5 km, with a width of 1–2 km. The deposit area is 450 km^2. The paleovalleys are made up of unlithified Cenozoic deposits and overlying basalt covers. The crystalline basement includes metamorphic formations of the upper Proterozoic, lower-middle Devonian volcanic rocks, and granitoids of the Egindavaa intrusive complex (T_3–J_1), which comprise the Chuluut massif with an area of 700 km^2. The complex is dominated by light pink leucocratic and biotite granite with characteristic round grains of smoky quartz. The presence of smoky quartz often indicates radioactivity. Their radioactivity varies from 30–35 to 45–60 μR/h, uranium content from 2 to 10 ppm, and thorium content from 14 to 50 ppm. The basement is dissected by a series of tectonic faults of northeast (40–45°) and northwest (315–340°) strike. The northeast striking faults are represented by zones of crushing, mylonitization, tectonic sutures with slickensides. Faults are often healed ("stitched") by dikes of microgranite and diabase with lengths generally from 2–3 km and up to 5 km. The thickness of the faults does not exceed a few tens of meters. They are often accompanied by zones of increased fracturing. In granites and in lower-middle Devonian volcanic rocks in tectonic zones, halos (alteration selvages) of epidotization, K-feldspathization, metasomatic, and veinlet silicification are widely developed. Some of the older structures have been rejuvenated by neotectonic movements and this neotectonic extension controls late Cenozoic architecture and deposition. The position of linear paleovalleys filled with Neogene-Holocene terrigenous and basalt formations is probably controlled by northwest-trending fault zones. At the base of the Cenozoic deposits, Pliocene sediments of piedmont-fan and lacustrine-alluvial facies are noted. They lie on the weathered surface of the basement or on buried weathered crusts (paleosols), marked by kaolinized, hematitized eluvium of granites up to 20 m thick. All types of rocks contain fine carbonized plant detritus, collomorphic humic organic matter, and, more rarely, small amorphous segregations of low-grade kerite (asphaltites). The color of the rocks varies from light gray clastic units to dark gray loam and clay. The background grade of uranium in the sediments is 2–5 ppm, and thorium is 8–12 ppm.

Butelinnuru Uranium Potential District

The district was identified within the radiogeochemical zone of the Proterozoic metamorphic Butelinnuru ledge. The ledge is an east–west trending fault-bounded-anticlinal uplift with a gentle dome in the center and asymmetric southern and northern limbs. There are a series of granite-gneiss and crystalline schist domes 3–7 km in diameter. The northern boundary of the zone is complicated by cataclasite zones associated with the development of the Jelter fault that in a structural sense, ties the (Mesozoic) occurrences of gold-bearing quartz-veins to uranium–thorium occurrences, in migmatites and pegmatites.

For the Butelinnuru zone as a whole, the leading minerals are uranium-containing rare and rare-earth mineralization in quartz-potassium feldspar-albite metasomatites and pegmatites. These bodies are widespread in the axial part of the zone.

Pegmatites and metasomatites with high grades of uranium and thorium are noted here. The most well studied occurrences of Bayan-Gol and Zuun-Under-Tsokh are located in the eastern part of the Butelinnuruu ledge, enclosed by a granite-gneiss dome. Biotite, amphibole-biotite gneisses, and granite-gneisses give rise to quartz-feldspar metasomatites, and locally grade into cataclasis zones. Metasomatite bodies strike northwesterly and are from tens to 200 m in length, and with thickness 1–1.5 m. Elevated levels of radioactivity (up to 250–1500 μR/h) are sporadically distributed. Uranium-bearing metasomatites consist of isometric quartz grains and large tabular microcline feldspar replaced by albite, biotite laths, and single almandine garnet crystals (in general terms, alumino-sodic alteration).

Mineralization is represented by euxenite and brannerite intergrown with titanium magnetite and rare-earth minerals. The uranium grade, according to channel samples, is 0.006–0.15% with a pronounced symmetric distribution of values centered on larger grade zones. The higher-grade samples contain (%): thorium—0.006–0.01, tantalum—0.03–0.2, niobium—0.8–5.6, yttrium—0.5, gadolinium—0.04–0.2, lanthanum—0.02. In the zone with a northeasterly strike, albite metasomatites developed along plagio-granites in bodies up to 240 m long and up to 10 m thick. The primary mineral is brannerite associated with tantalum-niobates. The region's prospects for uranium are not highly rated.

Bayankhongor Uranium Potential District

The Bayankhongor uranium district covers the Baidrag ancient metamorphic block, which is composed primarily of the Neoarchean Baidrag metamorphic complex and the Paleoproterozoic Bumbugur metamorphic complex which is overlain by Mesoproterozoic carbonate-terrigenous sediments of the Burdgol Formation. Paleozoic volcano-plutonic and Mesozoic intrusive complexes are also common in the district. The Neogene and Cretaceous terrigenous formations that developed here may be of interest as potential host rocks for exegenous (roll front type) uranium deposits. The regional uranium footprint was identified primarily on the basis of AGSM data; the ground-based follow-up and study is insufficient.

In the Bayankhongor radiogeochemical zone, the occurrence of elevated grades are of a mixed uranium–thorium–potassium nature and, most likely, due to the redistribution of NRE during hydrothermal-metasomatic processes that are also related to gold mineralization. The prospects for uranium mineralization in this zone have not been determined; though one can readily predict the possibility of uranium concentration due to the contact effect of granitoids on carbonaceous rocks. In terms of radiogeochemical data, the most interesting anomalies are in the southern and northwestern parts of the zone.

3.2.1.5 Khentii-Daur Uranium Zone (Province)

Geotectonically, the Khentii-Daur province corresponds to a wide, extended arcuate area of Triassic-Jurassic rift magmatism superimposed on the Precambrian metamorphic complexes of the Tarbagatai Craton in the west and greenschist metasediments of the Khangai-Khentei turbidite basin in the north. The radiogeochemical zone continues further northeast into Russia along the Zachikoi mountainous region and the Daurian zone. Note that earlier, Abramovich (2004) considered this structure a deep-water trench filled with turbidites related to subduction under the southern margin of the Siberian continent. Early Mesozoic rifting and maximum extension zones control the widest development of granitoids, including leucocratic varieties notably enriched in uranium and rare metals (W, Sn, Mo).

In radiogeochemical terms, the Khentei-Daurian province is characterized by high levels of uranium enrichment, primarily due to highly radioactive Triassic-Jurassic granitoids. Anomalous radiogeochemical zones (East Khangai, Ikhehairkhan, Nalaykh, Iluur) and blocks (Asakan-Shumilovsky, Ingodinsky, Daursky) (Mironov 2003, 2005, 2006a, b, 2007a, b, 2008), occupy about 35% of the area. These zones and blocks are associated primarily with Jurassic leucogranite massifs, themselves enriched in uranium, and to a lesser extent with late Mesozoic VTS developed along large northeast- and northwest-trending faults. Uranium grades in granite range from 8 to 12 ppm, thorium from 35 to 40 ppm, and potassium from 3.5 to 4%. The host metasediments and metacrystalline rocks, as a rule, are weakly radioactive (1.5–2.5 ppm uranium, 5–10 ppm thorium, 1.5–2% potassium). It has been established that local zones of redistribution and deposition of uranium are associated with hydrothermal-metasomatic (alteration) processes that are widely developed within the granite massifs and their immediate surroundings.

The Khentei-Daur province is characterized by a broad deposition of uranium mineralization, directly or indirectly associated with massifs of Mesozoic leucocratic granites enriched in uranium. In Mongolia, numerous uranium occurrences are attributed to vein type deposits in fault-sheared leucocratic granites and to sandstone type deposits in terrigenous deposits of the Mesozoic basins.

Janchivlan Potential Uranium District

This district was identified based on the work of the MGSE (Mironov 2003) and named at the time "Central". The district is characterized by uranium mineralization associated with granites.

The potential Janchivlan uranium district (PUD) is located in the northeastern section of the Khentei-Daur uranium province. In geotectonic terms, the Janchivlan region occupies the northeastern section of the Khentei turbidite basin, intruded and altered by Mesozoic (Triassic-Jurassic) granitic magmatism.

The Khentei basin is composed primarily of Devonian–Lower Carboniferous siliceous-terrigenous sequence and Lower-Middle Carboniferous flyschoid sand-silt formations. All sediments are intensively dislocated, crumpled into folds, and broken into blocks. Cross-linked lower Paleozoic massifs of diorite-plagiogranite are widely developed in the area, the largest of which, Bayanjargalan, has an area of 460 km^2.

The main fault structures that control the emplacement of the Triassic-Jurassic intrusions of the Janchivlan complex are the zones of the South Khentei and Tavansubeet regional faults with a northeast strike. Massifs with similar structure, radiogeochemical, and metallogenic features are concentrated in the Janchivlan complex. The largest of which, with an area of a few hundred square kilometers, are the Janchivlan, Bayandelger, and Togoot. All massifs are composed of porphyritic leucocratic and intensely greisenized granites with a pronounced rare-metal signature. In addition to greisenization, they are characterized by microcline-albite metasomatites (sodic alteration) associated with niobium and tantalum, and tungsten-tin mineralization is widely developed. The Jargalant, Yudeg, Shorootingol, and a number of smaller massifs are of the same type in terms of their structure and geochemical signature. In the apical (roof region) parts of igneous bodies, greisenization, and high-temperature veinlet silicification are common and generally accompanied by elevated concentrations of tungsten, tin, and molybdenum.

The main structural discontinuities of the Janchivlan District are large zones of the South Khentei and Tavansubeet (Kerulen) regional faults with a northeast strike, as well as tectonic sutures of the northwestern Choir-Khentei zone. These important foci of structural shear zones and the nodes of their intersection control the emplacement of almost all Mesozoic granite massifs and occurrences of minerals of Mesozoic age. The primary ones are Tsogt, Janchivlan, and Gurvankhukhii. The length of each of them is 20–25 km, and the width of the zones of tectonically disturbed rocks is 1.0–1.5 km. Among the north–south trending fault structures is the Batur fault, the largest submeridional fault, which was traced for 15 km with a width of 0.5–1.5 km. Discontinuities of a higher order are sublatitudinal (east–west trending) crushing zones that determine the position of most occurrences of uranium mineralization. These are the Elstiin, Arshaan, and Urt-Tamga zones, which are 4–6 km long and 100–300 m wide.

Within the Khentei arch in the Janchivlan district, the Upper Iluur molybdenum and Nariin tungsten deposits are known, as well as a significant number of tungsten and tin ore occurrences (Bayan-Undur, Kharaat, Tarkhat, Khonkor, Khaalga, etc.). A common feature of the occurrences of rare metals in the area is a close spatial and genetic relationship with the massifs of leucocratic granites of the Janchivlan and Jargalant complexes.

Occurrences of quartz-vein and greisen mineralization of cassiterite-wolframite-quartz mineral type, tantalum-niobate mineralization in albitites, placers of cassiterite, wolframite, tantalum-niobates, and gold are associated with the Janchivlan granite massif. Due to the small size of the mineralized bodies and the low grade of rare metals, the occurrences have no economic significance at present.

Fluorite mineralization is also associated with granite-leucogranite massifs of the Janchivlan complex. In the Janchivlan, Bayandelger, Togoot massifs, fluorite is present in pegmatites, rare-metal greisens, and also in the form of independent low-temperature quartz-fluorite veins. Fluorite is sporadically present in pegmatites, where its optical varieties are found in miarolitic cavities; fluoritization is noted in almost all occurrences of uranium mineralization. Even though there is significant areal extent of fluoritization, there is only one small fluorite deposit (Napchtai) and four prospects currently known.

The Janchivlan district is noted for numerous and various radiogeochemical occurrences, most of which are related to elevated uranium grades in Mesozoic granitoids. The primary prospects for higher-grade uranium within the district, are associated with mineralization in the crushing zones of rare-metal Mesozoic granites enriched in uranium. Five uranium occurrences, more than 50 ground-verified anomalies with established uranium mineralization, numerous primary and secondary heavy mineral placers and brine and gas uranium halos have been identified in the district.

The potential Janchivlan uranium district is well known in the area, and spatially coincides with the Janchivlan leucogranite massif and anomalous radiogeochemical zone controlled by tectonic sutures of the Khentei-Choir fault zone. It includes four uranium occurrences (Tamga, Urt, Elstiin, and Arshaan), numerous points of uranium mineralization, local anomalies with a uranium signature, and groundwater brine uranium dispersion haloes.

The Janchivlan massif is a composite intrusion with three phases of emplacement. The first phase includes porphyritic coarse-medium-grained biotite and leucocratic biotite-bearing granites of the Gorikh type. They form the main foundation, composing vast fields in the western, central, and eastern parts of the massif. Numerous bodies of pegmatites with crystal encrusted miarolitic cavities are associated with Gorikh type granites. Postmagmatic alteration of the granites of the first phase, are represented by early K-feldspathization and greisenization. The second phase includes predominantly equigranular or weakly porphyritic biotite and two-mica granites of the Modot type, which form two separate adjacent bodies in the center of the massif with a total area of about 25 km^2, and a third separate outcrop in the Elstiin block. Albitization and greisenization with wolframite-cassiterite mineralization are widely developed in the granites of the second phase.

The third phase is known as "Sharkhad type" and consists of leucocratic biotite-bearing granites, alaskites, and apogranites comprising a large body in the north section of the massif and two small outcrops in the southern section. They contain zoned bodies of microcline-albite with zinnwaldite, amazonite-albite with zinnwaldite, and albite-lepidolite granites, quartz-biotite and quartz-muscovite greisens. These formations are characterized by extremely variable but high grades of uranium and thorium.

As can be seen, according to a number of features, the Janchivlan granites may well be on-strike extensions of the rare-metal leucogranites, which are part of the Asakan-Shumilovsky complex of the Chikoi uranium district in the Russian section of the Khentei-Daur arch. This conclusion is also confirmed by geochronological data—the age of the Janchivlan and Asakan-Shumilovsky granites has been determined to be ~195 Ma (K–Ar and Pb–U methods).

Additional studies of uranium mineralization in the Janchivlan District, were undertaken jointly with experts from COGEMA, a French company. These studies showed that the Elstiin and Urt ore occurrences are similar to the economic uranium deposits of the "granite" type found in eastern China, the Central French Massif, and northern Kazakhstan. The similarities are expressed in the control of mineralization by fault zones, the development of sericitization ("episyenitization"), beresitization, hematization, intense chloritization, vein quartz-sulfide, and fluorite mineralization, and a marked nasturan (nasturan-coffinite) stage of mineralization.

One can also see similarities between mineralization in the Elstiin and Urt sites and the poorly studied hydrothermal uranium occurrences in the Chikoi and Daur regions of Russia. The ore occurrences of Anomalnoe (Lomovoe), Mergenskoe, and Onkoekskoe also show a spatial relationship with granites of the Asakan-Shumilovsky and Kharalga complexes enriched in uranium, tungsten, and tin. It is useful to compare details of the Janchivlan District of Mongolia with the fault-sheared basement hosted uranium in the Chikoisky, Daursky, Olovsky regions of the Russian Zabaikalia in order to estimate potential future uranium resources.

Intrusions of highly radioactive leucocratic granite and occurrences of uranium mineralization localized within them are unevenly distributed—most of them are located within the Khentei (Khentei-Daur) arched uplift (the Elstiin, Tamga, Urt, Arshaan occurrences in the Janchivlan massif and the Borzhi occurrence in the western part of the Bayandelger massif).

Commercially interesting uranium deposits (deposits in Gornoye, Berezovoe, etc.) in the Mesozoic leucocratic granites of the Khentei-Daur arch uplift in the region of Zabaikalia (a large region mostly east of Lake Baikal that transects Mongolia and Russia), cannot be unambiguously considered as occurrence models for Mongolia, since the occurrences of uranium established in Mongolia in connection with similar granite massifs, even within a single regional structure, have significant geological differences.

Occurrences at Elstiin (Fig. 3.9), Tamga, and other locations within the Janchivlan leucogranite massif are relatively well studied. Uranium mineralization in the Janchivlan massif is represented by pitchblende and coffinite in association with sulfides: marcasite, galena, and sphalerite deposited directly in veinlets with montmorillonite. The oxidation zone of sulfide and uranium ores is widely manifested on the surface, forming a dispersion halo that can be traced to a depth of 400 m. The minerals of the oxidation zone are manganese hydroxides, malachite, azurite, covelline, chrysocolla, pyromorphite, mimetesite, wulfenite, pharmacosiderite, metathorbernite, metaotenite, zeynerite, uranophane oh, casolite, and β uranotyl.

Mineralization is localized in zones of crushing, brecciation and intense mylonitization, in tectonic sutures with sheared clay gouge. At occurrences delineated by surface trenches and drilling, uranium mineralization has been traced to a depth of 400 m. It has been established that economic grades of uranium are formed in aureoles developed around reserve blocks in zones of sericitized, beresited, argillized granites. Particularly favorable for ore deposition are bends in faults and areas of intersection and development of multidirectional faults.

Economic mineralization occurs as clusters and short lenses with lengths of up to a few tens of meters and thicknesses of several tens of centimeters to a few meters. The uranium content

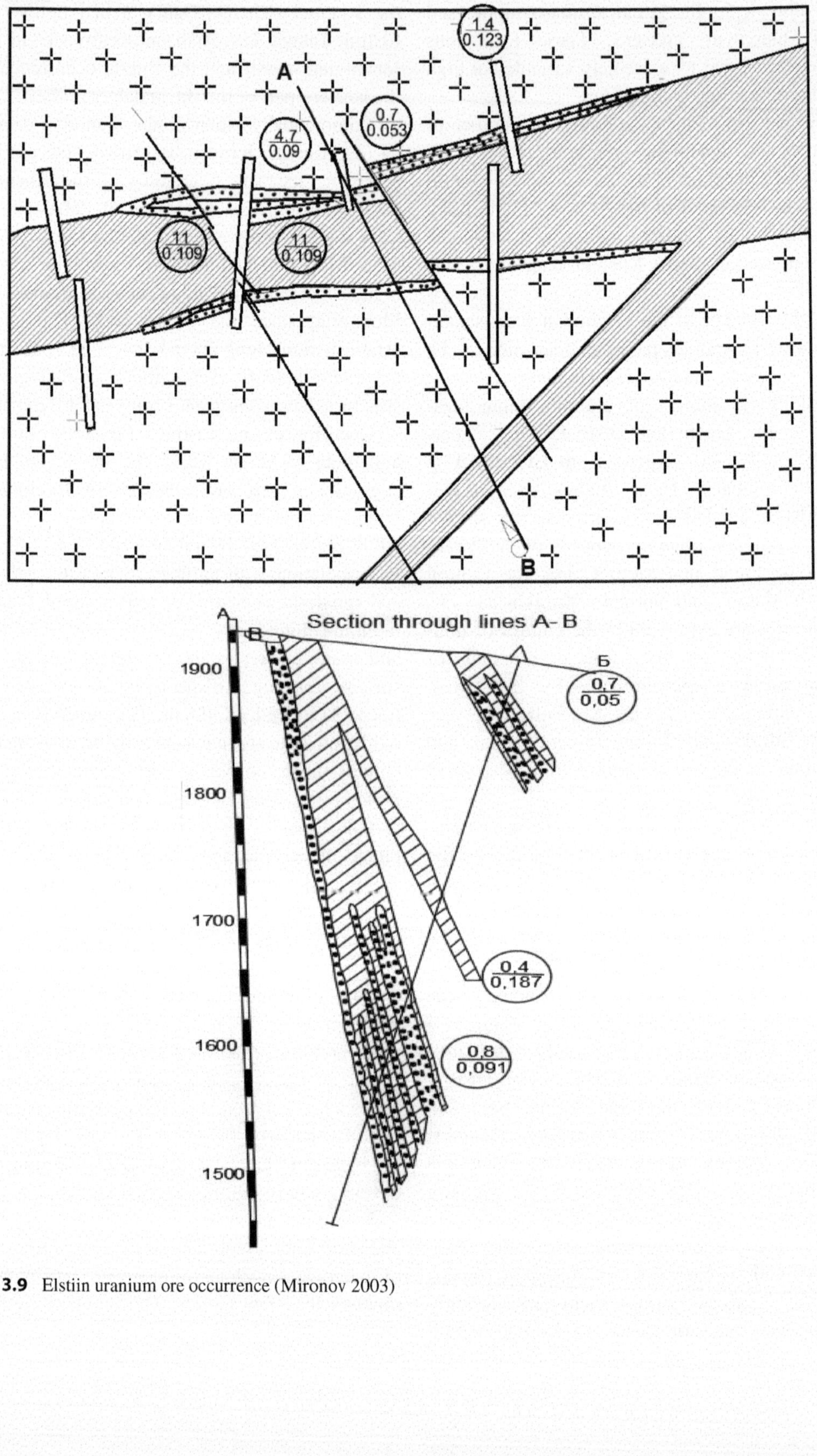

Fig. 3.9 Elstiin uranium ore occurrence (Mironov 2003)

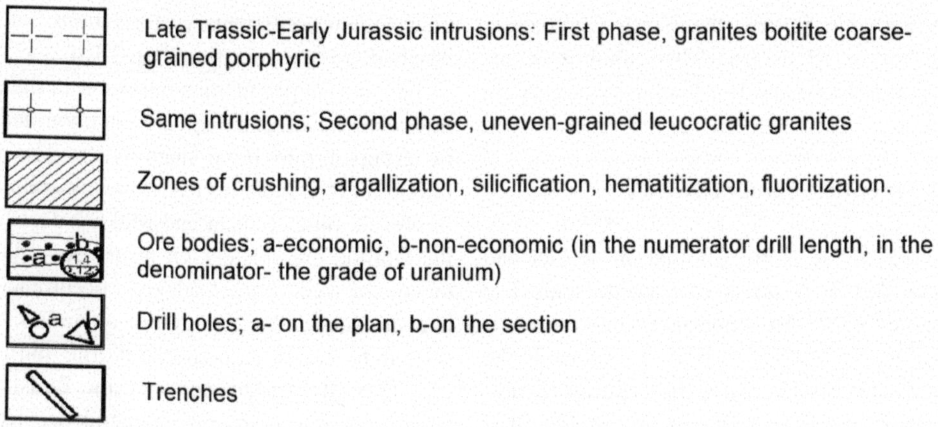

Fig. 3.9 (continued)

is usually 0.05–0.1, less often 0.15–0.25%, in some areas up to 0.325%.

Primary uranium mineralization is represented by pitchblende and coffinite, commonly associated with pyrite, arsenopyrite, chalcopyrite, galena, and sphalerite. Of the vein minerals, fine-crystalline violet and pink fluorite are consistently present, while carbonates and sometimes zeolites are less common. In the oxidation zone (100–400 m thick), uranium minerals are replaced by supergene minerals metaotenite and metathorbernite; casolite, uranophane, and less commonly zeinerite.

Malachite, azurite, chrysocolla, covellite, scorodite, wulfenite, pyromorphite, mimetesite, pharmacosiderite, and limonite are formed due to oxidation of sulfides. The outboard (of uranium) geochemical halo elements include lead, molybdenum, zinc, arsenic, and silver.

The age of mineralization has been estimated using the lead isotope method, in the laboratories of VIMS, IGEM, and VSEGEI, resulting in a bimodal distribution curve with a peak at 100–110 Ma apparently reflecting the deposition time of the bulk of uranium mineralization. Exploration and appraisal work conducted within the Central district did not identify significant mineralization with economic uranium grades. Identification of new targets that do not outcrop is of course possible, at depth in "structural traps", such as, for example, intersections of steeply dipping faults with relatively gentle plunge.

Higher-grade uranium mineralization can be found at the bends of the controlling fault zones, under the cover of gentle contacts and at the base of Devonian schists occurring as xenoliths in the granite. In such situations, the formation of small and medium-sized deposits is possible. The most prospective zones are along the Batur and Janchivlan faults in the area of the Tamga occurrence, as well as the zone of the Bayandavaa fault, which terminates the Janchivlan ore district in the north.

3.2.1.6 East Mongolian Uranium Zone

The nature of the distribution of uranium mineralization and the features of the geological structure of Mongolia made it possible to distinguish the East Mongolian uranium zone (Fig. 3.4) as a discrete district with endogenous (primary) and exogenous (secondary, mobilized, and redeposited) mineralization.

The geological structure and metallogenic appearance of the zone is very similar to the Khentii-Daur zone, therefore the zone is considered one of the most prospective for the discovery of new and potentially economically significant uranium deposits. In radiogeochemical terms, the zone is characterized by a high amplitude of anomalies in areally extensive blocks and zones, such as the Middle Gobi-Oshinnuur, Kharayrag, Berkh, and Choibalsan-Onon.

Within the zone, combined with the established uranium grade and the occurrence of other favorable factors, the North Choibalsan, Choir, Shivee, Ulziit, Khairkhan uranium occurrences, and Berkh and Bor-Undursky targets are noted.

Berkh Potential Uranium District

The Berkh District is located 100 km west-southwest of the North Choibalsan district and is controlled by a series of variable scale late Mesozoic VTS. An important ore-control is the transverse northwest-striking system of the Munkhkhaan-Binder faults. In addition to uranium mineralization, economically viable fluorite deposits (Berkh, Idermeg, and others) have been proven within the region.

According to the geological structure and the occurrence of uranium mineralization in the area, two potential uranium centers are highlighted in the western part of the Ikhbulag (Batnorov) cluster, and on the eastern flank-Ulzii-Saikhanuul (Tumenkhaan) cluster.

The first cluster is characterized by late Mesozoic depression structures filled with sedimentary rocks and volcanic rocks of basic and intermediate composition. Rhyolites are present as separate necks and subvolcanic bodies.

The second cluster is formed around a system of rhyolite volcano-dome structures, relatively highly enriched in radioactive elements (6 ppm uranium and 27 ppm thorium). They combine into the large Ulzii-Saikhanuul VTS, within which one other uranium occurrence (Tumenkhaan) and a series of uranium anomalies related to fluorine–molybdenum–uranium mineralization are known. Except for a few occurrences where moderately detailed work was conducted, the potential of the Berkh uranium district is poorly studied.

North Choibalsan Uranium District

The metallogeny of the North Choibalsan district is characterized by polymetals, uranium, fluorite, rare metals, and gold, the formation of which is thought to be associated with collisional rifting (Table 3.6). Economic deposits (Dornod, Gurvanbulag, etc.) occur in volcanogenic-sedimentary formations of the basalt-rhyolite association (Fig. 3.10).

For most types of mineralization in the area, multielement associations are common and can be grouped into three main types. The most important type includes lead–zinc mineralization in breccia pipes (Ulaan and Mukhar deposits in the Dornod ore cluster). Of nearly equal importance are vein type lead–zinc occurrences in beresite (quartz-sericite-pyrite) alteration zones related to small granodiorite-diorite intrusions (the Tsav, Bayan-Uul deposits, the Baitsiin ore occurrence, etc.). In some mineral occurrences of this type, silver is the dominant element.

Epithermal fluorite deposits are the primary type, with economic importance, in the district. Within the Dornod ore cluster, the large Kholoi deposit was discovered, and additional prospective ore occurrences were identified in the Turgen, Tsav, and other ore clusters. All prospects are characterized by the quartz-fluorite composition of mineralization and the vein type ore bodies. The identified reserves and potential resources of fluorite in the North Choibalsan district are comparable in size and grade with the reserves of the Berkh, Borundur, and Khar-Ayrag ore districts of the East Mongolian fluorite belt.

The mineral occurrences of the North Choibalsan district were formed in two main stages of the Mesozoic metallogenic epoch. In the early-middle Mesozoic stage, rare-metal mineralization formed in association with leucocratic granite intrusions, and gold, molybdenum and lead. Zinc paragenetically associated with granodiorite-diorite complexes. Formation of polymetallic, uranium, fluorite mineralization, and related mercury and zeolites is associated with the late Mesozoic stage.

The North Choibalsan uranium district is confined to the central part of the Choibalsan-Onon volcanic province, which includes a number of large depression-type volcanic-tectonic structures superimposed on the North Choibalsan uplifted block of ancient crust. The Dornod deposit, and Ugtam and Turgen prospects have been identified in the area. Of these, the Dornod cluster is of the greatest interest, within which economic deposits of uranium,

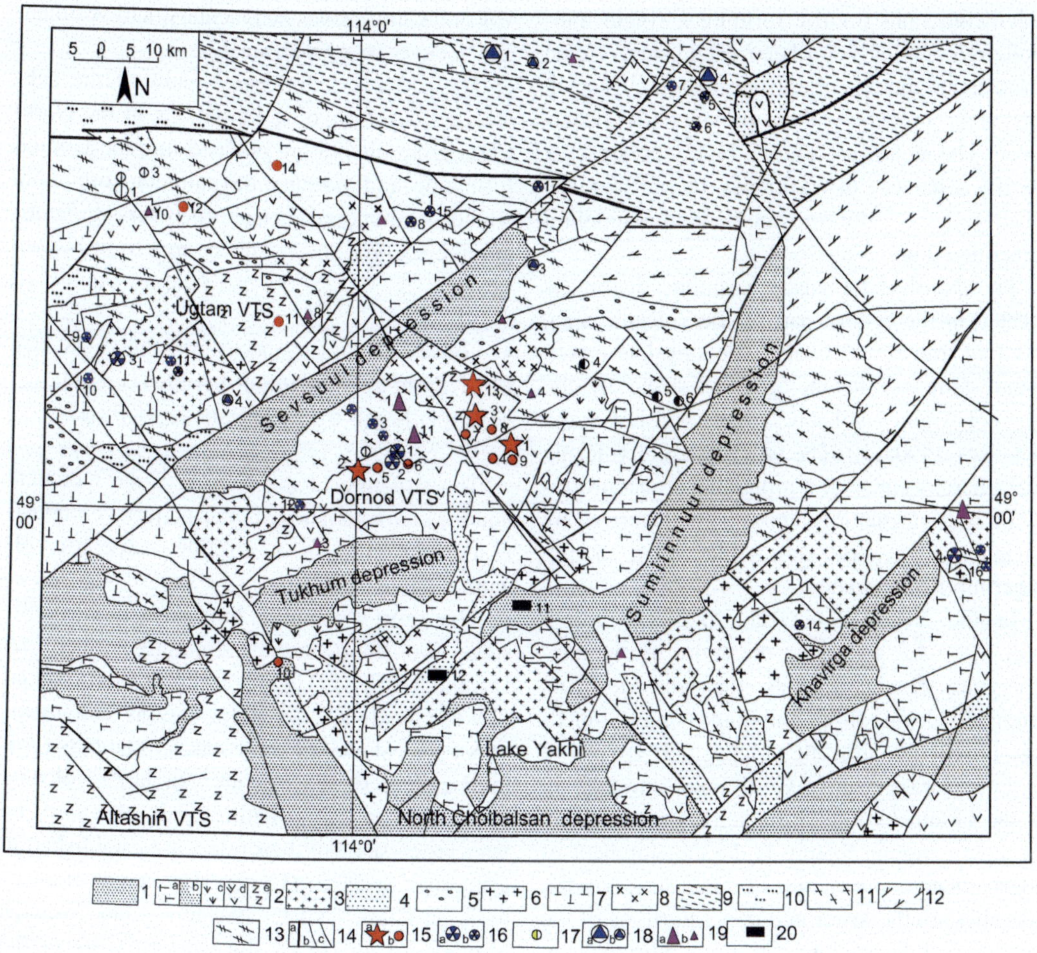

Fig. 3.10 Schematic geological map of the North Choibalsan uranium district (according to Mironov 2003): 1-2—Late Mesozoic formations: 1—terrigenous coal-bearing (K_1), 2—subalkaline basalt-rhyolitic (J_3–K_1) with three subformations: upper (A)—trachyandesites (a), sandstones, conglomerates (b), rhyolites (v); middle (B)—rhyolites, trachyrhyolites, their tuffs; lower (C)—basaltic andesites, dacites, trachyrhyolites, their tuffs, sandstones, conglomerates; 3-4—Early Mesozoic formations: 3—monzonite-granite-granosyenite (MZ_1), 4—coarse clastic terrigenous molasse (T_3); 5-11—Early and Middle Paleozoic formations: 5—terrigenous molasse (P_2), 6—leucogranite (P_1), 7—andesite-rhyolite (P) formation, 8—late orogenic diorite-granite formation, 9-12—accretionary-collisional: 9—terrigenous flyschoid (D–C_1), 10—carbonate-terrigenous (D_1), 11—granite-granodiorite (PZ), 12—spilite-diabase-shale formation (NP_3–E_1); 13—metamorphic complex of the Precambrian basement: undivided carbonate-volcanogenic-terrigenous formations; 14—faults: a—transregional faults, b—regional faults, mainly of Paleozoic age, c—late Mesozoic syn- and post-collision faults. Minerals: (a—deposits, b—ore occurrences): 15—uranium: 1—Dornod, 2—Gurvanbulag, 3—Mardaingol, 4—Khavar, 5—Davan, 6—Mukhar, 7—Tsever, 8—Ilrekh, 9—Khar, 10—Bagazosnuur, 11—Ugtam, 12—Baruunkhooloi, 13—Nemer, 14—Tsagaanchuluut, 15—Road; 16—polymetals: 1—Ulaan, 2—Mukhar, 3—Bayan—Dun, 4—Tsav, 5—Eltrud, 6—Lutaagiin, 7—Nomint, 8—Bolostyn, 9—Sogoot-Uul, 10—Sant-Uul, 11—Arbulag, 13—Butan, 14—Bayanbulag, 15—Modotur, 16—Bayan-Uul, 17—Erdene-Tolgoi, 18—Tsairyn, 19—North Ulaan; 17—gold: 1—Turgen, placer, 2—Khudag, 3—Chuluut—Khooloi, 4—Ugtam. 5—Urliynovoo, 6—East-Dagai, 7—Tokhoi; 18—tin, tungsten, molybdenum: 1—Chuluun-Khureet, 2—Chuluut-Bulag, 3—Dagai, 4—Zun-Naimt-Bulag, 5—Avdar-Tolgoi; 19—fluorite: 1—Baruunsuuj, 2—Khuvbulag, 3—Buduunbulag, 4—Delger, 5—Tsagaannuur, 6—Ozernoe, 7—Ganzhuur, 8—Ugtam, 9—Jaraakhai, 10—Baruunkhooloi, 11—Khooloi: 20—brown coals, including those with uranium mineralization (c): 11—Delgernuur, 12—Shinebulag

polymetals, and fluorite forming a single mineralized complex have been established. Some researchers include the Engershand cluster in this area, including Tsav polymetallic occurrence. The area is considered highly prospective for the discovery of additional, though perhaps small-scale, fluorine–molybdenum–uranium mineralization.

The North Choibalsan uranium district is located in the central part of the volcanic area, trending transverse (likely reflecting late extensional faulting) to the general strike of the Mongol-Argun volcanic belt (Mironov and Solovьev 1990a, b). The late Mesozoic volcanic rocks here occur with a pronounced unconformity on older formations and are characterized by an independent structural fabric and a sharply superimposed character.

In the North Choibalsan district (Mironov et al. 1989), as well as in the entire Mongolian-Priargunsky volcanic belt, the late Mesozoic eruptive volcanic flows and domes and sedimentary rocks of contrasting composition are as old as Late Jurassic-Early Cretaceous and belong to the basalt-rhyolite (subalkaline) assemblage. Its most common feature is the development of intermediate and basic rocks (andesites, basaltic andesites) in the lower and upper portions of the section, and the predominance of acidic rocks in the middle part. At the same time, important elements of lateral zonality were noted. Rhyolitic volcanism, intense in the axial part of the range, thins out on its periphery, where synchronous tuffaceous-sedimentary strata of acidic composition are more widely developed.

In the late Jurassic and early Cretaceous settings, thin layers of basaltic andesites and relatively thick rhyolitic tuffs, siltstones, sandstones, and conglomerates are typical and usually underlain by "lower" and overlain by "upper" basaltic andesites. The significant rhyolite composition of clasts in sediments, the significant role of acidic tuffaceous material in cement and clasts, the presence of pebbles of characteristic rocks of the rhyolite complex (sanidine and morion (quartz) rhyolites, trachytes, dacites, etc.) make it possible to confidently correlate volcanics, tuffaceous-sedimentary, and sedimentary rocks across broad areas.

The effusive-sedimentary rocks of the subalkaline basalt-rhyolite formations in the North Choibalsan district are grouped into the Dornod formation and younger Bayantumen formation, which is composed of three members. In stratigraphic section, the members are (from bottom to top):

- Lower basalt-trachyandesite-rhyolite (J_3–K_1bt_1).
- Medium trachyrhyolite-rhyolite (J_3–K_1bt_2).
- Upper basalt-trachyandesite (J_3–K_1bt_3).

The structure of the section of the members and the characteristics of their material composition are given in the description of specific volcano-tectonic structures.

The most "mature", long-term, and multistage volcano-tectonic structures, such as Dornod (DVTS), Ugtam (UVTS), and Turgen (TVTS), include all the formations and members listed above. In some cases, members form independent small structures. In large volcanic structures, earlier structures are obscured by later processes, but their relics are established in a number of cases (Upper Turgen volcano-tectonic structure—VTTS, Western Ulan, Western Mardaingol, in UVTS the western wing of the Zharaakhai trough, South Ugtam, Ukhagchin structure in TVTS, etc.).

The structures of the basalt-trachyandesite-rhyolite units are located along the periphery of large volcano-tectonic structures or locally outside of the main Choibalsan-Onon zone (Bagazosnuur, Jargalant, Khaistovoo, etc.).

The Ugtam caldera proper is the most distinct structure, composed mainly of rocks of basalt-trachyandesite-rhyolite affinity. The southern flank of the Ugtam structure and the Jaraakhai trough, the Ulaan and Dornod depressions, and other late structures of large volcanic domes are predominantly composed of rocks of the trachyrhyolite-rhyolite member.

The basalt-trachyandesite member developed both in the southern part of the Ulaan and Tsagaannuur depressions and in the central part of the Jaraakhai trough, gradually merges into (interfingers with sedimentary units of) the Early Cretaceous terrigenous depressions (Sumiinnuur, Sebsuul, etc.).

The Dornod Volcano-Tectonic Structure (Fig. 3.11), including a number of volcano-tectonic structures of the North Choibalsan block, is by far the largest and the most well-studied. It occupies an area of about 1800 km^2 and is elongated in the northeast direction. The Dornod VTS is confined to the area of large regional conjugate fault structures such as the northwestern Choibalsan-Onon and northeastern Ulaan and consists of a series of down-dropped blocks (volcanic and tectonic depressions) with the predominant development of late Mesozoic sedimentary-volcanogenic formations and uplifts separated by protruding blocks of the pre-Mesozoic basement.

Theories about the deep structure of the Dornod VTS are based on the interpretation of geophysical surveys carried out by MGSE. Careful analysis of their results revealed the complex block structure of the DVTS. The down-dropped and uplifted basement blocks are

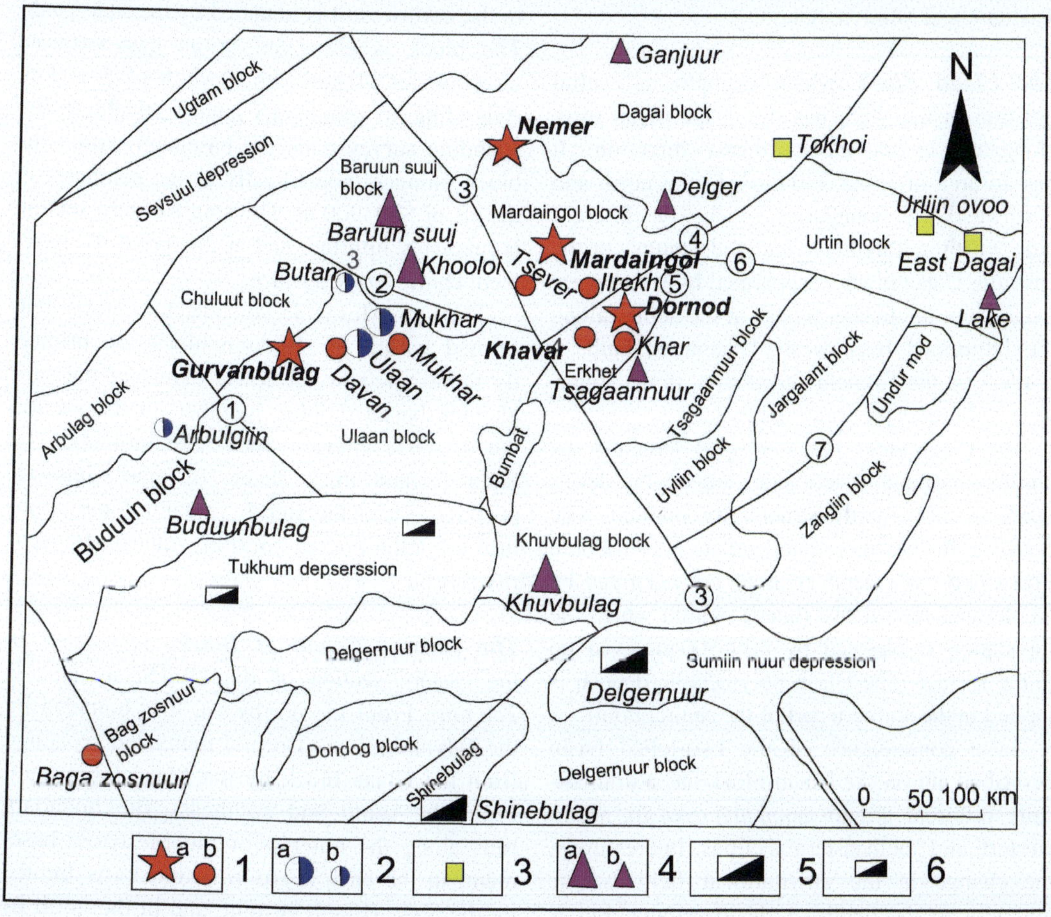

Fig. 3.11 Geologic block structure and mineralization of the Dornod volcano-tectonic structure region (according to Mironov et al. 2012). Deposits (a) and ore occurrences (b): 1—uranium: 1—Nemer, 2—Mardaingol, 3—Gurvanbulag, 4—Khavar, 5—Dornod, 6—Davan, 7—Mukhar, 8—Tsever, 9—Irlekh, 10—Khar, 11—Baga-Zosnuur; 2—polymetallic: 1—Ulaan, 2—Mukhar, 3—Butan, 4—Arbulgiine; 3—gold occurrences: 1—Tokhoi, 2—Urliinovoo, 3—East Dagai; 4—fluorite: 1—Baruunsuuj, 2—Khooloi, 3—Khuvbulag, 4—Ganjuur, 5—Delger, 6—Tsagaannuur, 7—Lake, 8—Buduunbulag; 5—brown coal deposits: 1—Delgernuur, 2—Shinebulag; 6—uranium bearing coal ore occurrences. The main interblock faults (numbers in the circle in the figure): 1—Tsakhildag, 2—Talynnuur, 3—Mardaingol, 4—Talyntolgoi, 5—Tsagaannuur, 6—Zuunurt, 7—Sharart

characterized by lower and higher gravitational fields, respectively, and can therefore easily be mapped with gravity surveys.

Large faults with long and complex history are identified by zones of linear gravity gradients with an intensity of 1–10 mG/km (horizontal gradient maps), zones of sharp magnetic field gradients, discontinuous but linear magnetic anomalies, field deformations, etc. Based on the interpretation of gravimetric, magnetic, and geological maps within the Dornod VTS, the following large subsided blocks were identified: Ulaan, Buduun, Dornod, Mardaingol, Urtyn, Tukhem, and Sumiynnuur.

The Ulaan Block, which occupies a central position in the VTS, has the shape of an oval, elongated in an east-northeast direction. It has an area of about 330 km^2. In the north and northwest, the boundaries of the block coincide with the Talynnuur and Side faults. In the east, the Ulaan block is separated by a basement ledge from the Erkhet block; in the south, along the latitudinal (east–west) Ulaannuur fault, it borders on the Tukhem depression; in the southwest, it borders the Buduun block.

The Ulaan block has a complex structure. Its northwestern and eastern sides are weakly developed, possibly partly structurally omitted. The depths to the basement interface are 600–1000 m. These two basin depth regimes are separated by an elevated central rise (horst), within which the foundation is elevated by 400–500 m, even up to the surface. The Ulaan polymetallic deposit is located in the northern part of the central uplift.

As a consequence of the two-lobed basin described above, the Ulaan block has a dramatically different gravity anomaly over its northwestern and southeastern sections. Interestingly, the change in the gravitational field is due not only to the depth of the basement, but is also influenced by a sharp change in its material composition—from crystalline schists in the northwestern section to granitoids in the southeast. The difference in measured gravity between the two basin lobes is over 10 mG.

Late Mesozoic volcanism (subalkaline basalt-rhyolitic and volcanoclastic formations) was intensely and most fully developed in the Ulaan block, culminating in the formation of a depression (collapse) structure with elements of caldera subsidence.

The Buduun Block is located in the southwestern section of the Dornod VTS and covers an area of about 150 km^2. In the east, it adjoins the Ulaan block along the system of the northwest striking Tsakhildag fault. In the southeast, along the Ulaannuur fault, the block borders on the Tukhum depression, in the west it is separated by a fault from the small Takhilgat block, and in the northwest it is bounded by the Side Fault. The block is elongated in an east-northeast direction for 20 km with a width of 7–10 km. According to gravimetry and vertical electric sounding survey data, the basement within the block plunges monoclinally to the southeast to depths of 800–900 m. The southwestern section is relatively uplifted, and at this point the basement comes to the surface.

The Late Mesozoic volcanic formations that formed within the Buduun block are primarily represented by the lower layers of the subalkaline basalt-rhyolite member, in the section where felsic volcanic rocks predominate. The latter formed thick sheets of quartz-feldspar porphyries and the Berkh volcano-dome structure of oligophyric morion (quartz)-feldspar rhyolites.

The Dornod Group of Blocks is located in the eastern section of the DVTS, its area is 230 km^2. From the northwest, it is bounded by the Talyntolgoi and Dornod faults and separated from the Ulaan block by the Mardaingol fault. From the south and southeast, the boundary runs along the outcrop of the Mesozoic basement, in the east, a group of blocks is limited by the Shavart fault system, and in the north by the Zuunurt deep faults. Along the Tsagaannuur fault, the group is divided into two blocks, Erkhet and Tsagaannuur, which are associated with depression structures controlled by tectonic faults.

The Erkhet block is divided along the Bagaerkhtet fault into a relatively uplifted northeastern section (average volcanic thickness 500–700 m) and a down-dropped southwestern section (average volcanic thickness 1000–1200 m). Structures of the northeastern section are narrow erosion-tectonic depressions ("paleovalleys"), elongated in a northeast direction, and the effect is to complicate the underlying structural fabric.

The Tsagaannuur block is poorly studied. According to gravimetry and electric survey (VES = vertical electrical sounding) data, the block has a simple structure. The maximum depth of the basement interface as imaged in the central section is around 1000–1200 m.

Late Mesozoic volcanic rocks within the Dornod group of blocks are represented by an almost complete section of the subalkaline basalt-rhyolite formation. Subvolcanic facies of rhyolitic and basalt-andesitic composition are widely developed. In the modern erosion section, the marginal parts of the block are composed of rocks of essentially felsic composition, and the central section, corresponding to the Tsagaannuur subblock, is filled with late basaltic andesites. The nature of the distribution of volcanic products, the vertical changes in stratigraphic section, and the analysis of the tectonic structure of the block allowed G. V. Zverev and G. A. Shatkov to interpret a probable caldera origin of the Tsagaannuur depression. The Mardaingol block is located in the northern section of the DVTS and occupies an area of about 130 km^2. In the south, it is separated by a basement high from the Dornod and Ulaan blocks. The rest of its boundaries coincide with the outline of the DVTS. The Mardaingol block is an isometric structural depression complicated by large displacement faults of various orientations. The northern section of the block is downdropped relative to the southern section along the latitudinal (east–west) Zuunurt fault, while the western section is uplifted relative to the eastern section along a meridional (north–south) fault, fragments of which are evident as basement outcrops in the northern and southern sections of the block. The greatest thickness of volcanic rocks, according to gravity and VES data, is noted in the eastern section of the block, and the thinnest, in the southwestern section. In terms of composition and completeness of the section of volcanic formations, the block differs from the Dornod block only in lesser thicknesses of volcanic rocks.

The Urtyn block is located in the northeast section of the DVTS, its area is about 120 km^2. The block is limited by outcrops of crystalline basement on the northwest and northeast. The southern boundary is the Zuunurt fault.

In the gravity survey, the Urtyn block is demarcated by two gravity lows separated by a relative high. This gravity gradient indicates the presence of the Shavart fault, striking northeasterly, dividing the block into two sections. The northwestern section of the block is characterized by deeper basement rocks.

Late Mesozoic volcanic rocks form a volcanic edifice of the central type (Urtyn paleovolcano) on the Urtyn block. The block is composed of vent and near-vent facies of dacites, rhyodacites, trachydacites, and trachyandesites, with the lower part of the section dominated by subalkaline basalt-rhyolite.

The Tukhum terrigenous basin (about 300 km^2 in area) is a depression elongated in the east-northeast direction, consisting of three separate troughs. The northwestern side, according to geophysical data, has a more, gentle basement dip, and the southeastern side is steeper, probably due to down-stepping normal faults. The basin is filled with deposits of the Zuunbayan formation (K_1), whose thickness, according to geophysical data, ranges from 500 to 1300 m.

The Sumiinnuur Early Cretaceous terrigenous depression limits the DVTS from the east and southeast. The depression has tectonic boundaries and is characterized by an asymmetric structure. The deposits of the terrigenous coal-bearing formation (the Zuunbayan formation) are in contact with the DVTS volcanic rocks along the northeastern marginal fault, extending beyond the study area. In some areas, we observed the Zuunbayan formation transgressively overlying the VTS volcanic rocks.

The Ugtam Volcano-Tectonic Structure in the northwestern section, of the North

Choibalsan region, occupies a central position among three large volcano-tectonic structures of the Choibalsan-Onon area. The regional faults with a NW direction (Erdenetolgoi, Ugtam, Tsogdelger, Ulyast), established within the UVTS, are linked, respectively, with the Tsakhildag, Mukhar, Mardaingol, and other faults of the DVTS. The Ugtam structure (as well as the DVTS) is confined to the junctions of these faults and the northeastern regional faults (Sebsuul, Jaraakhai, etc.). The combination of the NE and NW faults determines the block structure of the UVTS. From west to east, the Suuj, Ugtam, Tavnykhundi, and Jaraakhai blocks are highlighted.

The Suuj and Tavnykhundii blocks are composed primarily of pre-late Mesozoic formations, and only in the southern sections are the basement structures overlain by late Mesozoic volcanic and sedimentary rocks, as infill of the Ugtam caldera and Jaraakhai trough.

The Jaraakhai block of the UVTS and the Khailaast block of the TVTS are characterized by minor rhyolitic volcanism. The major feature is trough-like structures filled with tuffaceous-sedimentary formations, synchronous with rhyolitic volcanism. Whereas, to the west of the Ugtam (UVTS) and Sumtyn (TVTS) blocks, felsic volcanism is widely developed.

In the UVTS, two structural floors are distinguished—the lower one, which is the foundation, and the upper one, composed of upper Mesozoic sedimentary-volcanogenic formations.

The block structure of the UVTS and its bounding faults are clearly expressed in geophysical data, and, above all, in the gravimetric data. The magnetic survey refined the boundaries of the blocks and confirmed the interpretation of fault tectonics.

The Ugtam Block (caldera) occupies a central position in the UVTS and has a pentagonal shape in plain view, somewhat elongated in a northeasterly direction. It has a markedly different gravity field compared to the uplifted Suuj and Tavnykhundii blocks, expressed as low values of the Bouguer anomaly, minus 92–100 mG against a background of minus 75–85 mG.

In the southeast of the caldera, biotite-bearing dacites, corresponding petrochemically to trachyfelsite-porphyries, are light gray, massive, sometimes with flow textures, and often with biotite flakes. Dacite thickness is 40–200 m. At the bottom of this sequence, there is a horizon of essentially layered lavas and tuffs. The lavas are spheroidal, spheroidal with flow textures, to massive, and pinkish-gray in color. In the upper part of the sequence, there is a sheet-like body of polyphyric quartz-feldspar porphyries. The base of the tabular porphyry unit is composed of interbedded lithocrystal-lithoclastic tuffs, ignimbrites, quartz-feldspar porphyry volcanoclastics, and volcanic glasses (vitrophyre). Massive polyphyric quartz-feldspar porphyries occur conformably above. The thickness of the plate-like body, according to drilling data, is from 300 to 500 m.

In the final phase of the formation of the caldera along the northern and eastern bounding faults, subvolcanic intrusions of granite-porphyries and granosyenite-porphyries were introduced as ring fault resurgent intrusions, in the central part of the caldera and in the northwest of the UVTS, late-stage extrusions of coarse phenocryst quartz-feldspar porphyry were erupted.

The basalt-trachyandesite-rhyolitic member rests on rocks of the pre- and late Mesozoic basements in various parts of the section. Its average thickness, according to gravity and VES data, reaches 800–900 m. Along the sides of the Ugtam caldera and Jaraakhai trough rock layers dip at angles of 15–20° into the trough and become nearly horizontal in the central parts of these structures, essentially forming a concentric basin structure.

The Turgen Volcano-Tectonic Structure is located on the left bank of the Ulz River below the mouth of the Turgen River tributary, in a zone of conjugation of the North Choibalsan block with the Dashbalbar block. The area is about 450 km^2, elongated in an east–west direction.

The structure is controlled by regional faults of northwesterly strike and is divided into two large blocks by an apparent east–west trending uplift (horst) of the basement, resulting in the separate and distinct Sumtyn and Khailaast blocks. The western section of the Sumtyn block contains the Early Cretaceous Ust-Turgen depression, filled with terrigenous deposits of the Lower Cretaceous. Late Jurassic-Early Cretaceous rhyolitic volcanism was intense in the central part of the block, mainly as flow dome complexes, with a subordinate element of stratified formations. Within the Khailaast block, basaltic andesite flows are often interbedded with sedimentary rocks, the fragmental composition, which is dominated by acidic volcanic rocks showing both alluvial and air transport.

The total thickness of the upper Mesozoic deposits, according to the VES data, is more than 600 m. The dip angles of the rock layers do not exceed 10–15°. The basal deposits and basement in the center of the Khailaast block were not penetrated by wells. Bounded by prominent structures, pre-late Mesozoic formations are represented by Devonian chlorite-sericite and siliceous-chlorite shales and sandstones intruded by upper Paleozoic medium-coarse-grained leucocratic biotite granites.

Steeply dipping faults are widely developed and gentle interstratal accommodation failures are common. The steeply dipping Tsagaanchuluut fault, which controls uranium mineralization, has been well studied. The siltstone-filled Khailaast trough adjoins the NNE-striking (10–30°) Tsagaanchuluut fault. The fault is represented by a series of branching and en echelon-shaped faults (possibly anastomosing) with zones of intensely crushed, brecciated, and fractured rocks.

Especially intense steep fracturing is developed in the top of the upper horizon of trachyandesites and basaltic andesites, as well as near the base of unstratified siltstones and "pudding" gravelstones of the upper horizon. The rocks in the damage zone are hydrothermally altered. Beginning at the surface, rocks are limonitized and covered with carbonate and hyalite crusts. The total thickness of crushed, intensely fractured, and hydrothermally altered rocks reaches 115 m. The main suture has a northwesterly dip of 49–74°. The vertical component of movement, along it, reaches 130–150 m (the block is lowered to the northwest).

The Upper Turgen Volcano-Tectonic Structure (UTVTS) is located in the northwestern section of the Choibalsan-Onon area in the upper reaches of the Turgen and Duch-Gol rivers, where it occupies Mongolia (it extends northward into Russia), it covers an area of more than 450 km^2. It is elongated in a northeasterly direction and within its boundaries two late Mesozoic depressions are identified, the Upper Turgen trough-shaped depression and the Duchgol graben, both filled with sedimentary-volcanogenic formations of a subalkaline basalt-rhyolitic assemblage.

The pre-late Mesozoic basement of the UTVTS sharply differs from other volcano-tectonic structures of the Choibalsan-Onon area.

The enclosing geology of the Upper Turgen depression and the Duchgol graben are composed of a weakly metamorphosed flyschoid sequence, which we attribute to the terrigenous flyschoid formation ($D–C_1$), with intercalation of sandstone, chlorite and sericite shale, and less often limestone and quartzite.

The subalkaline basalt-rhyolite formation is represented by two members: (a) basalt-trachyandesite-rhyolite and (b) trachyrhyolite-rhyolite. The section of the basalt-trachyandesite-rhyolite member begins with a 50–70 m horizon of basal conglomerates, overlain by basaltic andesite lavas represented by a series of amygdaloidal bubbly lava flows. These are distinctive bimodal volcanic associations.

The thickness of the cover reaches 120–150 m. A thin (up to 50–60 m) cover of trachydacites occurs higher in the section, developed mainly in the western section of the Upper Turgen VTS and not found in other parts of the structure. This suggests its formation within the realm of a small and confined volcanic center, the vent, which was near Bor-Under. Acidic varieties of the formation are also represented by extrusive facies of quartz-feldspar rhyolites, the formation of which is associated with the action of several centers of volcanism of the central

(Lun and Khurmen paleovolcanoes) and linear (Duchgol paleovolcano) types.

The relationship between different volcanic facies is clearly observed in the example of the Lun paleovolcano, which has an annular structure. Vent and extrusive facies compose the central part of the volcanic edifice, while mantle facies compose the outer. The volcanic vent is a 2 km diameter concentrically dipping tubular body of crystallized rhyolites, developed within various breccias of quartz-feldspar rhyolite. The area of cover begins with a thin (1.2–4 m) horizon of red volcanic glasses (vitrophyre) and spheroidal lava, above which there is a cover of trachydacite and rhyolite up to 50 m thick.

The trachyrhyolite-rhyolite member, within the UTVTS, is a subvolcanic facies of limited extent. It forms linear dike-like bodies of quartz-sanidine rhyolite breccias or isometric granite-porphyry bodies intruding the basalt-trachyandesite-rhyolite member and basement rock.

Based on the content of radioactive elements, the UTVTS rocks are divided into two major groups. The lowest content of uranium and thorium are characteristic of the terrigenous flyschoid formation (U (1.7–2.5) $\times 10^{-4}$%; Th (9.1–12.7) $\times$ 10–^{4}%) of the basement and basaltic andesites trachyandesite-rhyolite formation (U (1.8–2.7) $\times 10^{-4}$%; Th (5.4–10.9) $\times$ 10–^{4}%). The uranium and thorium grades are notably higher in acidic volcanics of the basalt-trachyandesite-rhyolite formation, 5.8 $\times 10^{-4}$ and 22.9 $\times$ 10–^{4}%, respectively, while subvolcanic formations of the trachyrhyolite-rhyolite formation have average uranium grades of 9.3–11.0 $\times$ 10–4^{-4}%.

There are significant differences between the UTVTS and other structures of the Choibalsan-Onon structural-formational zone, for example the incomplete development of the subalkaline basalt-rhyolitic formation established for the DVTS, UVTS, and TVTS, the small thickness of volcanic rocks and their poorly expressed geochemical enrichments, and the fundamentally different type of pre-late Mesozoic basement.

The Engershand Volcano-Tectonic Structure (EVTS) with an area of about 3000 km^2 along the eastern closure of the North Choibalsan depression is mainly centered on the junction zone of the northeasterly striking Kerulen deep fault and the northwest striking system of Mangirt faults.

In the northeastern section of the EVTS, volcanic rocks from the lower part of the upper Mesozoic section (andesite-basalts and overlying felsic rocks) are predominantly developed. The "upper" andesite, trachyandesite and andesite-dacite are less important. In the uplifted northeastern section of the EVTS, the upper part of the section is apparently removed by erosion.

The early Mesozoic subvolcanic complex is developed in the northwestern section of the EVTS and shows highly complex intrusive relationships in the surrounding Paleozoic syenite, syenite-porphyry, granite-porphyry, and granosyenite. Those rocks and relationships are overlain by Upper Jurassic-Lower Cretaceous volcanogenic formations.

The late Mesozoic complex in the northwestern section of the EVTS is represented by dikes of fine grained diorite and diorite porphyrite with a northwesterly strike. The late Mesozoic sedimentary-volcanogenic complex of EVTS is represented by extrusive flow and myroclastic facies of acidic and intermediate composition, as well as continental sedimentary formations of Late Jurassic-Early Cretaceous age.

It should be emphasized that about 35% of the current uranium resources of Mongolia are concentrated within the North Choibalsan region.

Choir Uranium District

The Choir district is characterized by the large size of the Choir Valley. The Choir Valley is surrounded by highly radioactive Mesozoic granites. The Kharaat uranium deposit formed as a result of in situ oxidation. It was identified, explored, and prepared for development along with a large number of uranium occurrences. This ambitious development plan eventually stalled at an inchoate stage.

To the north, the district borders on the Nilga depression, where the promising uranium-bearing area of Shivee is located. The Choir depression, in which the Kharaat deposit is located, is a large valley 10–25 km wide and stretching for about 150 km from north to south filled with Cretaceous sedimentary deposits. The Kharaat uranium deposit is located in the southern part of the Choir depression. The deposit occupies an area of about 30 km^2. Exploration work revealed 11 potential ore zones, the largest of them, zones 1 and 2, were further explored.

Shivee Potential Uranium District

The Shivee district lies in the western section of the Shivee VTS and corresponds to a large radiogeochemical anomaly. Separate radioactive anomalies have been identified in the Mesozoic granites of the VTS basement, as well as in coal-bearing continental basin sediments that fill the superimposed Cretaceous-Neogene Nilga, Tavansuveet, and Bayanayrag depressions.

Six uranium occurrences were discovered in the Shivee area (Enger, Narangiin-Bulag, Elgen, Durtai, Tsaidam-Nuur, and Naidal). Additionally, about 50 uranium anomalies and a significant number of radioactive points were identified, including two thorium and rare-earth occurrences, six fluorite, one copper, and four tungsten occurrences. In recent years, the Enger uranium prospect has been explored and successfully upgraded to the category of a deposit, called Engershand.

The primary structural features pertaining to the occurrences are the Shivee VTS, the surrounding Mesozoic-Cenozoic depressions (Bayanairag in the west, Nilga in the south and southeast, and Tavansuveet in the north) and the Bayan-Ovoo Mesozoic granite massif. Elevated concentrations of uranium and thorium mostly occur in rhyolite (average grade of uranium 5.3 ppm, thorium 33 ppm) and particularly volcanic glass (uranium 17 ppm, thorium 53 ppm). In the main volcanics, the average grade of uranium is 2.5 ppm, thorium 9 ppm. The uranium grade in the basement gneiss and shale of the VTS ranges from 1.7 to 17 ppm and in the granite of the Bayan-Ovoo massif the background grade of uranium is quite high at 8.56 ppm.

The Shivee VTS is a dome structure, with a near-perfect annular shape, with a protrusion of pre-Mesozoic basement in the central section and a periclinal occurrence of volcanic-sedimentary formations around its flanks. Within a ring about 4–6 km wide, there are local centers of felsic magma eruptions. Basaltoids erupted along fissure channels controlled by deep faults of latitudinal (east–west) and submeridional (north–south) strike. The basalt-rhyolite association is fully developed only in the southern section of the VTS, where alternating, approximately equal thickness flows of basalt, basaltic andesite, and ultra-acidic rhyolite are preserved. The northern part of the VTS is dominated by basaltoids.

The rhyolites are characterized by various and alternating flow textures that are relatively favorable for ore deposition. Flow units 300–700 m thick alternate with horizons of lithoclastic tuffs and volcanic glasses, less often tuffites. The maximum thickness of felsic volcanic rocks, established by boreholes, is about 700 m; the volume of tuffs does not exceed 20%. The volcanogenic-sedimentary sequence in the Shivee VTS is capped by Lower Cretaceous conglomerates and an overlying package of basalt flows up to 450 m thick.

The complex geological structure at Shivee is formed by the suturing of two different age blocks at the junction of regional deep faults. A series of uplifted and subsided intensively dislocated blocks, with different history of geological development, is characteristic. Hydrothermal-metasomatic processes (alteration) strongly transformed the rocks and hosted hydrothermal uranium, fluorite, and polymetallic mineralization.

Exploration and evaluation work has established that the prospects for expanding the scale of known hydrothermal occurrences are limited. There are a number of factors unfavorable for the localization of economic uranium mineralization including the diorite-metamorphic composition of the VTS basement and its weak enrichment in uranium; dome rather than depression morphology of VTS; relative monotony of the volcanic section; lack of contrasting physical–mechanical properties and chemistry of the rocks and lack of intense alteration processes.

Uziit Uranium District

The Ulziit district spatially coincides with the Ulziit depression, one of the largest late Mesozoic depressions in eastern Mongolia. The northeast striking depression is both an erosional and tectonic graben-like structure, controlled by the zone of the Main Mongolian lineament. The length of the depression is 200 km and its width varies from 10 to 25 km.

Uranium mineralization of sedimentary units in the Ulziit depression has been known since the late 1950s. Since that time, exploration work including ground prospecting and evaluation has been conducted, and in some areas, an assessment of potential ore content was made by drilling to a depth of 100 m. As a result, several uranium ore occurrences and anomalies were identified and the Ulziit uranium deposit has been further explored.

Khairkhan Uranium District

The presently defined Khairkhan uranium district is a combination of the Gurvansaikhan and Khairkhan late Mesozoic depressions. These depressions are typically filled with terrigenous-sedimentary Lower Cretaceous coal-bearing deposits, which over vast tracts, are overlain by younger Upper Cretaceous and Neogene-Quaternary sediments. Both depressions are surrounded on all sides by outcrops of early Paleozoic granitoids containing xenoliths and roof pendants of shale-carbonate lithologies of Early and Late Proterozoic age. Early Mesozoic intrusions are locally present. The crystalline rocks of the margin are markedly faulted, characterized by crushing, brecciation, and increased rock fracturing, which contributed to the leaching of uranium and other associated components.

The uranium deposits of Gurvansaikhan and Khairkhan have been explored and largely prepared for exploitation.

Bor-Undur Potential Uranium District

The crystalline basement of the Bor-Under district is composed of lower Proterozoic metamorphites and lower Paleozoic granitoids. These are overlain by Permian acidic volcanic rocks, all of which are locally intruded by alkaline Mesozoic granitoids. According to geological mapping, electrical sounding, and drill data, the western section of the Alagtsav VTS is characterized by a complex subsurface basement topography, where uplifted and erosional ledges alternate with structures plunging toward the center such as paleo-troughs, paleo-bays, and paleovalleys. The axial sections are filled with basal volcanoclastics and polymictic breccias and conglomerates. Volcanites in the lower part of the section are represented by Late Jurassic-Lower Cretaceous alkaline basaltoids, while the upper part of the section is dominated by Early Cretaceous molasse.

Bor-Undur, Khongor, and Ulaannuur, potential uranium clusters, are combined into this unified uranium district. The Bor-Undur cluster in the western section of the Alagtsav volcano-tectonic structure and its margins spatially coincides with the Bor-Undur fluorite cluster.

Bor-Undur I, Bor-Undur II, and Zurkh uranium occurrences are located here, near or within large potentially economic fluorite deposits. The uranium mineralization was identified during exploration of the fluorite deposits, not as a uranium-focused effort. The major radiogeochemical anomaly is limited in the east by the Meridional fault, in the north by the Khajuulaan fault, in the south by the Noen fault, and in the west by the Western fault.

The occurrences of uranium in the district are divided into three groups: (1) vein, stockwork-like, and sheet-like in Jurassic-Cretaceous basaltic andesites and basal autobreccias; (2) vein and stockwork-like in granites of the Bor-Undur complex; and (3) vein and stockwork-like in Permian volcanics. Mineralization here is very similar to the Dornod district in the North Choibalsan. The Bor-Undur I mineral occurrence is challenged due to its small scale and limited prospects. The evaluation of the occurrence at Bor-Undur II has not been completed. The zone of the Meridional Fault, with direct and indirect signs of uranium mineralization, remains unexplored. Widespread development of uranium mineralization among the rocks of the Jurassic-Cretaceous volcanogenic-sedimentary strata is indicated by the presence of significant anomalies in 10 prospecting and exploration wells for fluorite. There are signs of in-situ uranium mineralization in the basal tuffaceous-sedimentary horizon and significant primary uranium halos found 6 km south of the Bor-Undur II occurrence, in the embayment structure of the Alagtsav VTS-Bor-Undriin-Khudag.

The possibility for the discovery of uranium deposits, in the crystalline basement, is indicated by the Zurkh occurrence with the mineralization of the Bayan-Uul fluorite deposit at the intersection of the submeridional (approx. north–south trending) Bayan-Ulaan fault. This area hosts an economic quartz-fluorite ore body about 3 km long, with the northeast striking Bayan-Zurkh fault. The host rocks are represented by Mesozoic leucocratic biotite granites and granosyenites, where a mineralized zone, 8.9 m thick with 0.098% uranium, was intersected in a borehole. Non-economic mineralization was found in several intervals.

Within the limits of the Bor-Undur cluster, it is recommended that additional prospecting be conducted. This should include preliminary structural and geophysical evaluation aimed at identifying paleovalleys, paleotroughs, and other negative paleostructures in the basement and margins of the VTS followed up by drilling.

The Khongor cluster covers the Dalanjargalan VTS, a semicircular caldera-type volcanic edifice formed by numerous coalescing volcanic centers. In terms of the composition and structure of the section, this VTS is very similar to the Dornod VTS in the North Choibalsan district. Its foundations are Proterozoic metamorphic schists, marbles, and limestones intruded by Paleozoic and Mesozoic granites. Acidic volcanic flows, tuffs, and volcanic glasses are characterized by high Clark (background) values of uranium and molybdenum; however, the basement rocks of the VTS do not have a clearly defined geochemical signature.

At the Khongor occurrence in the Khongor field of fluorite deposits, uranium mineralization was found in metamorphites of the VTS basement and in quartz-sanidine rhyolite-flows and domes. Identification of new uranium anomalies within the Khongor cluster will most likely be in gently dipping zones of intense crushing and fracturing in acidic effusive (flows and domes) rocks, sheeted fault zones, and at the intersection of faults with diverse geometry, accompanied by strong alteration of the host rocks. Common host units will include: the basal sedimentary horizons of the late Mesozoic volcanogenic-sedimentary sequence, enriched in pyrite and organic matter; and extrusions, necks, stocks, and other subvolcanic bodies and their contact zones, which may act as structural traps and screens (depositional sites) favorable for the formation of vein and stockwork mineralization.

The Ulaannuur cluster and VTS are transected by a large sublatitudinal (approx. east–west striking) fault. The cluster includes two occurrences and 22 interesting anomalies of uranium, with halos of primary and secondary uranium mineralization. Separate large-footprint uranium anomalies are found inside the cluster and include the Ulaannuur, Khur, and Khavtsal prospects.

The Ulaannuur VTS is an approximate east–west elongated volcanic-sedimentary depression 30 km long, superimposed on an ancient granite-metamorphic basement. It is composed dominantly of basalt-rhyolite formations of the Upper Jurassic—Lower Cretaceous. The thickness of the felsic volcanic rocks is 200–250 m, while the thickness of the underlying sediments and basic volcanic rocks is at least 1000 m.

The largest and most important structure of the cluster is the somewhat discontinuous Ulaannuur zone that can be traced along the southern margin as a series of down-stepped normal faults, with 10–100 m displacement, that rapidly drop geologic units to the center of the depression.

The distribution of radioelements in the formations of the Ulaannuur VTS is very complex.

The rocks of the crystalline base exhibit predominantly thorium mineralization. Quartz-sanidine rhyolite and volcanic glass of acidic composition are characterized by the highest concentrations, containing 6–10 ppm U and 27–40 ppm Th. In the same rocks, elevated grades of molybdenum, arsenic, cobalt, tungsten, lead, zinc, beryllium, and silver are noted.

Endogenous uranium mineralization of the Ulaannuur cluster is represented by volcanic-related occurrences Ulaannuur-I and II. These prospects remained underexplored at depth and on the flanks. Additional analysis of information on the geological structure and the results of an airborne gamma spectrometric (AGS) survey at a scale of 1:25,000 revealed new anomalies and radioactive assemblages, primary and secondary uranium halos and fields of hydrothermally altered rocks. These results require a new review of the Ulaannuur VTS and a complete re-evaluation of its potential.

Middle Gobi-Oshinnuur Potential Uranium District

The Middle Gobi district is located in the western section of the East Mongolian zone and stretches along the main Mongolian lineament on the northern side within the volcanic belt for approximately 250 km, and a width averaging 40 km and reaching up to 60–70 km. Within the district, aerogamma-spectrometric uranium anomalies and some potential mineral occurrences were revealed at Khashaat and Ger-Ovoo in the east, Ulziit-Uul in the west, and several additional points of uranium and uranium–thorium mineralization and radioactive anomalies. Subvolcanic bodies and dome structures play an important role here, as well as sediment-filled depressions. Volcanic rocks are only sparsely distributed within these basin fills.

The foundation of the area is characterized by widespread Late Proterozoic carbonate rocks. Alkaline igneous rocks are locally intruded into the carbonate basement and form rare-metal and rare-earth anomalies. Uranium mineralization was discovered in the region though occurrences are limited. In addition, several dozen local radiometric anomalies were discovered during the airborne gamma spectrometric (AGS) survey and these are good candidates for additional exploration work to determine their level of significance.

Taking into account the generally poor knowledge of the area and the significant parameters of the radiogeochemical zones, it can be assumed that within the Middle Gobi area, rather large blocks favorable for the formation of uranium deposits can be identified. One possible uranium deposit model analog is the Zaozernoye deposit in the eastern section of Northern Kazakhstan. Zaozernoye is localized in carbonate rocks and was identified at the surface as thorium anomalies and points of phosphate and rare-earth mineralization.

Of keen interest is the comparison of geological data of the Middle Gobi district with data from Eastern Mongolia (North Choibalsan district), adjacent territories of Transbaikalia (Urulyunguev district), and more remote regions (Northern Kazakhstan), including the analysis of previously established structural levels of localization of uranium mineralization. As is known, at the deposits of the Dornod district, ore deposits are noted at different levels within the VTS. At the deposits of the Streltsovskaya group in Transbaikalia, uranium mineralization is found at several levels in the VTS and also found in basement granites. Additionally, the Argun economic deposit, also related to volcanic rocks, is completely localized in the Proterozoic carbonate strata. So there are many points in common here with regionally known uranium districts that could effectively guide further exploration work.

3.2.2 South Mongolian Metallogenic Province

The South Mongolian uranium province includes folded structures formed in rocks of the paleo Neo-Asian Ocean and the Paleotethys Ocean. Lying between these oceanic structures are the South Gov and Nukhetdavaa-Ulyastai massifs of Grenville age. Uranium districts, zones, and clusters are distinguished both within folded structures and in Grenville-age massifs.

The eastern half of the province is characterized by a greater distribution of late Mesozoic taphrogenic (fault-bounded basin) sediments than the western section, where not only late Meozoic, but also Cenozoic taphrogenesis developed. This different age span of basin sediment fill makes it possible to divide the province according to the nature of potential uranium mineralization into the Gobi and Western Mongolian uranium zones.

3.2.2.1 Gobi Uranium Zone

This area is relatively well studied, and many uranium deposits and occurrences have been discovered. The volcanic-sedimentary rocks of the Proterozoic and Paleozoic zones are intruded by middle, late Paleozoic and Mesozoic intrusions resulting in a complex metallogenic setting including copper, lead, zinc, fluorspar, uranium, and rare-earth elements, of which uranium mineralization is of paramount importance.

Within the depressions, Cretaceous formations are the most widespread and well studied and large deposits of important minerals for Mongolia, such as coal, uranium, oil, and gypsum, are associated with these basin-filling sequences.

The lower part of the section in all depressions is composed of lacustrine (lake) sediments of the Lower Cretaceous, generally overlain by variegated platform and continental sediments of the Upper Cretaceous-Paleogene, and continuing up the section, variable molasse, and red beds of the Neogene-earliest Pleistocene. Additionally, as a general rule, permeable formations of the Cretaceous are uranium-bearing.

In the Gobi zone, the Zuunbayan and Sainshand uranium clusters, and prospects including Mandalovoo, Ail, Madat, and Nukhetdavaa are noteworthy.

Sainshand Potential Uranium District

Sainshand district is located in the central part of the Gobi uranium ore province and is marked by the Nars and Durbulj exogenous deposits, classified as a sandstone (roll front) type deposit, hosted in terrigenous deposits of the Mesozoic-Cenozoic depressions.

Geotectonically, the region straddles the Gobi subsidence zone and includes the late Mesozoic-Cenozoic taphrogenic (block faulted graben) Sainshand depression.

Radiogeochemically, the district is characterized by a well-defined distribution of radioactive elements.

In the Sainshand depression on the periphery of the Tugreg horst, is the Nars deposit, which consists of three primary occurrences: Inga, Mysovoy, and Nars (Afanasiev 1981, 1989).

The Ingin site, which is a small deposit in terms of uranium resources, is located along the southeastern side of the Sainshand depression, at the northeastern end of the Tugreg uplift, one element of the regional Saikhandulaan-Sainshand zone of neotectonic horsts.

The basic stratigraphy of the Nars occurrence is as follows (from top to bottom): red-colored Cretaceous-Neogene and various Upper Cretaceous platform units, unconformably lying on the lacustrine Lower Cretaceous formation, which in turn overlaps the Jurassic-Cretaceous lithified terrigenous formations and older rocks. The uranium-bearing unit is quite variable in litho-stratigraphic character. Mineralization is confined to the fourth unit up from the bottom consisting of a clayey-sandy member 70–130 m thick, lying between the variegated (40–60 m) and red-colored (40–50 m) clayey members. In the transitional mineralized unit, lower (up to 30 m) and upper (up to 90 m) aquifers are identified, separated by an aquiclude (5–14 m thick), within which mineralized intervals in drill holes are found at depths of 180–480 m.

Structurally, the Ingin site is confined to the northeastern slope of the horst indicated above, which is complicated by a zone of normal faults with minor displacement of the northeast striking Saikhandulaan fault system. The vertical displacement along individual structures range from several meters up to 70–90 m. Faults break the hinge line of the anticline into a series of extended linear, sometimes monoclinal, both positive and negative structures, all gently plunging to the northeast.

Linear structures, in turn, are complicated by a system of younger northwest-oriented faults with somewhat smaller vertical and horizontal displacements, which creates a complex structural nexus hosting the deposit. The hydrologic interpretation is that channels of exfiltration of deep circulating waters entered the Upper Cretaceous aquifer complex essentially preparing (creating favorable permeability, porosity, and chemistry) the necessary mineral-hosting zones of secondary reduction. This is thought to be confirmed by wide development of epigenetic changes in the reduction series observed in the sediments of the Upper Cretaceous unit, which controls uranium mineralization. The apparent supply of deep circulating solutions containing gas–liquid reducing agents, alternated in time, with the predominant environment of oxygen infiltration processes (i.e., a diachronous transgression of typical "redox" setting).

The most recent changes in the oxide series are various forms of late stage reducing alteration. The superimposition of a reducing environment upon the oxidized zones, is an uncharacteristic process, leading to multi-stage new mineral formation. This creates a large spread in the absolute age of mineralization, from 45 to 1.7–0.2 Ma, and a wide range in the radioactive equilibrium coefficient, from 20 to 1100%.

There are also late stage oxidation events operating over a variety of scales and times, with major variations from place to place. One example is the infiltration of oxygenated waters probably derived from the Central Mongolian tectonic block, which acted over a long period of time, and caused distinct zones of reservoir oxidation. This process probably transported mobile uranium to geochemical barriers, concentrating uranium at classic redox boundaries.

Another oxidation event, that developed along the periphery of neotectonic uplifts, for example at Tugrig, gave rise to the development of stratified oxide lenses, spread laterally for several kilometers as distinct limonitization zones. This process is interpreted to have caused the redistribution of previously formed uranium concentrations, often with near complete depletion, which is observed within profile II at the Mysovoye occurrence.

The complex changes in the reduction direction, which controls the ultimate deposition of the primary uranium mineralization, are expressed in a change in the color range of rocks along lateral dimensions. The central parts of the main reduced mineralized zones have a gray and dark gray color and are characterized by widespread bituminization and sulfidization. Further from the center of reduction is a zone of light green rocks, the color of which is the result of the development of hydromica, supergene pyrite, and often chlorite. This is followed by a zone of red rocks, which owe their color to hematite and hydrohematite. A series of changes ends with a whitish zone of sulfide-free reduction, where kaolinite and carbonates are formed. On the periphery, the last of these zones is composed of rocks that bear signs of in-situ oxidation, which is expressed in the field in a whitish color of cement as a rare veneer of limonite altered gravel and pebbles. Thus a typical concentric alteration sequence can be observed which is very helpful in the early exploration stages to understand what part of the system you are in.

Uranium mineralization is noted in all of the above zones, but the primary concentrations are confined to the contact of gray and light green rocks, at which point the physical and chemical properties of the host units change, and uranium is deposited from groundwater. Mineralization, in this case, is characterized by a shift in the radioactive equilibrium coefficient towards uranium, and its values range from 20 to 100%. In the remaining epigenetic zones, mineralization occurs sporadically and is of a relic nature, which is indicated by a significant shift of the radioactive equilibrium coefficient towards radium, reaching 1100%. In general, they are more typical of uranium concentrations in the near-contact parts of gray-colored aleuropelitic (silty claystone or argillite) interlayers, which are the outer halo of the concentrically zoned mineralization.

The Inga site was studied along a northeast striking strip about 6 km long and 1.5 km wide. The main uranium-bearing zone is a set of closely spaced en echelon-subhorizontal bodies

800–1200 m long, 100–400 m wide, 0.3–7 m thick with a uranium content of 0.016–0.67%. The cumulative thickness reaches 13 m with an average uranium grade of 0.052%. In most epigenetic zones developed in permeable horizons, the shape of mineralized bodies is lenticular and sheet-like, commonly formed at the contact of gray sulfidized sands, impregnated with hard carbonaceous minerals and with hydromicaceous light green loose sandstones, which is classic roll-front morphology.

Ore textures are highly diverse and include colloidal, collomorphic, grain edge coatings; disseminated, veinlet arrays, and rarely layered plate forms. The ores are represented by three mineral types: keritonic pitchblende, black pitchblende, and coffinite pitchblende. The most common pitchblende-kerite mineral is characterized by close intergrowth of pitchblende with a solid carbonaceous mineral, which fills gentle cracks and contact boundaries and unconformities of rocks of various permeability and forms spectacular pseudomorphs of plant remains. Interpretation of the infrared spectrum of the uranium-bearing solid carbonaceous mineral indicates affinity with the higher-order kerites. Nasturan-cherniy (black nasturan) mineral is equally common. Nasturan is noted in gray sulfidized poorly sorted sandy gravel in the form of sporadic vugs, as well as rim coatings on globular pyrite segregations. Uranium oxides form locally isolated black powdery earthy aggregates, enveloping sand grains, gravel and filling the pore space between them. The nasturan-coffinite mineral is of subordinate importance. Nasturan is observed in the form of rare vugs in association with coffinite, which stands out as highly visible microveinlets among a light green hydromicaceous-montmorillonite matrix containing octahedral pyrite and as cherry-red hydrohematite impregnations in fractured and porous zones. Typomorphic minerals are represented by solid carbonaceous substances (higher-order kerite), pyrite, hematite-hydrohematite segregations, large-scale hydromica-montmorillonite masses, and chlorite. Iron and manganese hydroxides, jarosite, and gypsum are developed in the hypogene (surficial oxidation) zone. There are also elevated grades of barium from 0.2 to 0.9%, lanthanum, chromium, vanadium, zinc, arsenic, tungsten, copper, germanium, and strontium from 0.02 to 0.09%, and molybdenum, gadolinium, cerium, lead, nickel, cobalt, antimony, and yttrium from 0.002 to 0.009%. In general, three multistage epigenetic processes are manifested here, of which the reducing one controls the main concentrations of uranium mineralization. The uranium supply to the geochemical barrier is largely controlled by the zone of strataform oxidation. It should be noted that the structure of this area is very similar to the deposits of the Central Kyzyl Kum (Uzbekistan).

The Mysovoe Ore Occurrence is located 15 km southwest of the Ingin site. Along the periphery of the Tugrig horst, a zone of stataform oxidation develops at a distance of 5 km from the edge of the lithified substrate and 1 km from the outcrop of the uranium-bearing member exposed at the surface. Weak uranium mineralization is present where the zone pinches out in the upper parts of the limnic (K_1, lake deposits) and at the base of the lacustrine-alluvial (K_2) formation in an interval 20+ m thick, at depths of 150–250 m. The front of the pinch-out of the limonitization zone was studied from well data along three drillhole profiles over a distance of 6 km. Uranium concentrations of up to hundredths of a percent stretching for hundreds of meters at a thickness of 0.3–1.3 and even 6.6 m are associated with the outer front and the upper lobe of the oxidation zone that is confined to sands and clays with carbonaceous organic matter. Individual carbonaceous-argillaceous layers, locally, have a higher content of uranium. Weak uranium occurrences of a similar nature were also noted at the Inginsky site, in particular, on profile II.

The Nars Occurrence is located 35 km southwest of the Ingin site. Along the contact zone between Lower Cretaceous lacustrine units and Upper Cretaceous conglomeratic units, a zone of increased radioactivity was traced on the surface for 11 km with a width of up to 10 m. Within this zone, pods with uranium concentrations of 0.01–0.34%, 1.5 m wide and 50–250 m long, are known from field surveys. Secondary uranium minerals are also noted: uranophane, otenite, and schrekingerite. The grade is dependent on the degree of strataform oxidation in conglomerates along the unconformity, with the deposition of uranium primary in gray-colored siltstones at the lower contact. Apparently, this mineral occurrence has a relic (fossilized) character, since no anomalous concentrations of uranium were found in the wells that would indicate an active uranium dispersal process.

At the Durbulj occurrence, uranium mineralization is associated with Aptian-Albian sediments. Here, a zone of soil oxidation was revealed with interfingering limonitization, probably developed from groundwater flows related to the Modot uplift. Mineralization is deposited by groundwater, mildly enriched in uranium.

The Sainshand district is considered prospective for exogenous (secondary dispersion) uranium mineralization, suitable for mining using the in situ leach (ISL) method.

Zunbayan Uranium District

This district includes the northeastern section of the Zuunbayan graben and the Tavankhar horst. The Zuunbayan graben is divided into the Unegt and Zuunbayan taphrogenic (fault bounded) depressions of Mesozoic age, separated by the Zuunbayan fault. The Tsagaanels and Zuunbayan oil fields, the oil fields which lie below the uranium-bearing deposits, are located in the Zuunbayan depression.

In recent years, the Dulaan-Uul uranium deposit in the Unegt depression and the largest uranium deposit in Mongolia, Zoovch Ovoo, in the Zuunbayan depression, have been explored and largely prepared for exploitation using the method of in situ leaching.

Mandalovoo Potential Uranium District

Proterozoic and Paleozoic volcanic-sedimentary formations in the Mandalovoo district are intruded by the Upper Paleozoic and Mesozoic magmatic complexes. Meso-Cenozoic sediments cover almost 80% of the area and are of interest for potentially hosting hydrogenous (water table) uranium deposits amenable to ISL extraction. The district includes the Mushgaikhudag radiogeochemical zone and the southern section of the Middle Gobi-Oshinnuur zone. A significant gamma spectrometric anomaly, detected in 2015–2018, effectively extended the known footprint of the Mushgaikhudag zone. The district is known for the Teg-Uul uranium occurrence, where there are many points of uranium mineralization associated with alkaline intrusions of the Mesozoic orogeny and the occurrence of Ongiyingol containing uranium–thorium mineralization. Alkaline varieties predominate among Mesozoic granitoids at Mushgai-Khudag and other massifs.

Mironov (2009) reported that in the Mandalovoo District there is good potential for discovering uranium deposits associated with carbonate rocks, similar to Zaozerskoye in North Kazakhstan and the deposits of the North Choibalsan region.

Undurshil Potential Uranium District

Undurshil is distinguished by a wide distribution of Cretaceous uranium-bearing sediments and the identification of several uranium mineralization points based on the results of an airborne gamma spectrometry survey. Otherwise, it remains practically unexplored.

Ail Potential Uranium District

The Ail district is located on the southwestern extension of the Zuunbayan district and includes the northeastern section of the Zuunbayan graben and the Tavankhar horst. The district is defined by Ail and Khanbogdin radiogeochemical zones. In the Tavankhar horst, uranium occurrences Khad, Ail, Dalai, Elgen, and others are known.

The Ail anomalous zone is composed of elevated radioactive element grades in pegmatites and pegmatoids (up to 30 ppm uranium and up to 50 ppm thorium), in Devonian and Permian subalkaline granitoids (8–13 ppm uranium, 30–45 ppm uranium, and 30–45 ppm uranium) and in quartz-sanidine rhyolites of the Late Mesozoic Ail VTS (15 ppm uranium and

12–18 ppm thorium). Unconsolidated deposits filling the depressions contain widely dispersed and localized uranium anomalies and broad fields of higher uranium concentrations that largely enclose and contain the known occurrences.

Matad Potential Uranium District

The Matad district is located in the extreme east side of Mongolia, in the northeastern section of the Gobi province. Geotectonically, it corresponds to the largest Late Mesozoic-Cenozoic sedimentary basin in the eastern half of the country, which combines a number of local taphrogenic (block faulted) and platform depressions filled mostly with terrigenous gray-colored and variegated, often coal-bearing deposits of the Cretaceous-Paleogene-Neogene age with a total thickness of up to 3000 m. The basin represents the eastern part of the large Sukhbaatar continental shelf (Byamba et al. 2018), which includes a Proterozoic basement covered with a 5000 m thick oceanic Silurian-Ordovician sequence and local rift-related limestone. Carboniferous, Permian, and Triassic granitoids are locally present. The Tamtsag artesian basin water source is from the mountain ranges surrounding it to the south and north; groundwater is discharged into Buir-Nuur and Dalai-Nuur lake systems, as a result of which the basin has a well-established infiltration setting. The sedimentary filling of the basin has undergone intense hypogene alteration resulting in widespread oxidation associated with surface, ground, and formation waters (i.e., meteoric) and reduction associated with deep-seated waters percolating up and along fault zones. Therefore, ideal redox settings are large and expansive in this region, conducive to roll front type uranium mobilization and redeposition.

The region includes the northeastern part of the Matad anomalous radiogeochemical zone, where Permian and Jurassic-Cretaceous felsic volcanic rocks enriched in uranium are emplaced. Late Mesozoic-Cenozoic largely unconsolidated sediments are characterized by moderate, usually poorly differentiated uranium grades (1–3 ppm) as a background component, upon which radioactive grades increase and significant uranium mineralization is present. Uranium anomalies and halos are associated with the addition and redistribution of uranium during superimposed epigenetic processes (supergene remobilization). Examples include radioactive anomalies in the sediments of the upper coal-bearing strata in the northern section of the Tamtsag depression, at the junction with the bounding Matad uplift. This area is composed of Lower Cretaceous pebbles and gravels with interlayers of medium-fine-grained polymictic sandstones. Permeable rocks are subject to surface oxidation, expressed in uneven whitening, patchy hematitization, and contain psilomelan oolites. The uranium dispersion halo is approximately 150×400 m in size with grades of $5–14\times10^{-4}\%$ largely confined to limonitized sandstone layers. The depth extent of the anomaly was not assessed since drilling was not conducted. Local hypogene enrichment of uranium has been identified at the Zuun-Bulag occurrence near outcrops of brown coal seams of the Zuun-Bulag coal deposit, where sooty oxidized coals contain fractions of a percent uranium. A number of anomalies are confined to the Hauterivian-Albian age sediments over a distance of 16 km approximating the marginal down-warped region flanking the Zuun-Art-Uul uplift. The uranium content of these sands and sandstones is confirmed by a significant group of anomalies. The Bayan-Sum anomaly, group No. 9 on the transverse uplift separating the West Tamtsag and Mogoit depressions, is confined to oxidized yellow conglomerates in the basal horizon of the coal-bearing strata which also includes clays and sandstones with interlayers of phosphoritized marls. Potential mineralization of the Cenomanian age deposits in this area is evidenced by the anomalous radioactivity of whitish-light yellow small-pebble conglomerates of the basal Cenomanian, which lie on the Aptian-Albian clayey member. On the southwestern subsidence of the Matad massif along the northeastern section of the Tamtsag depression and in the Khukh-Us area in the Hauterivian-Albian and Cenomanian sediments, formation-scale oxidation was established in mature sandy sediments with oxidation of clay interlayers in contact with sandstone. The presence of uranium in the basement, local concentrations of uranium in the bounding geologic structures, a favorable section of gray-colored coal-bearing sandy-argillaceous strata, strong signs of epigenetic changes in the oxidative and reductive states, and an abundance of areal and local radiogenic anomalies, all indicate significant accumulation of uranium in the terrigenous deposits of the depressions. These indications make it possible to consider the Tamtsag region as prospective for the identification of sandstone type uranium mineralization. However, negative factors including insufficient volume of crystalline rocks surrounding the depression and frequent facies variability in the Lower Cretaceous deposits which could limit ore body continuity must be considered. In the Matad district, it is recommended to conduct detailed exploration studies, including geological mapping, structural-geophysical surveys, trenching, and exploration drilling.

Nukhetdavaa Potential Uranium District

Structurally, the Nukhetdavaa potential uranium district corresponds to the Nukhetdavaa paleo continental margin (Byamba et al. 2018) structural basin. The lower part of the basin is filled with Ordovician and Silurian metaterrigenous and mixed metaterrigenous-carbonate sequences. Volcanic rocks and volcanic-sedimentary rocks of the Late Silurian-Early Devonian Modot Uul formation unconformably overlie the earlier strata. Throughout the Middle Jurassic, syntectonic granitoids were intruded, and flanking depressions (Unagadai and others) were filled with coarse clastic material and finally alkaline basalt flows. The region's prospectivity was identified by the presence of a distinct radiogeochemical zone and such uranium occurrences as Saikhanuul and Tsagaanuul associated with granitoids.

3.2.2.2 Southwest Mongolian Uranium Zone

The southwestern zone is composed of volcanogenic-sedimentary formations of the Silurian, Devonian, and Carboniferous age, intruded by Paleozoic monzonites, syenites, gabbro, and ultramafic rocks and has a metallogenic signature of gold, silver, lead, zinc and copper. An AGS survey conducted in 2015–2018 revealed radiogeochemical zones and anomalies. The uranium content of the zone, however, is practically unknown due to lack of follow-up studies.

On the basis of radiogeochemical anomalies and other geological prerequisites, the Baruunkhurai, Atasbogd, and Nemegt potential uranium district are identified in the zone.

Baruunkhurai Potential Uranium District

The Baruunkhurai potential uranium district is located on a natural continuation of the structures of the Zuungar uranium district of China. It is identified on the basis of geological and geophysical prerequisites for uranium mineralization. In 2015–2018, the Uvur-Altai radiogeochemical zone was identified based on AGS surveys. Structurally, the Baruunkhurai district consists of several structures of the Southern Mongolian province: the western end of the Baitag paleo-continental margin, the central section of the Bar accretionary prism, and the eastern section of the Tseel block. The geological structure of the area involves Cambrian-Ordovician metamorphites, Devonian, and Carboniferous volcanogenic-sedimentary strata, and Carboniferous and Permian intrusions.

Sedimentary deposits of the Paleogene, Neogene, and Quaternary were widely developed within the region.

Nemegt Potential Uranium District

The district is located within the western section of the Noen-Tsokhiot late Mesozoic-Cenozoic taphrogenic (extensional) structure. Radiogeochemically, the area is characterized by reasonably well-defined radioactive anomalies. The geological structure of the region includes sedimentary-volcanogenic formations of the Ordovician, Silurian, Devonian, Carboniferous, Permian and cut through by intrusions of the Carboniferous and Permian. Mesozoic-Cenozoic rock or sediments cover about 70% of the area. The western half of the region corresponds to the South Gobi oil-bearing Nemegt basin.

Atasbogd Potential Uranium District

Atasbogd has recently been identified as a potential uranium target at the western end of the Noyon-Tsokhiot structure based on results of an airborne gamma spectrometric (AGS) survey conducted in 2015–2018, which revealed a radiogeochemical anomalous zone of the same name. Detailed follow-up work has not been conducted. This target is an excellent opportunity to continue with early-stage exploration to determine if economic potential is present.

Conclusions: Metallogenic zoning of Mongolia in terms of uranium was initiated by VSEGEI employees in 1985, and later modified and refined by Mironov (1993, 2003). Accordingly, Mongolia was divided into four uranium ore provinces: Mongolian-Argun, Gobi, Khentei-Daur, and North Mongolian. Later (2009), this division scheme was further refined based on a much better metallogenic understanding by incorporating new results of terrane analysis. In this work, the metallogenic uranium zoning of Mongolia is made on the geodynamic (tectonic) basis of the long-lived and complex development of the continental crust of Mongolia. As such, Mongolia is now divided into the North Mongolian and South Mongolian metallogenic provinces. Within these provinces, uranium-bearing zones are identified, usually occur in areas that combine interconnected structures of rifting, collision, and accretion heritage, and are commonly associated with events of the same two to three geologic ages. For a more robust identification of continental structures throughout geologic history of Mongolia, a classification of the rift structures of Mongolia (Fig. 2.1) was made, as the most important for metallogeny, and genetically linking them with recognized geodynamic (tectonic) stages. The dominant uranium-bearing structures turned out to be granitoids of marginal-continental arcs and back-arc rift zones, marked by alkaline-bimodal volcanic rocks of especially basic, intermediate, and felsic compositions. Such back-arc tectonic settings occurred at least three times in Eastern Mongolia (in the Late Permian, at the end of the Triassic-Early Jurassic, and at the end of the Mesozoic).

Based on the compilation of resources and occurrences above, it is clear that for uranium mineralization, taphrogenic structures (block-faulted graben) are the most prospective setting. They are formed after the recession of the syncollisional uplift and coincide with the transitional stages between the riftogenic (extension) and convergent plate stages (compression) leading to development of continental block-fold structures (nappes). Taphrogenic structures developed within the continental shelf and lagoonal-marine terrigenous-sedimentary deposits formed in a redox environment, particularly favorable to subsequent uranium mineralization.

Though, as has been pointed out above, an enormous amount of work exploring for and discovering uranium resource was conducted during Soviet and post-Soviet times, establishing the fact that Mongolia contains significant, world-scale uranium resources and relatively little modern era exploration has taken place. There is an exceptional opportunity to apply state-of-the-art best exploration methods, combined with a modern understanding of the genesis and economic potential of uranium resources, to both advance known resources and prospects and to discover entirely new deposits.

3.3 Types of Uranium Deposits in Mongolia

For the first time, the genetic types of uranium deposits in Mongolia were identified by VSEGEI and MGSE in the 1970s–1980s. Twentieth century, and later by Yu. Mironov in 2003 [Uranium of Mongolia].

In recent years, the IAEA has developed a classification of uranium deposits in the world, which is recommended for all IAEA member countries.

According to this classification, uranium deposits are divided into the following types:

1. Deposits related with intrusions.
2. Deposits related with granites.
3. Deposits related with breccias with complex gold-uranium-copper ore.
4. Deposits related with volcanics.

5. Metasomatite deposits.
6. Metamorphic deposits.
7. Deposits of unconformity.
8. Deposits in collapse breccia.
9. Deposits of sandstone type.
10. Deposits in ancient conglomerates.
11. Near-surface deposits.
12. Deposits related with lignite and coal.
13. Carbonatite deposits.
14. Deposits associated with phosphorites.
15. Deposits in black shales.

There was a difference of opinion on the genesis of uranium deposits in Mongolia. For example: Mironov (2003) believed that the Dornod, Gurvanbulag, Mardaingol deposits were hydrothermal, while the Kharaat and Nars deposits were hydrogenous. Bayanmonkh and Debabrata considered the occurrence of uranium deposits at Khar-Tolgoin syngenetically hydrothermal, N. Sarangua compared the Baganuraat occurrence with the Kharaat deposit, and G. Jamsrandorj preferred to consider the ore bodies 2B, 2V, 3A, and 3B of the Dornod deposit sedimentary genesis (2008).

According to the classification of uranium deposits (IAEA 2018) by the IAEA, the following types of uranium deposits (Jamsrandorj and Altankhuyag 2015) have occurrences in Mongolia:

1. ***Sandstone Deposits***. This group includes deposits of sedimentary genesis, formed in terrigenous deposits of taphrogenic basins of Eastern Mongolia, associated with the processes of formation and surface oxidation and secondary reduction. Examples of such deposits are Kharaat, Khairkhan, Gurvansaikhan, Ulziit, Dulaan Uul, Zoovch Ovoo, and others (Figs. 4.32–4.66). Some researchers include ore bodies 2A, 2B, 2C, 3A, and 3B of the Dornod uranium deposit, which are localized in thick sedimentary rocks of the Upper Jurassic-Lower Cretaceous, among the sandstone deposits. These types of deposits are characterized by a uranium grade of 0.01–0.02% and a relatively uniform distribution. Uranium mineralization is located both in gray-colored rocks on the border with red-colored limonitized deposits and in the zone of oxidized gray-colored rocks. Within the oxidation zone, it is characterized by low uranium grades and is confined mainly to areas of sandy-clayey deposits enriched in carbonaceous material. The thickest ore accumulations are located along the boundary of the ground oxidation zone, less often among gray-colored rocks, where they are often associated with longitudinal tongues of limonitization. The uranium grade in such ores reaches tenths of a percent. The richest and most contrasting mineralization is confined to sandy, sandy-clayey, silty, and clayey gray-colored rock varieties. The maximum concentrations of uranium tend to deposit in areas with a large amount of carbonaceous detritus (Afanasiev et al. 2004; Budunov 2002). Mineralization is associated with the sequential manifestation of multi-stage epigenetic processes. At the same time, the reduction process controls the main concentrations of uranium mineralization, while the source of uranium to the geochemical barrier can initially be endogenous concentrations, which are destroyed and transported in the zones of reservoir oxidation.

2. ***Deposits Associated with Volcanics***. This type of deposit is defined by Russian geologists as "hydrothermal deposits of fluorine–molybdenum–uranium formation". According to the IAEA classification, these deposits are associated with basalt-rhyolite volcanism in the VTS. The deposits of the North Choibalsan uranium district belong to this type of deposition. These deposits include medium–low-temperature hydrothermal uranium and complex deposition localized in Late Mesozoic (Late Jurassic-Early Cretaceous) volcano-tectonic structures and their crystalline basements. Deposits related with volcanics are closely associated with contrasting rhyolite-basalt rock types, which are widespread within the Mongolo-Priargunsky volcanic belt. The most important regional magma- and ore-controlling structure of the Late Mesozoic

collisional rifting. The closeness in time of rhyolite-basalt volcanism and uranium mineralization has been documented by an isotope-geochemical data (Mironov et al. 1986, 1989; Mironov and Solovyov 1990; Mironov 2003).

In the composition of these deposits, two primary minerals are present, fluorite–molybdenite–nasturan and uranium–fluorapatite. Fluorite–molybdenite–nasturan is present at the Dornod uranium deposit, as well as numerous ore occurrences in the northeastern and central parts of Mongolia. Uranium-fluorapatite is present at numerous ore occurrences and mineralized locations. In the North Choibalsan ore district, these are the Tsagaan-Chuluut, Tsagaan nuur, and Ust-Turgen ore occurrences in the Turgen VTS and in the Ikhkhet district-the Ikhkhet ore occurrence on the western outskirts of the rhyolite ring VTS of the same name.

3. ***Deposits Related with Granites***. This type of uranium deposition is quite widespread in Mongolia typically found in highly radioactive massifs of the granite-leucogranite-alkaline rocks of early Mesozoic collisional rifting. These include uranium occurrences localized mainly in the leucogranites of Khentei, Khangai, Eastern Mongolia, and Altai. The most representative of them are the Elstiin and Urt deposits in the Janchuvlan district. Intrusions of highly radioactive leucocratic granites contain uranium mineralization which is inconsistently distributed—Most of the occurrences are located within the Khentei (Khentei-Daur) arched uplift (the Elstiin, Tamga, Urt, Arshaan ore occurrences in the Janchuvlan massif and the Borj occurrence in the western part of the Bayandelger massif). Similar occurrences are found in the Nukhetdavaa arch-block uplift (Saikhanuul and Tsagaanuul occurrences), the Mongolian-Priargunsky volcanic belt (Tumentsogt and Baruuntsogt occurrences), as well as in Central Mongolia on the Khangai uplift. Uranium occurrences in Elstiin, Tamga, and other areas within the Janchuvlan leucogranite massif seem to be relatively well studied. Uranium mineralization in the occurrence at the Janchuvlan massif is pitchblende and coffinite in association with sulfides: marcasite, galena, and sphalerite deposited directly in veinlets with montmorillonite. The zone of oxidation of sulfide and uranium mineralization is widely manifested on the surface and can be traced to a depth of 400 m. The minerals of the oxidation zone are represented by manganese hydroxides, malachite, azurite, covelline, chrysocolla, pyromorphite, mimetesite, wulfenite, pharmacosiderite, metathorbernite, metaotenite, zeynerite, uranophane, casolite and β-uranotyl.

Economic significance of mineralization in connection with highly radioactive leucocratic granites, for Mongolia, is quite relevant. Recent geological studies indicate grounds for revising the assessment of the significance of this type of uranium mineralization. Examples of economic uranium occurrences, associated with greisenized granites, are found in the Kattasai-Alatanga ore field, the Charkasar fluorite–uranium deposit (the eastern part of the Kurama Range), the Kiiktal uranium deposit (the southern part of the Mogol-Tau Ridge), and in a number of deposits in the Ore Mountains (Germany). The ore at these facilities consists mainly of uranium black and minerals of hexavalent uranium.

Greisenized granites are also mined at industrial facilities in Chukotka and the Chikoi uranium ore region in the Khentei-Daur district in Russian Transbaikalia. The fissure-filled deposits of Gornoe, Berezovoe, and others, have rich and contrasting mineralization, represented exclusively by minerals of hexavalent uranium-β-uranotyl and uranophane. The mineralization is localized in steep dipping, crushing zones with clay-zeolite alteration superimposed on greisenized leucocratic granites.

All these examples testify to the value of uranium mineralization associated with leucocratic granites, and to the expediency of continuing to study favorable geological settings for uranium within geoblocks, which demonstrate leucogranite magmatism and postmagmatic processes.

4. ***Deposits Related to Intrusions***. The IAEA references uranium mineralization related to anatectic and deep intrusions, like the Rossing deposit in Namibia. Mongolia has possible examples of uranium-bearing pegmatoids in the Buteliin Nuruu, Khubsugul, and Altai ranges, which have not been well studied.
5. ***Deposits Related with Metasomatites***. The occurrence of uranium mineralization in the Kholboo Maikhan, Alag-Erdene, and other areas associated with potassium metasomatism. Metasomatites are formed in highly permeable Precambrian metamorphosed rocks and are associated with cataclasis and crushing zones. In the field, development of deeply metamorphosed rocks of Precambrian age (Buteelnuruu, Totoshan crystalline massifs, Kholboo-Maykhan, Onon, Gurvanzagal blocks of Precambrian, etc.), numerous occurrences of uranium–thorium mineralization were noted in metasomatites, migmatites, and pegmatoids of various compositions. The best-studied occurrences are located in the eastern part of the Buteelinnuruu ledge, framed by a granite-gneiss dome, where a system of mutually conjugated northeastern and northwestern local faults are mapped among crystalline schists and granite-gneisses, which are well manifested on space photographs. Biotite, amphibole-biotite gneisses, and granite-gneisses give rise to quartz-feldspar metasomatites, trending toward cataclasis zones. The length of metasomatite bodies with a northwestern strike is from tens to 200 m, the thickness is 1.5 m with an uneven distribution of increased radioactivity (up to 250–1500 μR/h). Ore dissemination is represented by euxenite and brannerite intergrown with titanomagnetite and rare-earth minerals. Incidental elements of uranium (%) are: thorium, tantalum 0.03–0.2, niobium 0.8–5.6, yttrium 0.5, gadolinium 0.04–0.2, lanthanum 0.02. Albitic metasomatites develop over palingenic granites and compose bodies up to 240 m long and up to 10 m thick. The main ore mineral is brannerite associated with tantalum-niobates. Occurrences of uranium and thorium minerals are quite common for metasomatites associated with alkaline and subalkaline rocks. In the Khubsugul district, as in the Eastern Sayan and in the south of Eastern Tuva, a series of massifs and dikes of subalkaline granitoids and syenites rocks, of presumably Early Paleozoic or Late Triassic-Early Jurassic age, are known. Zones of metasomatic and hydrothermal alterations with uranium- and thorium-bearing minerals develop in the marginal endocontact parts of alkaline granite massifs. Stock-shaped, nested, lenticular, veined mineralization with disseminated uranium mineralization is localized in zones of intense albitization, K-feldsparization, silicification, chloritization, hematitization and is accompanied by sulfide, fluorite and less often beryllium mineralization. Uranium mineralization is represented by disseminated orthite, thorite, orangite, zircon, monazite, and, more rarely, uraninite. These are the types of occurrences at Ujiggol, Aragol, Balbrig, Alag-Erdene, and others. Ore bodies contain the following minerals (%): uranium from thousandths to 0.23, thorium from hundredths to 13, zirconium 0.1–0.2 to 10, niobium 0.2–0.5, tantalum 0.03–0.2, beryllium 0.01–1, lithium 0.08–0.2, lanthanum 0.05–2, cerium 0.1–2, yttrium 0.1–2, gadolinium 0.5–0.06, and ytterbium 0.02.

A wider range of rare and trace metals is observed in uranium- and thorium-bearing metasomatites associated with alkaline and nepheline syenite massifs. In the areas of Khanbogd, Mushgaa Khudag, and Lugiyngol, linear and nesting bodies of metasomatites, ekerites, and pegmatite with superimposed mineralization were established in connection with the Lugiyingol and Khanbogd massifs of alkaline rocks (Kovalenko 1981). The metasomatites contain sphene, zircon, astrophyllite, britholite, monazite, tantalum-niobates, davidite, fluorite, apatite, and pyrite. The grades of rare earths reach 0.8, niobium 0.062, tantalum 0.033% (by weight of the rock). In individual fluorite and carbonate veins, the content of rare earths is a whole percent. According to the mineral composition and conditions of formation, they are considered as "carbonatites, of a small erosion cut", or as formations of the potassium branch of the specialization of metasomatites of alkaline rocks.

6. ***Deposits Related with Lignite and Coal***. This type of deposit is recognized by the IAEA as independent, although it has similarities with the sandstone type deposit. It was first established in Mesozoic uranium-bearing coals of the Choibalsan, Shive Ovoo, Ereen I and II, and Jargalant Nuur occurrences. This group combines uranium concentrations associated with brown coal, less commonly hard coals, and gray-colored terrigenous rocks containing carbonized plant detritus. It is assumed that uranium mineralization was formed due to the processes of uranium oxidation by groundwater and reservoir infiltration. These waters carry products of destruction and leaching of reduced rocks specialized for uranium. The zones of soil oxidation are most clearly manifested at the Baganuur, Aduunchuluun deposits, and the Durbulj occurrence, localized in coal-bearing terrigenous deposits of the Lower Cretaceous Zuunbayan formation, 100 km east of the city of Sainshand. Over an area of 4 × 2 km, at depths of about 100 m, uranium is concentrated in dark gray carbonaceous siltstones at the lower contact of the ground oxidation zone that mineralizes the sand strata. The thickness of the ore bodies does not exceed 0.5–1.5 m, and the uranium grade is 0.01–0.015%.

Zones of seam oxidation are present in the Choibalsan, Choir, Bulgan, Ulziit depressions, which are filled with coal-bearing strata of the Lower Cretaceous. At the Ulzit deposit in the depression of the same name, uranium concentrations are confined to the oxidized section of the coal seam over a distance of 2 km with a width of the uranium deposition zone 300–1500 m and with a thickness of 1 m. Numerous occurrences of uranium of this type are found in the Choibalsan depression and smaller structures adjoining it to the north.

7. ***Deposits Related with Phosphorites***. Aerial gamma spectrometric studies have established gamma anomalies in connection with bedded phosphorites in the Khubsugul region. None of the occurrences of this type of uranium deposition have been studied in detail. For a number of reasons, these deposits deserve further study.
8. ***Deposits Related with Black Shales***. In the Proterozoic and Vendian-Cambrian black shale deposits of the Khubsugul, Ereendavaa, Bayankhongor, and other ancient structures, various rocks enriched in uranium, vanadium, and molybdenum have been established. Elevated uranium grades are confined to areas of siliceous, siliceous-carbonaceous, and vanadium-phosphorus-containing shales. An example is the Zaanar ore occurrence in the Khubsugul District, which is localized in the metamorphosed phosphate-bearing Dood-nuur subformation of the Khubsugul formation. As part of the ore-bearing strata, according to Osokin (1998), limestones, dolomites, black carbonaceous-siliceous-clayey, siliceous, siliceous-phosphate and carbonate-argillaceous shales, horizons of phosphorites and sedimentary breccias are present. The length of carbonaceous rock horizons ranges from several tens to 1.5 km, with a thickness of a few to several tens of meters. Such horizons often have a lenticular structure. More interesting are the increased concentrations of uranium, primarily in the outcrops of carbonaceous-siliceous shales, which are controlled by zones of intense tectonic and hydrothermal development. The distribution of increased radioactivity (up to 100–400 μR/h) in tectonic zones is of a point, nesting or lenticular nature and uranium concentrations vary $(10–200) \times 10^{-4}\%$. In uranium-bearing formations, elevated concentrations (%) of nickel 0.01–0.15, vanadium 0.01–0.1, molybdenum 0.001–0.02, copper 0.01–0.3, zinc 0.01–0.15, and silver $(1.1–2.0) \times 10^{-4}$ are present.
9. ***Near Surface Deposits***. The Chuluut ore occurrence, now the Dald deposit, was assigned by Yu. Mironov to paleovalley type

of deposit. According to the IAEA classification, G. Jamsrandorj classifies it as a near-surface type deposit because the uranium-bearing deposits are exposed at the current surface, and indicator minerals such as calcite and gypsum are present in the reservoir.

3.4 Uranium Ore Epochs in Mongolia

The uranium mineralization in Mongolia has been formed since non-Archean-Proterozoic time (Table 3.7).

Uranium deposits and numerous occurrences located in Mongolia were formed during various uranium-mineralization epochs—Neoarchean-Proterozoic, Ediacaran-Paleozoic, Mesozoic, and Meso-Cenozoic. At present, two uranium ore epochs (Mesozoic and Meso-Cenozoic) account for economic uranium mineralization and two uranium-bearing epochs (Neoarchean-Proterozoic, Ediacaran-Paleozoic) resulted in the accumulation of uranium-enriched rocks, which typically don't become economic deposits. The earliest uranium ore epoch, which is of great importance, is the Mesozoic. During this epoch, most of the deposits related with volcanics that have been discovered to date were formed spatially and paragenetically related to the Late Mesozoic volcano-tectonic structures, such as the Dornod, Gurvanbulag, Nemer, and other deposits.

Isotope-geochronological studies, conducted primarily within the Dornod district (VSEGEI, IGEM), included determining the age of the main igneous complexes, hydrothermal-metasomatic ore processes, establishing relationships between mineralization and magmatism, and solving ore-genetic issues. The age of the Bayantumen formation volcanic rocks was determined using potassium-argon and rubidium-strontium methods. Reliable consistent dates have been obtained only for felsic volcanic rocks of the middle subcomplex (150–155 Ma). The age of the lower (160 ± 9 Ma) and especially the upper (144 ± 10 Ma) complexes

Table 3.7 Age of uranium mineralization in Mongolia by zone and regions (Jamsrandorj 2019)

Region (o), zone (h)	Districts	Age of uranium mineralization
Tuva-Mongolian (late Baikalids)	Khubsugul	Ediacaran—Paleozoic
	Khan Khukhii	Middle Paleozoic
Lake-Egiingol (early Caledonides)	Egingol	Jurassic
	Lake	Devonian
Mongolian-Altai (late Caledonides)	Tsagaan Shiveet	Carbon
		Silurus
Buteel-Khangai	Buteel	Carbon, Proterozoic
	Chuluut	Jurassic, Pleistocene
	Bayankhongor	Late Proterozoic
Khentii-Daur	Janchuvlan (Central)	Triassic-Jurassic
East Mongolian (Mongol-Priargunskaya)	North Choibalsan	Jurassic-Cretaceous
	Berkh	Jurassic-Cretaceous
	Choir	Cretaceous
	Shivee	Jurassic-Cretaceous
	Ulziit	Cretaceous
	Middle Gobi	Cretaceous
	Khairkhan	Cretaceous
	Bor Undur	Jurassic-Cretaceous
Gobi (early Hercynides)	Mandalovoo	Carbon, Cretaceous
	Nukhetdavaa	Devonian
	Sainshand	Cretaceous
	Madat	Carbon, Cretaceous
	Zuunbayan	Cretaceous
	Ail	Cretaceous
Southwestern Mongolian (early Hercynides)	Nemegt	Neogene, Cretaceous
	Atasbogd	Neogene
	Baruunhurai	Neogene, Cretaceous

Buteel-Khangai, and Lake-Egingol potential

require clarification. The time of manifestation of the main stage of hydrothermal-metasomatic activity in the DVTS was determined from hydromicas from near-ore argillisites using the potassium-argon method (140+3 Ma) and from uranium minerals using the uranium-lead method (135–138 Ma).

The Meso-Cenozoic uranium mineralization era is characterized by the appearance of ore concentrations from a number of different formations (Fig. 3.12). The most significant deposits of uranium are in terrigenous deposits of the Meso-Cenozoic depressions associated with soil and reservoir redox processes (Kharaat, Khairkhan, Nars) (Radiometric research …, 1988).

During this era, additional mineralization took place:

- Numerous uranium occurrences in the Mesozoic coal-bearing strata associated with zones and seams of soil oxidation.
- Economic deposits within the Daursky arch in Russia-Gornoye, Berezovoe, localized in the zones of crushing of leucocratic granites.
- Occurrences of a similar type in Mongolia in the Khentei-Daur uranium ore zone.
- Large-scale uranium occurrences, which, upon further study, will probably be transferred to the classification of deposits. These occurrences are related to the uranium mineralization in terrigenous deposits of the Cenozoic paleovalleys. A characteristic feature of this uranium ore epoch is the formation of deposits in connection with exogenous diagenetic processes.

The earliest uranium-bearing epoch—Neoarchean-Proterozoic—in Mongolia is associated with the formation of such ancient structures as the Buteeliin nuruu (Byamba et al. 2018). The oldest uranium-bearing epoch is associated with the formation of ancient palingenic granites, gneiss-granites, accompanied by widespread silicic-alkaline metasomatites, pegmatites, and pegmatoids. Small uranium occurrences associated with areas of silicic-alkaline metasomatism in these formations are most widely developed in the Khubusugul, uranium ore deposits. However, there is little reason to expect the discovery of economic uranium ore deposits in these formations.

The second uranium-bearing epoch, the Ediacaran-Paleozoic, is associated with the accumulation of Ediacaran-Early Paleozoic carbonaceous-siliceous shales. The wide age range is explained by insufficient knowledge of the stratigraphic position of carbonaceous-siliceous deposits. The uranium grade, most likely associated with primary sedimentation within water basins, ranges from 0.001 to 0.03%. The role of uranium-bearing formations of these epochs, in the formation of economic uranium deposits during subsequent processing by endogenous and exogenous processes, is significant.

Figure 3.12 shows that starting from the Paleogene to the end of the Neogene, potential areas of uranium mineralization have not been identified. This period of time has not been studied in relation to uranium mineralization. In recent years, airborne gamma spectrometric anomalies of uranium have been established in Southwestern Mongolia within the area of distribution of the Lower Cretaceous and Paleogene deposits, which indicates the need to study potential uranium mineralization.

3.5 Assessment of the Prospects for Uranium Mineralization in Mongolia

3.5.1 Uranium Reserves of Mongolia

Geological data indicates that the prospects for uranium resources are not limited to the reserves and resources of uranium in known regions but can be increased with new discoveries. In 1994, international experts of the IAEA, when preparing to publish the map "Uranium Deposits of the World" at a scale of 1:30,000,000, quantified the reserves and probable resources of uranium in Mongolia to 1100 thousand tons (IAEA 1995). The total figure was reduced by excluding from the general estimate of estimated resources (P_3) industrial types of deposits that are not taken into

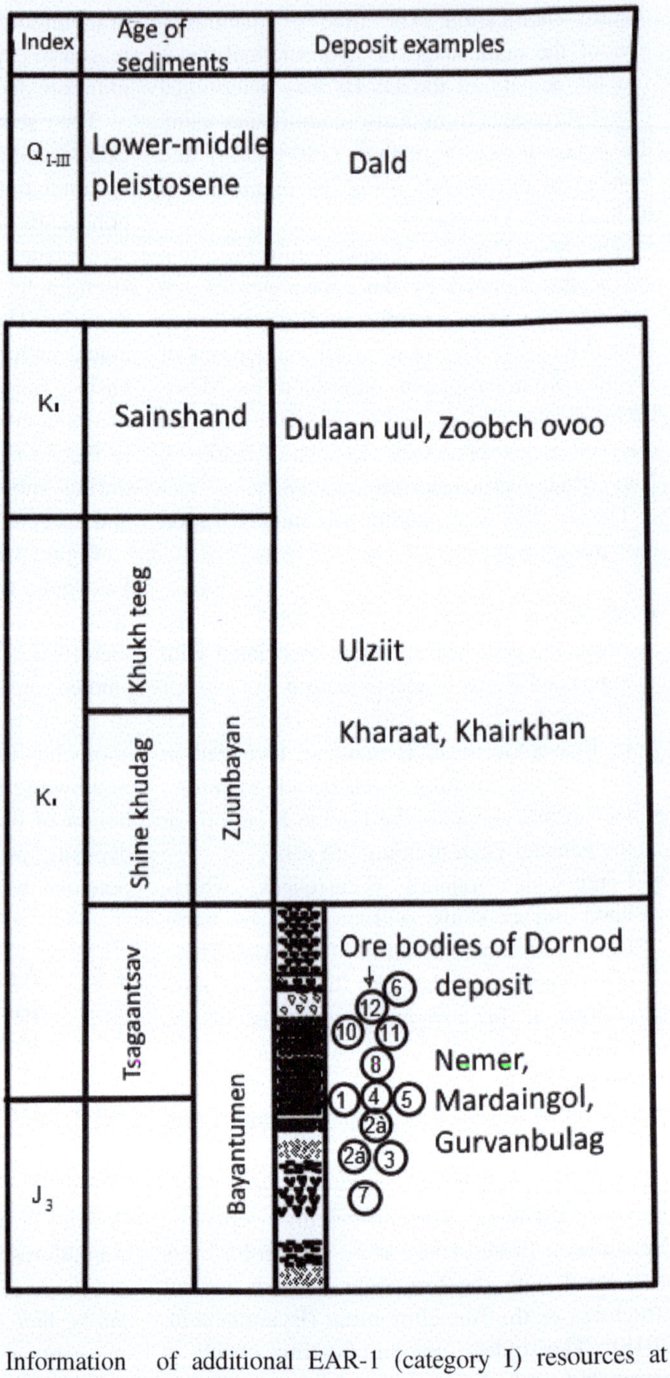

Fig. 3.12 Stratigraphic levels of the Meso-Cenozoic uranium deposit in Mongolia (Jamsrandorj 2019)

account by the IAEA classification. Information on the quantitative assessment of Mongolian resources was first published in the International Red Book in 1995. Based on materials from the IAEA technical committee symposium, 83 thousand tons of reliably recoverable estimated additional resources (RAR) and 11 thousand tons of additional EAR-1 (category I) resources at a price of less than \$40/kg. At the same time, the overall assessment of the resource potential of Mongolia was 1070 thousand tons, which is seven times higher than the coal resources in Mongolia, published in the global volume of coal resources. In 1994–2003 prospecting

Table 3.8 Distribution of uranium reserves of Mongolia by deposit (Jamsrandorj 2019)

No.	Deposits	Reserves	
		Ore (tonnes)	U (kg)
1	Mardaingol	3705.01	1948.1
2	Nemer	5338.21	5937.62
3	Dornod	29646.9	27221
4	Ulaan	241.8	270
5	Gurvanbulag	5305.76	13058.3
6	Haraat	28086.35	7288
7	Khairkhan	11791.8	8406.5
8	Gurvansaikhan	12650.3	4250.1
9	Ulziit	7843.9	3075.7
10	Dulaan Uul	32407.7	7054.71
	Southeast extension of Dulaan Uul	25710.108	4841.73
11	Zoovch Ovoo	288698.74	93290.6
12	Enger Ar	188824.1	80.52
13	Dald	51538.2	15518.11
	Total		192241.02

for uranium in Mongolia focused on identifying economic mineralization in sedimentary deposits of Late Mesozoic and Late Mesozoic-Cenozoic depressions within the Mongolo-Argun and Gobi provinces. As a result of work by foreign and Mongolian companies Gurvan-Saikhan JV and Kozhegob Emeelt Mines, Khan Resources of Mongolia, Nuclear Energy Mongolia and Adamas Mining, the reserves of the Kharaat deposit in the Choir depression, Dald deposit in the Chuluut district, the Gurvanbulag, Dornod, Mardaingol, Nemer deposits in the North Choibalsan district, the Zoovch Ovoo and Dulaan Uul deposits in the Gobi province were discovered and explored. Additionally, a prelimi nary assessment of potential areas was obtained in the Tamtsag and Sainshand regions (flanks of the Nars field), in the Nyalga, Undurshi, Ulziit, Oshiinnuur, Uneget, and other depressions. It should be emphasized that the quantitative estimate of uranium mineralization did not fundamentally change.

As of 2016, the IAEA reported that Mongolia has 141.5 thousand tons of proven uranium reserves, and these figures have remained virtually unchanged since 2013. However, Mongolian's geologic data for 2020 indicates that Mongolia's uranium reserves have slightly increased compared to 2016 to 166.6 thousand tons of uranium contained in 13 deposits, of which 48,435 thousand tons of uranium are contained in deposits related to volcanic rocks, and the remainder are related to sandstone deposits (Table 3.8).

Table 3.8 shows that (66.69%) are located in the deposits of East Mongolian and the Gobi uranium district, which are sandstone deposits. Volcanic deposits in Eastern Mongolia make up 25.19% of Mongolia's uranium reserves. At the end of 2020, Mongolia's uranium reserves represent about 2% of the world's total uranium reserves which ranked 12th largest in the world at the end of 2023.

3.5.2 Uranium Resources of Mongolia

Since 1991, the assessment of uranium mineralization in Mongolia, based on IAEA criteria, has not changed. As of January 1, 2022, there has been a new estimate of uranium resources in Mongolia based on the revision of the regularity in the distribution of uranium deposits (Table 3.3). In 2019, due to new exploration data and re-evaluation of past data of uranium deposits in Mongolia, a new theory of metallogenic zoning for uranium was developed. J. Byamba, Yu. B. Mironov, and G. Jamsrandorj made changes to the estimated resources for 1991 and introduced some changes in connection with the results of airborne gamma spectrometric studies in Southwestern Mongolia and the possibility of identifying new uranium resources—Nemegt, Baruunkhurai, and Atasbogd. As of January 1, 2022, the estimated resources dropped slightly, due to the transfer of resources to reserves, to 1004 thousand tons of uranium (Table 3.9).

Conclusions

1. According to the classification of uranium deposits by the IAEA, the following types of uranium deposits are found in Mongolia: sandstone deposits; deposits related with volcanic rocks, deposits related with granites intrusions metasomatites, lignites and coal, phosphorites, black shale, and near-surface deposits, of which the first 2 types are the most widely studied. The other types of deposits are identified by direct prospecting and need detailed studies in order to increase Mongolia's uranium reserves.
2. Uranium deposits and numerous occurrences in Mongolia were formed by various uranium ore epochs—Neoarchean-Proterozoic, Ediacaran-Paleozoic, Mesozoic, and Meso-Cenozoic. For the formation of economic uranium ore deposits in Mongolia, there are two uranium ore epochs that owe their manifestation to global tectonic events: (1) Late Mesozoic, due to rifting that accompanied the closure of the Tethys, resulted in the deposition of uranium in Late Mesozoic volcano-tectonic structures; (2) Meso-Cenozoic arch-block movements in the global orogenic belt stretching from the Carpathians to the Pacific Ocean.
3. The majority of Mongolia's uranium reserves (66.69%) are located in deposits in East Mongolian and the Gobi uranium district, which are sandstone deposits. Volcanic deposits in Eastern Mongolia make up about 25.19% of Mongolia's uranium reserves. At the end of 2020, Mongolia's uranium reserves represent about 2% of the world's total uranium reserves which ranked 12th largest in the world.
4. In 2019, due to new exploration data and re-evaluation of past data of uranium deposits in Mongolia, a new theory of metallogenic zoning for uranium was developed. J. Byamba, Yu. B. Mironov, and G. Jamsrandorj made changes to the estimated resources for 1991 and introduced some changes in connection with the results of airborne gamma spectrometric studies in Southwestern Mongolia and the possibility of identifying new uranium resources—Nemegt, Baruunkhurai and Atasbogd. As of January 1, 2025, the estimated resources dropped slightly, due to the transfer of resources to reserves, to 1004 thousand tons of uranium.

Table 3.9 Distribution of uranium reserves and resources in Mongolia by provinces and districts (Jamsrandorj & Mironov 2023)

Name of metallogenic subdivisions	**Reserves and predicted resources (thousand tonnes)**	**Reserves, (thousand tonnes)**	**Predicted resources (thousand tonnes)**
1	2	3	4
1. TUVA-MONGOLIAN POTENTIAL REGION	**80**	**0**	**80**
1.1. Khubsugul potential district	50		50
1.2. Khankhukhi potential district	30		30
2. LAKE-EGIYNGOL POTENTIAL ZONE	**50**	**0**	**50**
2.1. Egiingol potential district	30		**30**
2.2. Lake potential district	20		**20**
3. MONGOLO-ALTAI POTENTIAL ZONE	**45**	**0**	**45**
3.1. Tsagaan Shiveet potential district	35	0	35
3.1.1. Tashant-Deluun potential uranium ore cluster	5		**5**
3.2. Kharhiraa-Turgen potential uranium ore cluster	5		**5**
3.3. Kharus-Ulaat potential uranium ore cluster	5		**5**
4. BUTEL-KHANGAI POTENTIAL ZONE	**145.5**	**15.5**	**130**
4.1. Buteel potential district	**45**		**45**
4.2. Chuluut uranium district	60.5	15.5	45
4.2.1. Chuluut uranium ore cluster	30.5	15.5	15
4.3. Bayankhongor potential district	**40**		**40**
5. KHENTIY-DAUR ZONE	**25**	**0**	**25**
5.1. Janchuvlan potential district	**25**	**0**	**25**
5.1.1. Janchuvlan potential uranium ore cluster	**10**		**10**
6. EAST MONGOLIAN ZONE	**445.5**	**71.5**	**374**
6.1. North Choibalsan uranium district	87.4	**48.4**	**39**
6.1.1. Dornod uranium ore cluster	67.4	48.4	**19**
6.1.2. Ugtam potential uranium ore cluster	5		**5**
6.1.3. Turgen potential uranium ore cluster	5		**5**
6.1.4. Engershand potential uranium ore cluster	5		**5**
6.1.5. Sumannuur potential uranium ore cluster	5		**5**
6.2. Berkh potential district	**27**		**27**
6.2.1. Ulzii-Saikhanuul potential uranium ore cluster	15		**15**
6.2.2. Butnorov potential uranium ore cluster	12		**12**
6.3. Choir potential district	**79.3**	**7.3**	**72**
6.4. Shivee district	**35.1**	**0.1**	**35**
6.4.1. Nyalga potential uranium ore cluster	12		**12**
6.4.2. Shivee uranium ore cluster	3.1	0.1	**3**
6.4.3. Tavansuveet potential uranium ore cluster	20		**20**
6.5. Ulziit uranium district	**13**	**3**	**10**
6.6. Middle Gobi potential district	**151**		**151**

(continued)

Table 3.9 (continued)

Name of metallogenic subdivisions	Reserves and predicted resources (thousand tonnes)	Reserves, (thousand tonnes)	Predicted resources (thousand tonnes)
1	2	3	4
1. TUVA-MONGOLIAN POTENTIAL REGION	**80**	**0**	**80**
1.1. Khubsugul potential district	50		50
6.7. Khairkhan district	**17.7**	**12.7**	**5**
6.8. Borundur potential district	**35**		**35**
7. GOBI ZONE	**370.2**	**105.2**	**265**
7.1. Sainshand district	**30**	**0**	**30**
7.1.1. Narst cluster	30		**30**
7.2. Undurshil district	**30**	**0**	**30**
7.3. Zuunbayan district	**135.2**	105.2	**30**
7.3.1. Unegt cluster	21.9	11.9	**10**
7.3.2. Zuunbayan cluster	**113.3**	**93.3**	**20**
7.4. Mandalovoo potential district	**111**	**0**	**111**
7.4.1. Mushgaikhudag potential uranium cluster	**16**		**16**
7.4.2. Oshinnuur-Ongigol potential uranium cluster	95		**95**
7.5. Ail potential district	**10**	**0**	**10**
7.5.1. Ail potential uranium cluster	10		**10**
7.6. Madat potential district	**49**		**49**
7.6.1. Tamsag potential uranium ore cluster	4		**4**
7.7. Nukhetdavaa potential district	5		**5**
8. SOUTHWESTERN POTENTIAL ZONE	**35**	**0**	**35**
8.1. Nemegt potential district	15		**15**
8.2. Baruunkhurai potential district	15		**15**
8.3. Atasbogd potential district	5		**5**
TOTAL	**1196.2**	**192.2**	**1004**
8.3. Atasbogd potential district	5		
Total	1196.6	192.2	0

References

Afanasiev AM, Budunov AA, Mironov YuB (2004) New data on hydrogen deposits of Mongolia. In: Proceedings of the II Int. Conf. Tomsk, pp 63–65

Budunov AA (2002) Hydrogenous uranium deposits of Mongolia. Materials on the geology of uranium deposits: Inform. Sat. KNTS. Issue. 144. VIMS, pp 84–89

Byamba J (ed) (2012) Mongolian geology and mineral resources. Second edition. In: Book 4. Plit tectonic of lithosphere, p 487 (in Mongolian)

Byamba J, Mironov YuB, Jamsrandorj G (2018) Geodynamics and uranium metallogeny of Mongolia. Ulaanbaatar, p 302 (in Mongolian)

Gundsambuu Ch (2007) Metallogeny of Mongolia, p 32 (in Mongolian)

IAEA (1995) Uranium Resources, Production and Demand (Red Book)
IAEA (2018) Geological classification of uranium deposits and description of selected examples. Vena, p 430
Jamsrandorj G (2019) Uranium deposits of Mongolia, p 244 (in Mongolian)
Jamsrandorj G, Altankhuyag D (2015) Uranium metallogeny of Mongolia. Khaiguulchin 53:163–176
Jamsrandorj G, Baatartsogt B, Altankhuayg D (2016) Resent discoveries and resources of uranium Mongolia. J East China Univ Technol 39(Suppl)
Jamsrandorj G, Dandar S (2010) Genesis of uranium mineralization of Mongolia. Khaiguulchin 1:43 (in Mongolian)
Jamsrandorj G, Mironov YuB (2023) Mineral resource base of uranium in Mongolia. Ulaanbaatar, St. Petersburg, p 298
Korobeynikov AF, Rikhvanova LP (2002) Geochemical features of riftogenic magmatism of the Altai-Sayan region and metallogenic consequences. In: Proceedings of the international symposium, Mantle plumes and metallogeny. Petrozavodsk-M. pp 113–116
Mironov YuB (2003) Uran Mongolii. Izd-vo Anatoliya, Spb, p 326
Mironov YuB (2005) Uranovye mestorojdeniya Mongolii. In: Geologiya i metodiка izucheniya uranovych mestorojdenii: Mejdunarodnaya shкolaseminar. Spb, p 15 (in Russian)
Mironov YuB (2006a) Uran Mongolii. Izdatelstvo VSEGEI, Sanкt-Peterburg, p 199
Mironov YuB (2006b) Uranium of Mongolia. London, pp 229
Mironov YuB (2007a) Uranium deposits in Mongolia: emplacement patterns and potential. In: Proceedings of the ninth biennial meeting of the society for geology applies to mineral deposits, Dublin, pp 67–74
Mironov YuB (2007b) Uranium resources and potential in Mongolia. In: General rules, regulations and procedur for "Discaveri Mongolia 2007". Fifty intern, mining investors' forum, Ulaanbaator, Mongolia, p 15
Mironov YuB, Afanasьev AM, Bulosev AV, Buzovкin SV, Vysoкoostrovsкaya EB, Gerel O, Karpunin AM, Korshunov AM, Tamiraa A, Shor GM, Shuvalov YuM (2012) Uranovye mestrojdeniya Mongolii. Sanкt-Peterburg, p 304 (in Russian)
Mironov YuB, Naumov SS, Jamsrandorj G, Chuluun O (1993) The uranium resource of Mongolia. In: Resent development in uranium resources and supply. TECDOC 823. IAEA, pp 177–192
Mironov YuB, Oleynikov RM, Lemeshchenko GV (1986) Principles of forecasting uranium deposits in volcano-tectonic structures (using the Far Eastern Tectonic System as an example) // Materials on the geology of uranium deposits: Inf. collection. Issue 116. M. VIMS. pp 23–29
Mironov YuB, Samovich DA, Buzovкin SV (2007) Uranovyi пotentsial Zabaiкalya i Mongolii. In: Uran: resursy i proizvodstvo: Tez. Doкl. Vtorogo mejdunar. Simpoz. VIMS, Moscow, p 76 (in Russian)
Mironov YuB, Samovich DA, Buzovкin SV (2008) Metallogeniya urana yuga Vostochnoi Sibiri Region. Geol Metall 32:75–88 (in Russian)
Mironov YuB, Solovьev NS (1990a) Geologo-geneticheskaya model endogennych rudnych formatsi Severo-Choibalsanskogo raiona MNR. Geneticheskie modeli endogennych rudnych formatsi: Tez. III Vsesoyuz. Soveshaniya. Novosibirsк, p 2 (in Russian)
Mironov YuB, Solovьev NS (1990b) Osobennosti geologicheskogo stroeniya i metallogenii Severo-Choibalsanskogo raiona (Vostochnaya Mongoliya). Materialy пo geologii uranovych mestorojdenii: Inf. sb. Vyp. 126. VIMS, Moscow, pp 11–29 (in Russian)
Mironov YuB, Solovьev NS, Lvov VK, Pecherкin YuN (1989) Osobennosti geologicheskogo stroeniya i rudonosnosti Dornodsкoi vulкano-teкtonicheskoi struкturi (Vostochnaya Mongoliya). Geol Geofiz 9:22–32 (in Russian)
Mironov YuB, Solovyov NS (1990) Geological and genetic model of endogenous ore formations of the North-Choibalsan region of the Mongolian People's Republic. Genetic models of endogenous ore formations: Abstract of the III All-Union Conference. Novosibirsk, p 2
Osokin PV, Tyzhikov AV (1998) Precambrian tilloids of the Oka-Khuvsugul phosphorite-bearing basin (Eastern Sayan, Northern Mongolia). Lithology and minerals. No. 2. pp 162–176
Visokoostrovskaya EB, Krasnov AI, Samovich DA (1999) Rol kainizoiskikh basaltov v prognozirovanii uranovikh mestorojdenii Mongolo-Okhotskogo skladchadtogo poyasa. Ros Geophis J 15:58–64 (in Russian)
Visokoostrovskaya EB, Mironov YuB, Chuluun O (2007) Radiogeochemicheskaya kharakhteristka geologicheskikh obrazovanii Mongolii. Razvedke Okhrana Nedr 4:32–44 (in Russian)
Visokoostrovskaya EB, Zubov EI, Mironov YuB (2006) Effektivnost aerogammaspectrometricheskogo methoda na territorii Mongolii. Materiali po geologii mestorojdeniyam urana, redkikh i redkozemelnikh metallov. Info. Sb. Vip. 149. VIMS, Moscow, pp 131–156 (in Russian)
Volsky AS, Shatkov GA, Krasny LI, nar, (2004) Tectonics, deep structure and minerageny of the Amur region and adjacent territories. In: Shatkov GA, Volsky AS (eds). St. Petersburg, VSEGEI Publishing House, p 190

Websites

http://www.iaea.org/OurWork/ST/NE/NEFW/Technical-Areas/NFC/documents/uranium/tr-darwin-2012/Presentations/07.Rogers_(ERA)_-Geological_setting_of_the_Ranger_uranium_deposits.pdf

http://www.maptek.com/pdf/vulcan/case_studies/vulcan_olympicdam_casestudy.pdf

https://www.unece.org/fileadmin/DAM/energy/se/pp/unfc/ws_UNFC_Mongolia_2016/005_Hocquet_CogeGobi_UB.pdf

World distribution of uranium deposits (UDEPO) with uranium deposit classification, 2009 edn. IAEA

www.unece.org/...IAEA...July2013/...July/7_Bruneton_UDeposits_Class....

Uranium Deposits of Mongolia

4

4.1 Dornod Uranium Ore Cluster

The Dornod uranium ore cluster encompasses an area of about 2.1 thousand km^2 and is located in the central part of the Choibalsan-Onon zone. It is confined to the Dornod volcano-tectonic structure.

The uranium deposits are located in blocks, Mardaingol (uranium deposits Nemer and Mardaingol), Ulaan (uranium deposit Gurvanbulag), and Erkhetiin (uranium deposit Dornod) in the Dornod volcanotectonic structure (Mironov et al. 1995; Jamsrandorj et al. 2013). The geological structure of the district is relatively complex and includes the Paleoproterozoic Khaichingol metamorphic complex, Early Paleozoic intrusive rocks of the Modokhudag and Mesozoic Imalkh complexes, and volcanogenic-sedimentary rocks of the Bayantumen Formation of Upper Jurassic-Lower Cretaceous age (Jamsrandorj 2009, 2013; Gerel et al. 2021) (Fig. 4.1).

The foundation of the structure and its framing (boundaries) are composed of Proterozoic and Early Paleozoic metamorphic complexes with a predominance of sodic granodiorites and diorites.

The metamorphic rocks of the Khaichingol complex (PPhg), which make up the crystalline basement, were formed by metamorphism of rocks of different composition. They are composed of crystalline schists, gneisses, marble limestones, amphibolites, migmatites, granite-gneisses, and quartzites.

Paleozoic granitoids of the Modkhudag intrusive complex (δ, γδ, γPZ_1mh) have a hybrid composition and consist of granodiorites, plagiogranites, and quartz diorites. They often have a gneissic texture and numerous gneiss and schist xenoliths. The granitoids are intersected by dike-like bodies of leucocratic granites, aplites, and pegmatites. Dikes of basic rocks are noted much less frequently in granites. All rocks of the lower stage are affected to varying degrees by the processes of sodium and potassium metasomatism. The paleosurface of the basement is sharply dissected due to the alternation of erosion-tectonic swell-like uplifts and subsidence (paleovalleys), which had a significant impact on the section of the host strata, the development of fault tectonics, and the distribution of minerals.

Imalkh Complex (γδ, $\gamma^2 J_{2\text{-}3}i$). Massifs of Middle-Upper Jurassic granites of the complex, distributed in the northern, northwestern and northeastern parts of the area, were assigned to the Imalkha intrusive complex. The massifs are dominated by rocks of the second phase. The first phase is represented by fine- and medium-grained biotite-hornblende granites, and the second phase is represented by medium-grained biotite-hornblende porphyritic granites.

Granodiorites are distributed over a very small area along the edges of the massifs, while

The original version of the chapter has been revised.
A correction to this chapter can be found at
https://doi.org/10.1007/978-3-031-86481-0_5

J. Galsan, *Uranium Deposits of Mongolia*, Mineral Resource Reviews,
https://doi.org/10.1007/978-3-031-86481-0_4

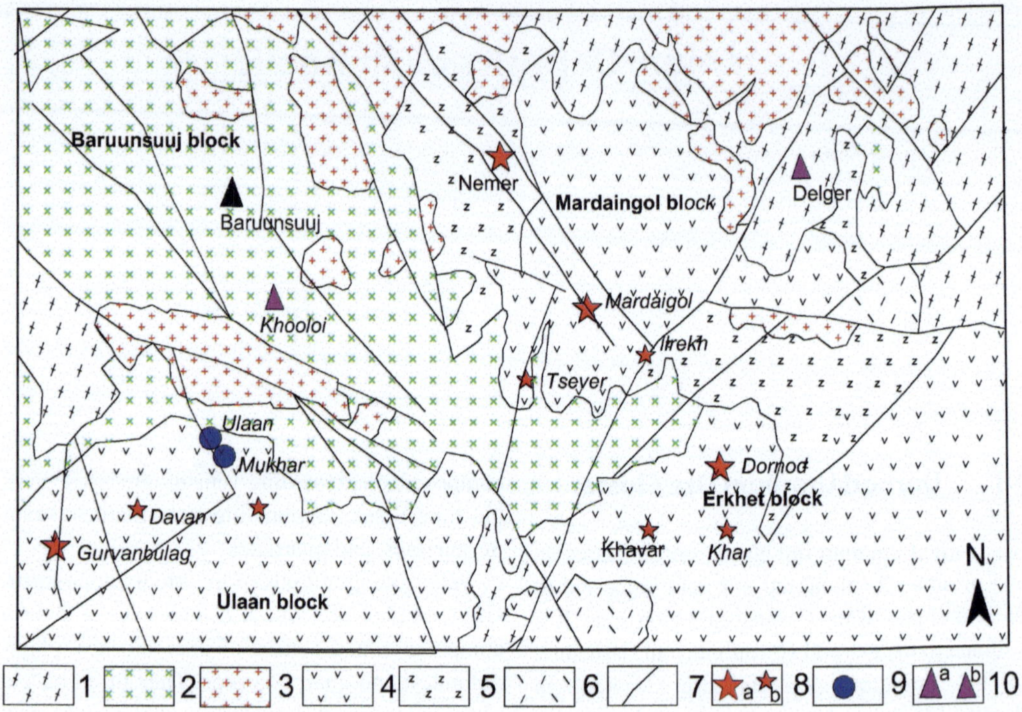

Fig. 4.1 Scheme of the geological structure and mineral deposits of the Dornod uranium cluster (Mironov 2003). 1—Paleoproterozoic. Khaichiingol complex: gneisses, amphibolites, quartzites, marbles, migmatites; 2—Paleozoic. Modokhudag intrusive complex: diorites, granodiorites, granites; 3—Middle and Late Jurassic. Imalkh intrusive complex: grandiorites, granites, granosyenites. Bayantumen formation: 4—upper subformation: basalts, trachybasalts, andesites; 5—middle subformation: rhyolites, tuffs, conglomerates, mudstones; 6—lower subformation: trachyrhyolites, rhyodacites, tuffs, conglomerates, sandstones; 7—faults (a), thrusts (b); 8—deposits of uranium (a), occurrences (b); 9—deposits of polymetals; 10—fluorite deposits (a), occurrences (b)

granites are the main component of the massif. Granite rocks are highly alkaline rocks of the alkaline type with a predominance of potassium. The absolute age of the granites is 164+5 Ma, or the middle-late Jurassic. Granites are dissected by thin aplite and alkali granite dikes. Previous researchers have established that silver-bearing polymetallic mineralization is associated with the rocks of this complex, and with a magmatic control criteria.

The sequence of volcanic-sedimentary rocks (Bayantumen formation J_3–K_1bt, 160–144 Ma), which fills the Dornod VTS, has a thickness of 1000–1500 m (Fig. 4.2).

It is divided into three subformations:

1. Lower basalt-trachyandesite-rhyolite (J_3–K_1bt$_1$).
2. Medium trachyrhyolite-rhyolite (J_3–K_1bt$_2$).
3. Upper basalt-trachyandesite (J_3–K_1bt$_3$).

The subformation section begins with a horizon of basal coarse clastic conglomerates, leveling out the dissected paleorelief of the basement.

It is represented by a sequence of sedimentary-volcanogenic rocks, heterogeneous in composition and physical and mechanical properties, formed in continental rift conditions, and gently dipping to the southeast at an angle of 5–10° (rarely 25°). The lower subformation (J_3–K_1bt$_1$) is composed of sedimentary, sedimentary-tuffogenic, and volcanogenic formations of basic and intermediate composition. The section of the subformation begins with a horizon of basal coarse clastic conglomerates leveling the dissected paleorelief of the basement. The thickness of conglomerates reaches 100 m in erosion-tectonic paleovalleys and sharply pinches out toward the sides. Higher in the section, there is a member

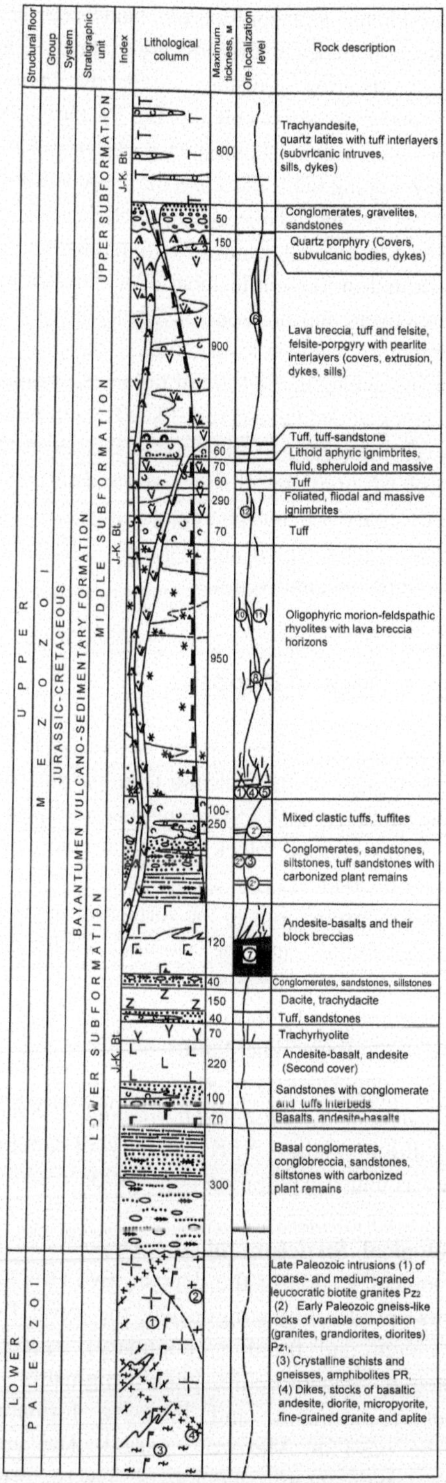

Fig. 4.2 Position of uranium ore bodies of the Dornod deposit in stratigraphic section (Mironov 2003; Jamsrandorj 2013)

of fine-clastic sedimentary rocks (sandstones, siltstones, mudstones) with unified plant detritus. Basal conglomerates and members of fine-grained sandy rock overlie alternating covers of basaltic andesites, trachyrhyolites, trachydacites with thin horizons of sedimentary and tuffaceous-sedimentary rocks of variable thickness.

These rocks are overlain by olivine andesite basalts, in the bottom part of which globular-block lava breccias are developed. The thickness of the cover reaches 100 m in low areas inheriting the paleorelief of the basement, and decreases or pinches out above the swell-like uplifts. Conditional uranium mineralization was noted in the lower and middle parts of the subformation (Fig. 4.2). A complete section of the subformation is noted within the Dornot and Mardaingol blocks.

The section of the lower subformation is completed by a thick sequence of fine and coarse clastic sedimentary rocks formed in lacustrine conditions. The lower part of the sequence, up to 100 m thick, is represented by a thin-layered alternation of fine-grained sandstones with two horizons enriched in carbonized organic matter. Ore bodies 2B, 2V, and 7 of the Dornod deposit and ore body 3 of the Gurvanbulag deposit gravitate toward them.

The upper part of the sequence is coarse clastic rock, filling individual isolated troughs. Its composition is dominated by conglomerates, conglobreccias, lenses and interlayers of gravelstones and uneven-grained sandstones. Tuffites with a low concentration of charred organics are observed sporadically. The maximum thickness of coarse clastic rocks reaches 50 m.

The middle subformation (J_3–K_1bt_2) is represented in the lower part of the section by a consistent sequence of felsic litho- and vitroclastic tuffs, with erosion overlying the terrigenous-sedimentary sections of the lower subformation. The tuffs contain interlayers of tuffites with unified organic matter. The thickness of the sequence is 20–40 m, and only on the southeastern flank, increases to 150 m. Higher up in the section, tuffs are replaced by covers of oligophyric rhyolites, lithoid ignimbrites, and massive and fluidal felsites. The thickness of the middle subformation is 600–750 m. Ore bodies 1, 4, 5, 8, 10, and 12 of the Dornod deposit, ore bodies 1 a, b, 2 a, b, and 4 of the Gurvanbulag deposit and the Mardaingol deposit are associated with it.

The upper subformation (J_3–K_1bt_3) is represented by covers of trachyandesites and basaltic andesites up to 300 m thick.

The Dornod VTS with uranium deposits is located at the intersection of faults with northeast and northwest directions. The latter, together with submeridional feathering faults, are ore-controlling and ore-bearing.

Within the Dornod ore cluster, the Gurvanbulag, Dornod, Nemer, and Mardaingol uranium ore deposits, the Ulaan and Mukhar polymetallic deposits, the Urliinobo and Tokhoi gold ore deposits, the Khooloi fluorite deposits and the Bumbat zeolite deposit were discovered and studied. Uranium deposits and ore occurrences are located in the northern marginal part of the Dornod volcano-tectonic structure and are combined into the Gurvanbulag, Dornod, and Mardaingol ore fields.

Uranium mineralization is localized in deposits of the Mesozoic sedimentary-volcanogenic complex in the basement rocks. The main ore deposits are localized in a narrow (150 m) stratigraphic range, covering the bottoms of the middle and upper parts of the lower subformation of the Bayantumen formation (Fig. 4.2). In this interval, the rocks are most tectonically disturbed, which, combined with the presence of permeable tuff-sedimentary rocks at this level and a thick screen composed of covers of massive acidic volcanic rocks, created favorable conditions for the localization of mineralization.

The main uranium ore deposits are located at depths from 30 to 700 m and do not have access to the surface; only individual vein-like ore bodies are exposed in the modern relief.

At the uranium deposits of the Dornod ore cluster, the structural and lithological control of mineralization is well manifested. The position of deposits and individual ore bodies is clearly controlled by large zones of steeply dipping submeridional and northwestern faults and their intersection nodes with northeastern faults. The main ore-bearing structures are gently dipping faults and zones of steeply dipping fracturing. As a rule, economic mineralization is absent in large steeply dipping faults that cross the entire volcanogenic-sedimentary sequence and reach the surface.

The lithological control of mineralization is manifested in the confinement of mineralization to the covers and horizons of permeable porous sedimentary and volcanic rocks enriched in reducing agents in the form of organics, sulfides, etc., stockwork-like and vein-like. The combination of structural and lithological factors in the control of uranium mineralization determined the diversity of morphological types of ore bodies, among which are plate-like, sheet-like, stockwork-like, and vein-like.

4.2 Dornod Deposit

The Dornod uranium deposit is located in the northern part of the volcano-tectonic structure of the same name (VTS), in the marginal part of the Erkhet tectonic block. The total area of the deposit is 20 km^2. From the north, the deposit is limited by protrusions to the surface of the basement granitoids and the Dornod northeastern deep fault, from the east and southeast by the Tsagaannuur and Fluorite faults, and from the west by the Bagaerkhet fault.

The deposit was discovered in 1973 by geologists of the Mongolian Geological Survey Expedition of the USSR Ministry of Geology and was under exploration until 1990.

Geological Structure of the Deposit. In the geological section of the deposit, two structural floors are clearly distinguished: the lower floor (basement and framing (boundary)) is represented by metamorphic, gneiss-migmatic, palingene-metasomatic and intrusive formations of the pre-Late Mesozoic period. The upper floor—sedimentary-volcanogenic—was formed by processes of collisional rifting with a sharp structural and stratigraphic unconformity and covers the foundation rocks.

Sedimentary-volcanic formations of the Bayantumen formation prevail in the lower structural stage at the deposit site (Fig. 4.3).

In the western and southern parts of the deposit, the mantle volcanics are intruded by numerous subvolcanic dikes and necks of felsite-porphyries, quartz porphyries, and trachyandesites. Dikes are concentrated near tectonic faults with northwestern and submeridional directions.

Economic uranium mineralization is localized in the rocks of the upper structural stage. In the basement of the deposit, only individual small ore bodies with low grade mineralization are noted.

The ore-bearing rocks of the Dornod deposit are terrigenous-sedimentary deposits of the horizon separating volcanics of the lower and middle subformations, specific globular-block lava breccias of olivine andesite basalts, oligophyric rhyolites, and vitroclastic tuffs of acidic composition.

Tectonics of the Deposit. The structural plan of the Dornod deposit is determined by its confinement to nodes of the intersection of differently oriented steeply dipping faults in the marginal part of the local Erkhet volcano-tectonic depression. The leading role in the formation of the ore-bearing tectonic knot belongs to the faults of the northeastern, northwestern, and submeridional directions. Most of them are hidden structures. They are distinctly fixed in the basement of the deposit and rocks of the lower subformation, but gradually fade in the formations of the middle and upper subformations, turning into discontinuous thin seams and bands of increased fracturing.

The faults of the northeastern 10–20 and 40–60° directions—Dornod, Tsagaannuur, Bulag, Fluorite, and others—are the most ancient. They divide the deposit area into a series of tectonic blocks 1.5–2.0 km wide and control the position of the main morphostructural elements (paleovalleys and swell-like protrusions) of the basement. In terms of the nature of their movements, they mostly belong to the type of fault-shifts with amplitudes of tens and a few hundreds of meters. On the surface, northeastern disturbances are expressed by zones of crushed and brecciated rocks 10–30 m thick.

The meridional faults are the most pronounced and ore-bearing: Western, Central, Eastern, etc. As a rule, these are zones of concentrated linear fracturing 2–8 m thick, increasing to 10–30 m at junction nodes with disturbances in other directions. Inside the zones of fractured rocks, central seams with friction clay 0.1–1.2 m thick develop. In the immediate vicinity of the seams, the rocks are intensively crushed to form small-clastic breccias (1–2 cm). These faults are accompanied by a series of closely spaced subparallel structures of higher orders, forming tectonic faulted areas 200–800 m wide and 4–8 km long. Vertical displacements along them is absent or does not exceed a few meters.

Submeridional faults control the distribution of all significant economic ore deposits at Dornod (Fig. 4.3). Among the steeply dipping faults within the deposit, faults with a northwestern (310–330°) strike are clearly manifested. The regional inter-block structures of the first order include the subvertical Bagaerkhet fault, which limits the deposit from the southwest. The fault is represented by a powerful tectonic suture, which is accompanied by a series of small subparallel echelon-shaped faults in a band of fractured and crushed rocks up to 100–150 m wide. The main seam is expressed by a zone of intense crushing and flaring, 35–40 m thick, with a wide development of friction clay.

Tectonic disturbances with a sublatitudinal direction are not widely distributed. In addition to the main steeply dipping fractures, small non-extended seams and cracks with a northeast and northwest strike, which have the character of thin spalls, have been identified in the field.

An important feature of the deposit is the intense development of gently sloping tectonic faults (Fig. 4.4). They are developed in the form of interlayer faults at the contacts of almost all horizons and covers of sedimentary-volcanogenic rocks, as well as at the junctions of differently oriented steeply dipping faults. As a rule, the maximum disturbances of this type occur in

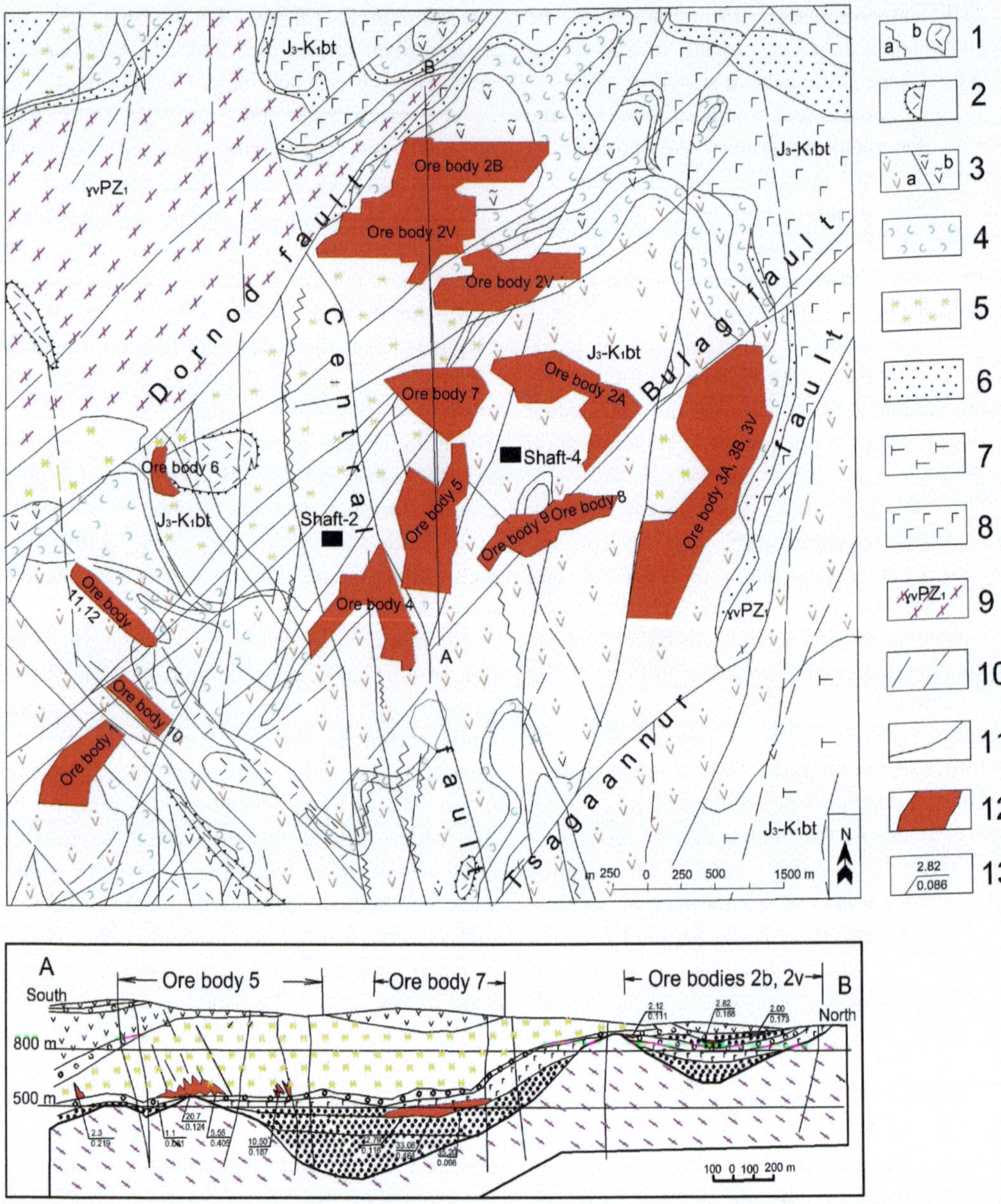

Fig. 4.3 Schematic geological map and section of the Dornoth deposit (according to Mironov et al. 1995): 1–8—sedimentary-volcanogenic formations of the Bayantumen formation (J_3–K_1, bt): 1—dikes (a) and extrusions (b) of basaltic andesites and trachydacites; 2—dikes and extrusions of rhyolites; 3—rhyolites (a), incl. h. coarse fluid (b); 4—felsite tuffs; 5—oligophyric rhyolites; 6—sandstones; siltstones, conglomerates; 7—trachyandesites; 8—basaltic andesites; 9—Early Paleozoic gneissic granitoids; 10—main faults; 11—gentle falls; 12—ore bodies with standard mineralization (projection on the day surface) and their numbers; 13—mineralization parameters (numerator—thickness, m; denominator—grade, %)

the most differentiated parts of the stratigraphic section, which are characterized by different physical and mechanical properties. Large layer-by-layer failures are usually represented by distinct seams up to 1.0–1.5 m thick, made by rocks ground to clay.

In the stratigraphic section of the area, five levels of development of layer-by-layer

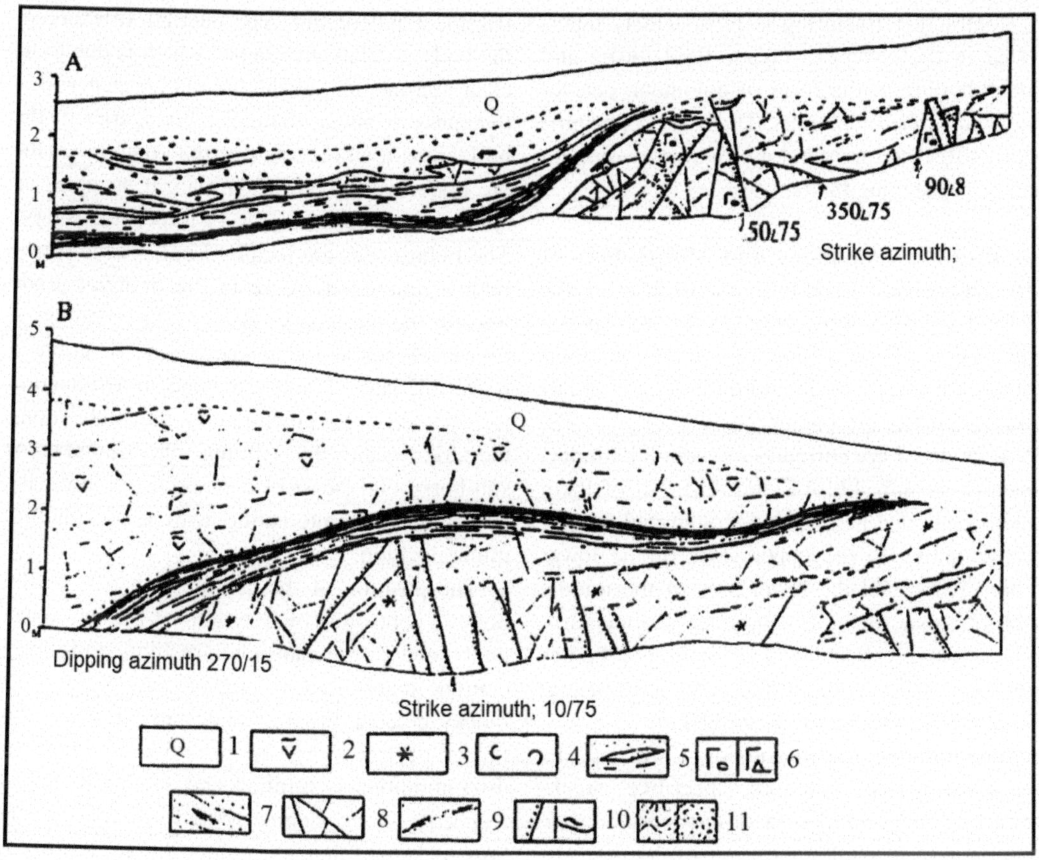

Fig. 4.4 Morphology of gently sloping faults at the contacts of sedimentary-volcanogenic rocks: **a** trench 638, northeastern wall; **b** trench 214, northeast wall; 1—deluvium; 2–5—middle subformation: 2—lithoid ignimbrites; 3—oligophyric rhyolites; 4—felsic tuffs; 6, 7—lower subformation: 6—massive and globular breccia andesite basalts; 7—sandstones and siltstones with coalified organics; 8—small fracturing; 9—gently sloping zones of contiguous fractures, intense crushing and flaring of rocks; 10—friction clay along tectonic seams and along rock contacts; 11—microfolding of sandstones and tuffs

ore-controlling faults were established (Mironov et al. 1986). The primary layers are the base of the cover of trachyrhyolites of the upper subformation, the base of oligophyric rhyolites of the middle subformation, and horizons of sedimentary-tuffaceous rocks with the upper cover of olivine andesite-basalts of the lower subformation. Uranium mineralization is localized within the established levels in gentle sheared zones within the areas of their conjugation with steeply dipping faults of the Central, Western, and Eastern tectonic zones.

Within the faults are areas of cleavage cracks, separation cracks and blocks of crushed rocks that are intensely manifested. For the most part, these cracks have a different direction, a steep dip, and are often limited in height by the contacts of the host rocks. The dimensions of the stockwork zones in plan reach 1500 × 400 m. Near the contacts, especially as layer-by-layer is shed, a flat-dip fracture is manifested, which is involved in the structure of the stockwork zones. Areas with high specific fracturing (40–80 fractures/m), as a rule, are confined to the junctions of systems of steeply dipping faults and blocks limited by them, and often develop over swell-like basement ledges and structurally complicated stratigraphic boundaries. The localization of small deposits is associated with stockwork fracture zones in the field.

Contraction fracturing is most widely manifested in the covers of coarse-fluid felsite and trachyrhyolite in the form of columnar, coarse-platy, and spheroidal units. These cracks significantly affect the morphology of steeply dipping and gently sloping faults.

Localization Conditions and Morphology of Ore Deposits. Uranium mineralization at the Dornod deposit is represented by 14 ore deposits localized in all rock complexes of the geological section (Fig. 4.5), including basement granitoids. Of these, ore bodies 2A, 2B, 2V, 3A, 3B, 3V, 4, 5, and 7 are currently of economic importance (Figs. 4.5, 4.6, 4.7 and 4.8). All of them are concentrated in a narrow stratigraphic range, corresponding to the upper sides of the lower subformation and the lower side of the middle subformations of the Bayantumen sedimentary-volcanogenic formation. The main ore bodies are not found on the surface and are concentrated at a depth interval of 30–650 m.

According to the morphology, ore deposits are divided into plate-like, sheet-like, stockwork-like and vein-like. In recent years, as new data became available due to re-exploration, G. Jamsrandorj proposes to revise the genesis of ore bodies 2A, 2B, 2C, 3A, 3B, 3C, that occur in the sedimentary strata of the upper part of the lower subformation of the Bayantumen formation, as sedimentary (Fig. 4.7).

The largest plate like ore deposit 7 (Fig 4 6), containing 49% of the total explored reserves of the deposit, is located in the central part of the deposit. It is localized in the horizon of specific spherical lava breccias, which occurs at the base of the third cover of olivine andesite-basalts of the lower member at depths of 400–500 m from the surface.

The structural position of the body is determined by its confinement to the node of the intersection of the submeridional Central zone of tectonic faults with the southern branch of the northeastern Kharaat fault, as well as its location above the most submerged section of the linear northeastern depression in the roof of the underlying sandstone and basal conglomerates, spatially combined with a large trench in basement paleorelief. The internal structure of the body is relatively simple, which is due to its clear confinement to the breccia horizon with a thickness of about 40 m. The boundaries of the horizon as a whole serve as the contours of the ore body in the section. In plan, it occupies an area of 800 × 400 m and has a trapezoidal shape. The richest ores are localized in the central part with a gradual decrease in the uranium grade toward the periphery. Behind the contour of the development of standard mineralization, a powerful halo of non-economic mineralization is observed. The pinch-out of the body along the strike occurs by splitting into a number of sub-horizontal lenses, composed of ordinary and poor ores, alternating with non-economic mineralization.

The position of the central, rich part of the body is controlled by a system of flat and diagonal zones of expansion, crushing and brecciation, manifested throughout the section of blocky-breccia basaltic andesites. Toward the flanks of the body, the network of diagonal discontinuities is rarefied and gradually fades, respectively, the degree of deformation of the host rocks and their ore saturation are removed. In the marginal parts of the cover, only separate extended gentle ruptures remain, which control the position of small sheet-like orezones.

Rich and ordinary veinlet-disseminated mineralization develops mainly in zones of maximum crushing and brecciation of rocks and along contraction cracks in the periphery of the blocks. In the areas of development of ordinary and poor ores, disseminated and dispersed ore mineralization develops exclusively along the cementing mass.

Steeply dipping discontinuities of visible connection with mineralization are not found. A statistically significant relationship between mineralization and variations in the thickness of the cover and the thickness of the productive horizon of blocky-breccia basaltic andesites has not been established.

Tear-like ore bodies 2 and 3 are located on the northern and eastern flanks of the deposit and spatially gravitate toward the marginal parts of the Erkhet local depression. In plan view,

Column	Ore localization setting	Morphological types of deposits	Number of ore zones	Ore body number	Persentage of reserves
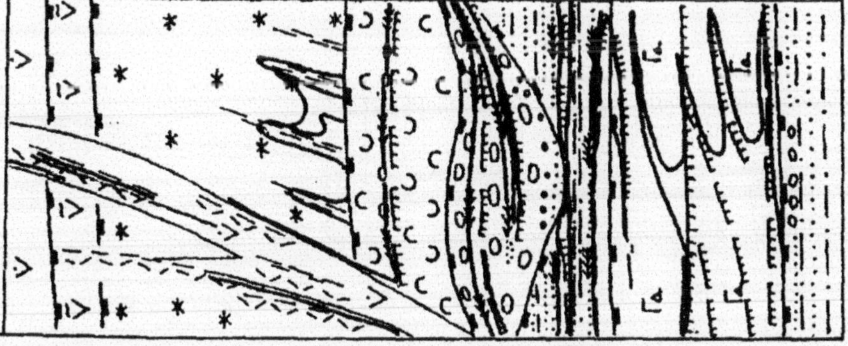	Contact zones of intruding bodies in volcanics of the middle subformation	Complex vein-like and stockwork-like	6 10 11 12		
	Steeply dipping fracture zones at the base of the oligophyric rhyolite cover	Vein-like and stockwork-like	1 4 5 8		4 13 1
	Zones of permeability and failure in carbon-enriched organic matter in tuffs, conglomerates, and fine clastic rocks at the boundary of the middle and lower subformations	Sheet-like	2 3 9	2A 2B 2V 3A 3B 3V	4 10 12 3 1 2 1
	Zones of permeability and failure in the horizons of large block breccias at the bottom of the basaltic andesite cover	Slab-shaped with complex pinching out on the flanks		7	49

Fig. 4.5 Localization conditions and morphological types of ore bodies of the Dornod deposit (according to Mironov 2003)

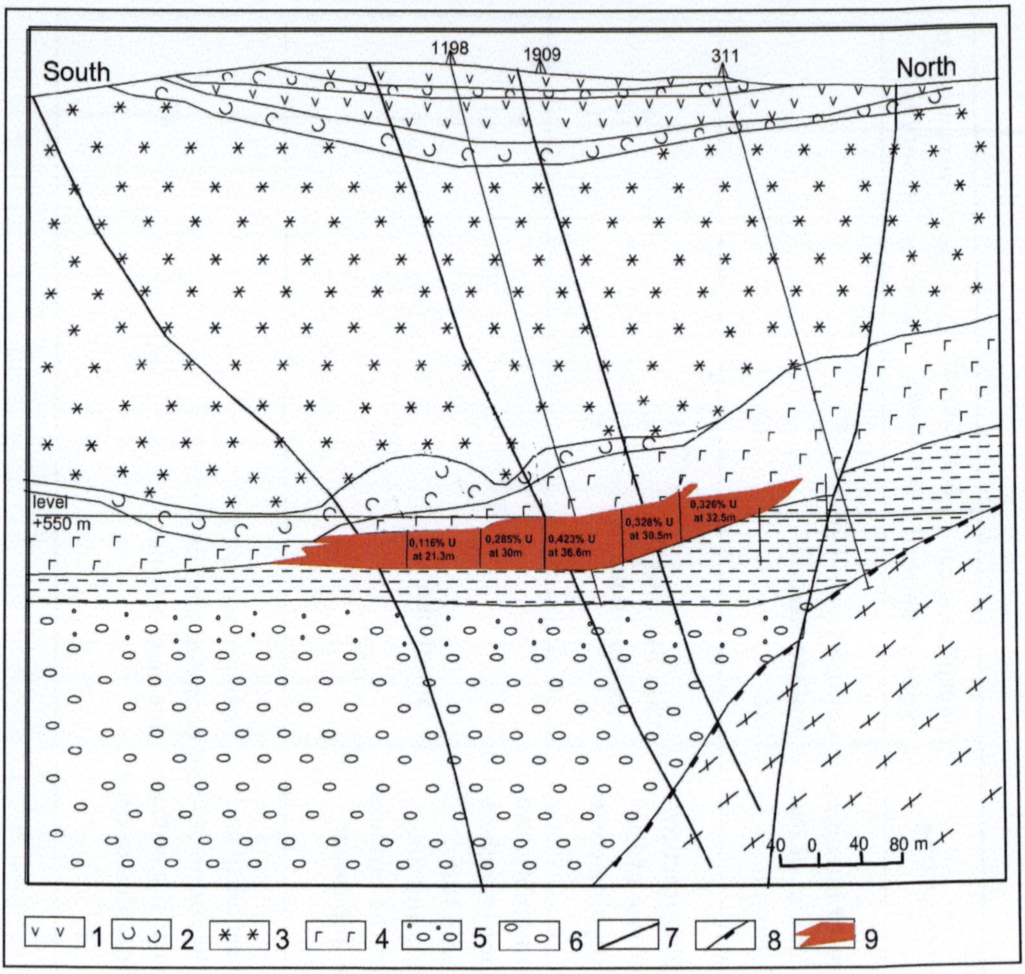

Fig. 4.6 Geological section of ore body 7 of the Dornod deposit (according to Mironov 2003): 1-6—sedimentary-volcanogenic formations of the Bayantumen formation (J_3–K_1bt): 1—felsite and their breccias; 2—felsic tuffs; 3—oligophyric rhyolites and their blocky breccias; 4—basaltic andesites; gravelstones, conglomerates; 7—tectonic faults; 8—gentle falls; 9—ore bodies; 10—parameters of mineralization (in the numerator—thickness, m; in the denominator—grade, %)

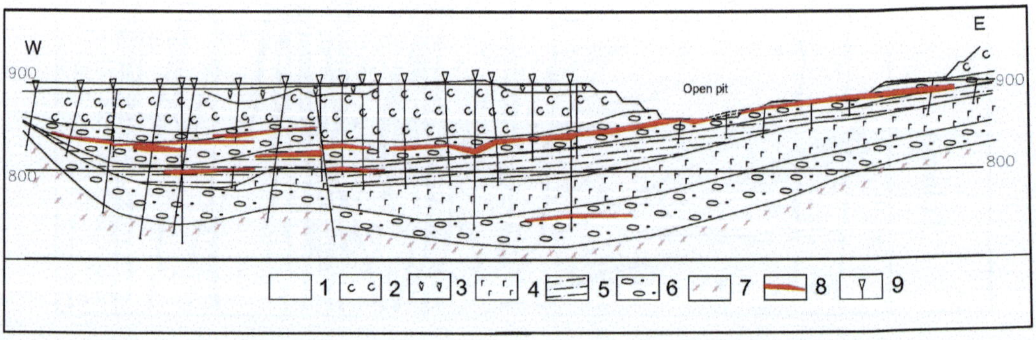

Fig. 4.7 Section through exploration line 179+50 of ore body 2B (Jamsrandorj 2009, 2013): 1—quaternary deposits; 2—rhyolites; 3—tuffs; 4—basalts; 5—siltstones; 6—conglomerates; 7—gneisses; 8—uranium ore bodies; 9—boreholes

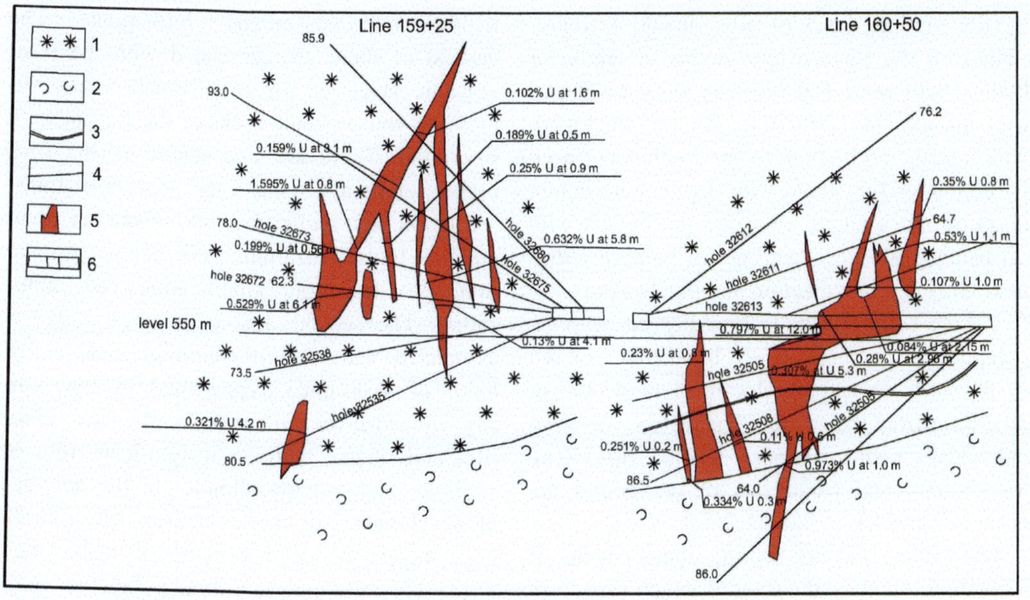

Fig. 4.8 Geological sections of ore body 5 of the Dornod deposit (according to Mironov 2003): 1—oligophyre rhyolites; 2—tuffs; 3—main faults; 4—gentle falls; 5—ore bodies; 6—underground mine workings

the ore bodies are elongated in a northeasterly direction and gravitate toward paleovalleys and slopes in the basement paleorelief. Uranium mineralization is localized in bodies of layer-by-layer fracturing, developed in a horizon of tuffaceous-sedimentary rocks 60–110 m thick at the boundary of the lower and middle subformations and is controlled by lenses and interlayers of fine clastic varieties enriched in unified remains. At the same time, the richest ore concentrations spatially gravitate toward discontinuous faults of near-meridional direction or nodes of their intersection with northeast-trending faults.

Several levels of development of uranium mineralization with a multi-tiered structure are found in the section. Some of the ore bodies are associated with fine clastic sediments, while others are confined to coarse clastic deposits. The distance between the ore bodies does not exceed 40 m. Economic ores are accompanied by wide halos of noneconomic mineralization. Sheet-like ore beds contain 34% of all uranium reserves of the Dornod deposit.

Ore-bearing zones 1, 4, 5, 6, 8, and 10 with fissure-vein mineralization, containing about 17% of the reserves of the deposit, are developed in the zones of the near meridional Central and Northwestern Bagaerkhet faults. The main part of mineralization is localized in the bottom part of the cover of oligophyre rhyolites and is controlled by a series of closely spaced steeply dipping tectonic sutures of near meridional and northwestern directions. The vertical range of mineralization usually does not exceed 50–80 m. The internal structure of ore-bearing zones is characterized by an extremely uneven distribution of economic mineralization, due to the presence of many small separated and differently oriented ore bodies.

One of the relatively large deposit of this type is ore body 6 of the Dornod deposit. It is confined to the western flank of the laccolithic extrusive felsite-porphyry. On the surface, the extrusion has a shape close to isometric with a diameter of about 500 m. With depth, its dimensions decrease to 200 m.

The main ore body, in the form of a small columnar stockwork, is controlled by a junction of steeply dipping submeridional and northwestern faults. Uranium mineralization was exposed by the modern level of erosion and is represented on the surface by a series of lenticular bodies with poor and ordinary ores developed over an area of 120×50 m. An ore stockwork with a steep declination to the south was traced by boreholes for 300 m with depth. With depth, it splits into separate vein-like ore bodies with off-balance (do you mean out of balance radiometrically?) mineralization and pinches out.

All ore bodies of the Dornod deposit, with the exception of ore body 6, are blind (not exposed to the surface) and are localized at depths of 30–600 m from the surface. The distance between individual bodies varies within 50–350 m. Morphology and parameters of ore bodies are given in Table 4.1.

Ore bodies and ore-bearing zones consist of separate ore bodies, the sizes of which vary from a few meters to 1 km.

Ore Mineral Composition and Host Rock Alteration. The uranium mineralization of the Dornod deposit was formed in the aureoles of polychronous, multistage hydrothermal-metasomatic formations. The earliest of them are associated with the post-volcanic stage of the transformation of sedimentary-volcanogenic rocks of the Bayantumen formation. There are two stages, in the first, higher temperature stage, the process of transformation of rocks was expressed in their areal K-feldspathization, pyritization, seritization, propilitization, silicification and albitization. The second, low-temperature stage, is associated with the formation of wide zoned halos of hydromica, represented by hydromicas, mixed-layer minerals, carbonate, chlorite, and quartz with an insignificant role of kaolinite and montmorillonite. In the aureoles of development of these minerals, the following parageneses are distinguished: in the outer zone—kaolinite-montmorillonite, in the middle zone—clayey-hydromicaceous, in the inner zone—quartz-hydromicaceous-chlorite with hematite. Hydromicaceous metasomatites often go beyond the deposit, being located along the main steeply dipping faults. A clear confinement of uranium mineralization to the most developed

Table 4.1 Parameters of ore bodies of the Dornod deposit (Mironov et al. 2012)

Number of ore bodies	Dimensions (length) (m)			Azimuth and angle fall
	To continue	To dip	Thickness	
Plate-like body				
7	800	480	11.2	Subhorizontal
Sheet-like bodies				
2A	1300	60–380	1.7	Same
2B	1000–1200	140	1.7	Same
2V	850	250–900	2.0	Same
3A	1700	200–650	1.2	Same
3B	600	600	1.2	Same
3V	1500	250–400	1.2	Same
9	990	70–300	1.3	Same
Ore-bearing zones with fissure-vein mineralization				
5	1100	80–400	–	Consists of sub-vertical small ore bodies of various orientation
4	380	80–230	–	
1	950	200–480	–	
8	450	50–200	–	
10	540	130	–	
6	500	80–280	–	

halos is not observed, but ore deposits in all cases were localized in previously hydromica rocks.

Ore-accompanying processes take place at the pre-ore, ore and post-ore stages. First, quartz-fluorite-sulfide veinlets with galena, sphalerite, and pyrite were formed. This mineral association is most widespread in ore body 7. Here, along with the above sulfides, marcasite, chalcopyrite, fahlore, and greenockite are present. In the ore stage, the formation of uranium mineralization proceeded with the formation of two mineral associations: jordisite-nasturan and uranium-titanate-nasturan. The first of them is present in all ore bodies, except for 7, and is accompanied by sulfides of lead, zinc, molybdenum, iron, and florite. In ore body 7, the range of productive minerals is wider. It includes pitchblende, branerite, uranium-titanium minerals, radioactive leucoxene, and anthraxolite.

Post-ore metasomatic veinlets are widely distributed in diagonal and subconformant discontinuities flanking them, and in the cover of globular-breccia andesite basals (body 7), in broken contacts and through steeply dipping discontinuities and rhyolite covers (body 5), as well as in small fissures. There are three mineral associations that successively replace each other: smectite kaolinite, coffinite-quartz-fluorite, and calcite.

The kaolinite-smectite mineral association includes rectorite, smectite, zeolites, fluorite, kaolinite, and calcite. Of these, smectite, kaolinite, and fluorite are the most common; in basaltic andesites, the leading mineral of this association is most often calcite. Often there are smectite-fluorite and kaolinite-fluorite veinlets. The formation of this mineral association, as a rule, was accompanied by brecciation and clarification of rocks and ores, replacement by smectite and kaolinite of the original rock-forming minerals and previously formed hydromica, chamosite and pitchblende, and in body 7, brannerite (Fig. 4.9). In this case, the redistribution and removal of uranium and iron and the formation of coffinite took place both in the form of pseudomorphs and independent segregations. For ore body 7, coffinite and phosphorus impurities are usually present. Characteristically, in

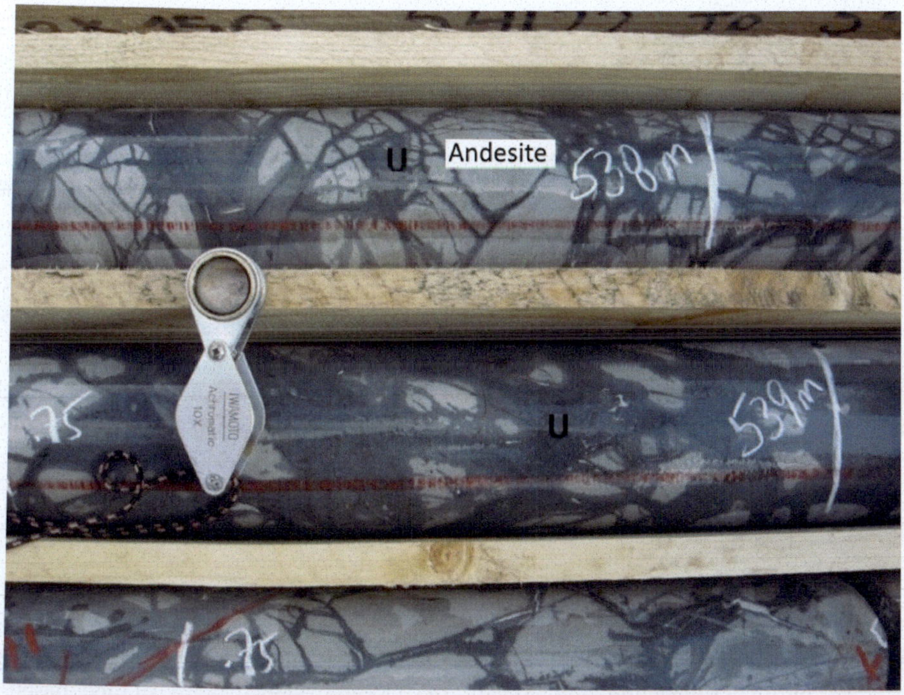

Fig. 4.9 Breccia texture of ores of ore body 7

the hydromicas preserved in the aureoles of the development of post-ore alteration, an increase in the swelling interlayers from 5–8 to 10–14% was noted.

The coffinite-quartz-fluorite association of minerals is common in all bodies, but to varying degrees. In ore body 2, the deposition of fluorite was accompanied by intense replacement of pitchblende by coffinite and even by the formation of not pseudomorphic, but independent segregations of the latter. The coffinite of this body contains an admixture of arsenic. In ore body 5, fluorite was followed by pyrite, sphalerite, chalcopyrite, fahlore, galena, and coffinite with an admixture of niobium (up to 1.5%) and yttrium, as well as phosphorus.

The oxidation zone at the deposit is poorly developed. Uranophane, otenite, beta-uranotyl, rarely welsendorphite, casolite, and amorphous uranium hydroxides have been identified only from tectonic faults. In addition to uranium minerals, there are hydroxides of iron and manganese containing a small amount of uranium.

According to G. B. Naumov (IGEM) and other researchers, the formation temperature of uranium ores at the Dornod deposit is 120–170 °C.

The age of hydrothermal uranium mineralization according to geological data is Early Cretaceous. The absolute age of uranium ores, established in the IGEM laboratory by the lead isotope method, is 136–153 Ma (J_3–K_1).

In the economic ores of the Dornod deposit, the main productive minerals are pitchblende, coffinite, uranium titanates; uraninite and uranium-bearing anthraxolite are less common. According to the peculiarities of the mineral composition, the following types of ores are distinguished: pitchblende-coffinite with molybdenite, pitchblende-coffinite with sulfides, pitchblende-coffinite with uranium titanates.

Nasturan-coffinite ores with molybdenite compose ore bodies in felsic volcanics, vein and stockwork in covers and extrusive bodies of rhyolites, sheet-like ores in lithoid rhyolites, tuffs and volcanic glasses. The amount of sulfides in them is 1.5–2%; pyrite, molybdenite, and galena predominate among them.

Nasturan-coffinite ores with sulfides form bedded ore bodies in sedimentary, tuff-sedimentary rocks that make up the basal horizon of the acid volcanic sequence. Coffinite here sharply prevails over pitchblende, the share of uranium associated with uranium-bearing anthraxolite is 2.5%. The amount of sulfides in ores reaches 5–7%. In addition to arsenopyrite, there are realgar, orpiment, and native arsenic.

Nasturan-coffinite ores with uranium titanates form plate-like ore bodies in the lower part of the third (upper) cover of basaltic andesites. Along with pitchblende and coffinite, brannerite, uranium-bearing leucoxene, and anthraxolite are present in significant amounts. Of sulfides, 2–3% are marcasite, galena, sphalerite, chalcopyrite, and greenockite. Uranium ores are mainly composed of an aggregate of fine and fine-grained, to a lesser extent finely dispersed uranium minerals with sinuous and rounded boundaries. The sizes of aggregates are from 0.001 to 2 mm. The texture of the ores is disseminated, dispersed-disseminated, spotty-disseminated, nested-disseminated, vein-disseminated, less often banded-disseminated and earthy. Rich ores contain breccia and cement textures. The structure of the ore is collomorphic, corrosive, emulsion, granular, looped, reticulate, spherulitic, tangled fibrous and lamellar (Fig. 4.10).

Chemical Composition of Ores. The chemical composition of the ores of the Dornod deposit were studied using metallurgical tests, on groups and individual samples (Table 4.2). From the data it follows that molybdenum (0.01–0.03%), lead (0.01–0.2%), zinc (0.5–2.0%), and zirconium are present in uranium ores in high concentrations. Arsenic (0.01–1.0%), copper and silver have no industrial value. According to the chemical composition, the ores of the Dornod deposit are of the aluminosilicate type with a low content of acidic components.

Geochemical Features of Ores. The geochemical characteristics of the layer-like deposits of the Dornod deposit were studied based on the results of analyzes of 6320 samples, including

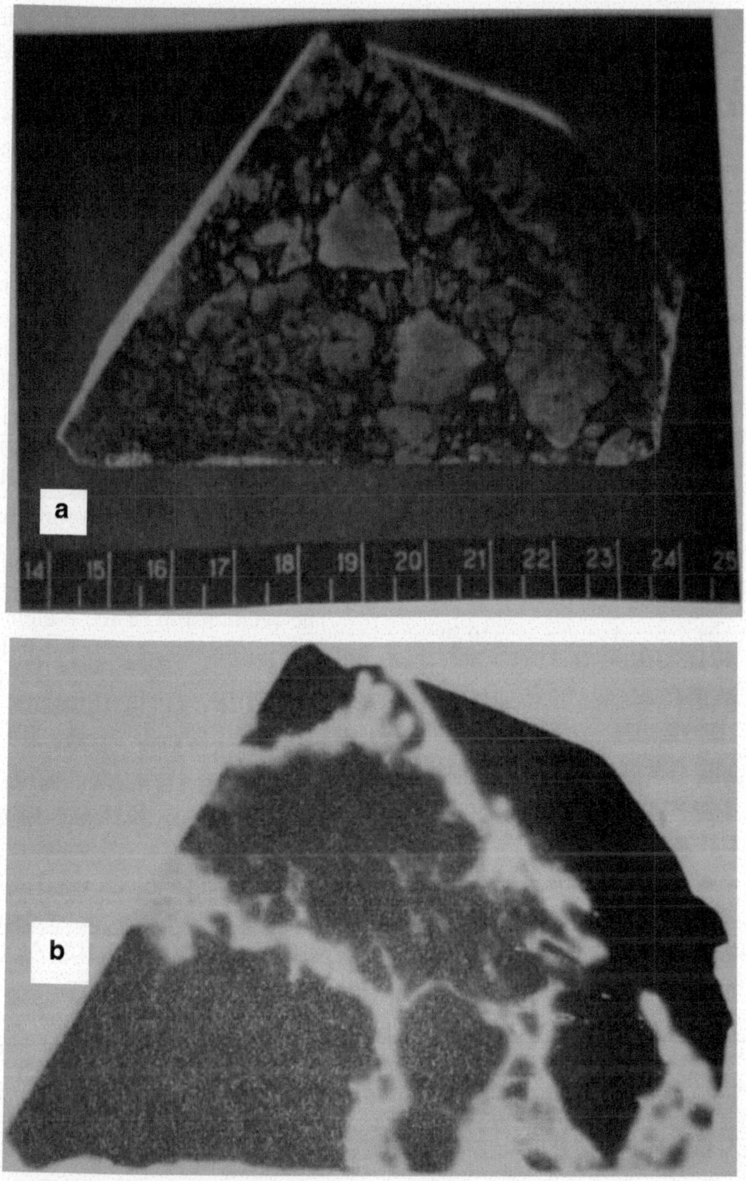

Fig. 4.10 Uranium minerals (nasturan, coffinite) are cemented and overgrown with rock fragments from the Dornod deposit. **a** Photo of the sample (natural size); **b** radiography (exp. 3 days)

320 mineralogical, core and channel samples. A wide range of satellite elements characterized by varying degrees of concentration and correlation with uranium have been identified in rocks and ores. The specialization of ore-bearing rocks (carbonaceous siltstones, sandstones, conglomerates) is determined by three associations of elements.

The first association includes elements with a high background concentration (Кк>3) (%): U—0.001, As—0.0028, Pb—0.0025, Zn—0.01, Th—0.0021, Nb—0.0034. Higher correlations (Кк>0.4; Кв—30–60 to 120%) have (%): Mo—0.0005, Cu—0.0015, Be—0.0006, Zn—0.0005, Li—0.0058, Co—0.0005. They are characterized by a weak correlation with

Table 4.2 Ore chemical composition (by group samples) (Mironov et al. 2012)

Chemical indicators	Layer like bodies 2A, 2B, 2V	Layer like bodies 3A, 3B, 3V	Plare shaped body 7
SiO_3	68.84	67.59	56.32
Al_2O_3	15.34	14.45	16.28
Fe_2O_3	0.71	2.23	1.38
FeO	1.30	1.15	3.53
TiO_2	0.46	0.45	2.96
P_2O_5	0.10	0.22	0.95
CaO	0.47	2.54	3.60
MgO	0.75	0.5	T
MnO	0.05	0.07	0.08
K_2O	3.26	3.47	2.01
Na_2O	3.02	2.56	3.38
LOI[a]	4.59	5.07	7.68
H_2O	2.14	1.64	1.48
a*	0.08	–	0.29
CO_2	0.77	0.78	3.26
S_{total}	0.187	0.184	0.31
S_{sulf}	–	0.177	–

a* *LOI* loss on ignition

uranium (Kcor 0.1–0.3), a uniform distribution (Kv 20–40 to 80%) and a low level of background concentrations (K 1.5–3).

Secondary indicator elements—Co, Mn, V, Cu, Be, B, Y, and Yb (group of input with concentrations in ores 2–3, and in halos up to 1.5 backgrounds) and Sn, Ga, Mg, Ca, Fe, Na, and Al (removal group with minimum grades in ores less than 0.5 background) have a weak ability to form halos. The connection with uranium is both positive (input group) and negative (output group), the correlation coefficient is up to ±0.5. The distribution is non-uniform (Kv 80–120%). The third association of elements, represented by Ba—0.043, Sc—0.0008, Ni—0.0008 and Sr—0.043%, has no association with uranium. The distribution of elements is homogeneous (Kb 40–60%), background concentrations are high (Kb>3).

Layer-like deposits in tuff-sedimentary rocks are accompanied by halos of indicator elements, the main of which are U, As, Zn (the level of accumulation in ores is 10–100 geochemical backgrounds and more, in halos 3–5 backgrounds). Pd, Mo, Cd also form halos, but have a lower level of accumulation (in ores up to 10, in halos less than 3 backgrounds).

The relationship of these elements with uranium is positive (Kcor—0.5–1.0, less often 0.3–0.5), the distribution is extremely heterogeneous (Kv—150–250, 300%).

Analysis of the average grades by ore classes shows that the maximum concentrations of elements-indicators of the input group are recorded in rich ores (%): As—0.6, Zn—0.237, Pb—0.017, Mo—0.005, Cd—0.002, Cu—0.003, Co—0.0013, Mn—0.074, V—0.0083, Be—0.0013, B—0.0017, Y—0.0118 and Yb—0.0015. At the same time, their contents are on average 2–4 times higher, and for the main elements 10 times higher, than in non-economic ores. Kv and Kcor also have higher values (by 1.5–3 times).

In a collection of 145 samples, an analysis was made of the occurrence of anomalous

Table 4.3 Occurrence of anomalous samples in the ores of the Dornod deposit (Mironov et al. 2012)

Element graes (%)	Occurrence (number of anomalies samples in the selection)
More 1.0	Arsenic (15), zinc (15), manganese (1)
0.1–1.0	Boron (7), strontium (4), barium (3), zirconium (3), titanium (3), molybdenum (3), lead (2), phosphorus (2)
0.01–0.1	Copper (9), cadmium (9), chromium (6), vanadium (6), gallium (5), cerium (5), lanthanum (4), yttrium (4), ytterbium (3), nickel (2), hafnium (2), niobium (2), tantalum (2), gadolinium (1)
0.001–0.01	Beryllium (11), scandium (9), tungsten (7), thallium (6), bismuth (3), germanium (3), tin (2)
Less 0.001	Silver (5)

values in ores (Table 4.3). The occurrence of anomalous values in ore range from 10 to 12, averaging 2–4%. The majority (60–80%) of anomalous values are found in industrial ores with economic uranium grade.

An analysis of the spatial distribution of elements showed that primary aureoles are significantly (2–5 times or more) larger than ore bodies. Their configuration is determined by the morphology of the ore deposits, ore-bearing rocks, and faults. The halos of all satellite elements are elongated along the bodies and are controlled by gently dipping, to a lesser extent, steeply dipping faults. In screening rocks, halos are developed mainly along steeply dipping structures, and in permeable (tuffs, tuff-sedimentary and sedimentary) formations—along the strike of horizons.

The most extended and contrasting (more than 2 backgrounds) are the halos of uranium (radioactivity). The thickness ranges from a few meters to 40–60 m, occupying the entire volume of productive rocks in some areas. Along the strike, the aureoles can be traced throughout the productive strata. Contrasting halos (more than 2 backgrounds) exceed the size of ore deposits by 1.5–3 times. Weakly contrasting (less than 2 background) halos are developed everywhere within productive rocks. The area of their distribution is more than 15 km^2. The haloes of dispersion of satellite elements of deposits 2B, 2C merge with the haloes of other deposits and form a common anomalous geochemical field with an area of about 80 km^2.

The halos of each satellite element individually are inferior in all respects to the halos of radioactivity.

Halos with maximum concentrations of satellite elements (arsenic 0.1–1, zinc 0.5–2, lead 0.01–0.2, molybdenum 0.01–0.03 and cadmium 0.01–0.02%) gravitate toward the western and central parts of the ore deposits, while they are more developed in coarse clastic sediments (ore-bearing rocks of deposit 2B). The length of aureoles varies from 109–150 to 250 m for the main ones, and 50–100 m or less for secondary satellite elements. The thickness of the halos is from a few meters, less often up to 10–15 m. It should be noted that the contrasting halos of the main satellite elements (arsenic and zinc) are recorded mainly in ore-bearing rocks, and of minor indicator elements beyond them.

Multiplicative scattering halos are more effective. Their sizes are comparable to the sizes of uranium scattering haloes, and in contrast they exceed them by 5–10 times or more. The multiplicative halos of satellite elements with maximum intensity are mapped in areas with rich ores.

4.3 Gurvanbulag Uranium Deposit

The Gurvanbulag uranium deposit is located in the northern part of the Dornod volcanotectonic structure (DVTS), in the northwestern marginal part of the Ulaan block, in the junction of differently oriented systems of steeply dipping faults and is localized in volcanic glasses and tuffs among acid volcanic rocks of the middle subformation of the Bayantumen formation, which are nowhere to be found marked (Figs. 4.11 and 4.12).

Discovered in 1973 by geologists of the Mongolian expedition of the USSR Ministry of Geology. According to the results of detailed exploration, it was determined as a large economic facility with ordinary ores. It consists of three ore-bearing zones Central, Intermediate and South-Western.

Gurvanbulag Deposit-Geological Structure. The geological structure of the Gurvanbulag deposit involves the formation of two structural stages. The lower level forms the foundation of the deposit and is represented by rocks of the pre-Late Mesozoic period. Two complexes of different ages stand out among them: Proterozoic metamorphic rocks of the crystalline substrate and Paleozoic intrusive anatectoid granitoids.

Proterozoic rocks include amphibolites, biotite-hornblende and hornblende schists, biotite-muscovite, biotite-sillimanite gneisses intercalated with quartzites, calciphyres, and marmorized limestones. Among the Paleozoic

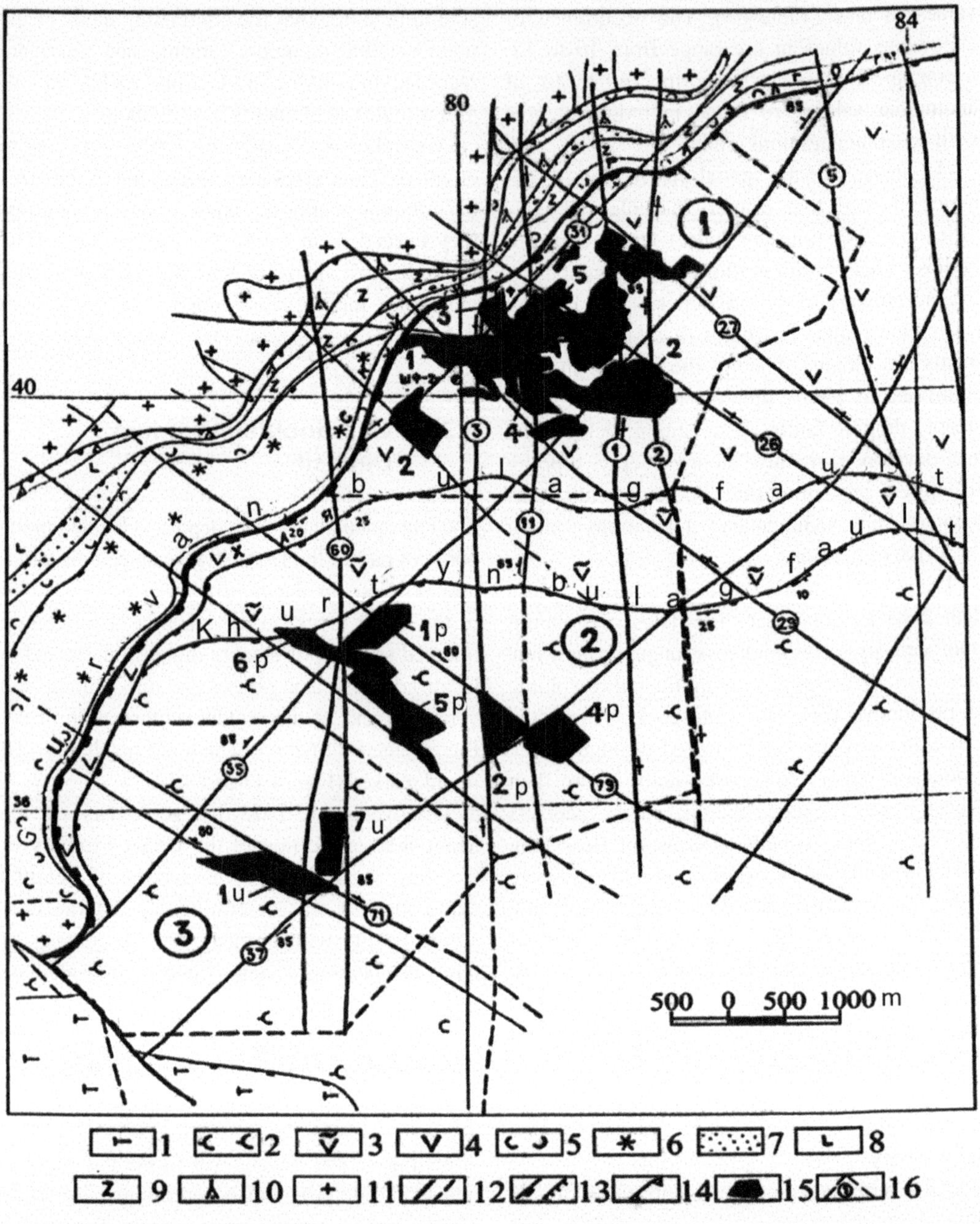

Fig. 4.11 Distribution of mineralization in terms of the Gurvanbulag deposit (according to Mironov 2003): 1-10—sedimentary-volcanogenic formations of the Bayantumen formation (J_3–K_1bt): 1—trachyandesites; 2—sanidine trachyrhyolites; 3—coarse fluid felsite; 4—felsite; 5—felsic tuffs; 6—oligophyric rhyolites; 7—sandstones; 8—basaltic andesite; 9—trachydacite; 10—quartz-feldspar porphyry; 11—undivided basement rocks; 12—steeply dipping tectonic faults; 13—gently dipping tectonic faults; 14—boundary of the Dornod volcano-tectonic structure; 15—projections of ore bodies localized at different lithological-stratigraphic levels, and their numbers; 16—contours of the ore zones: 7—central; 2—intermediate; 3—south-west

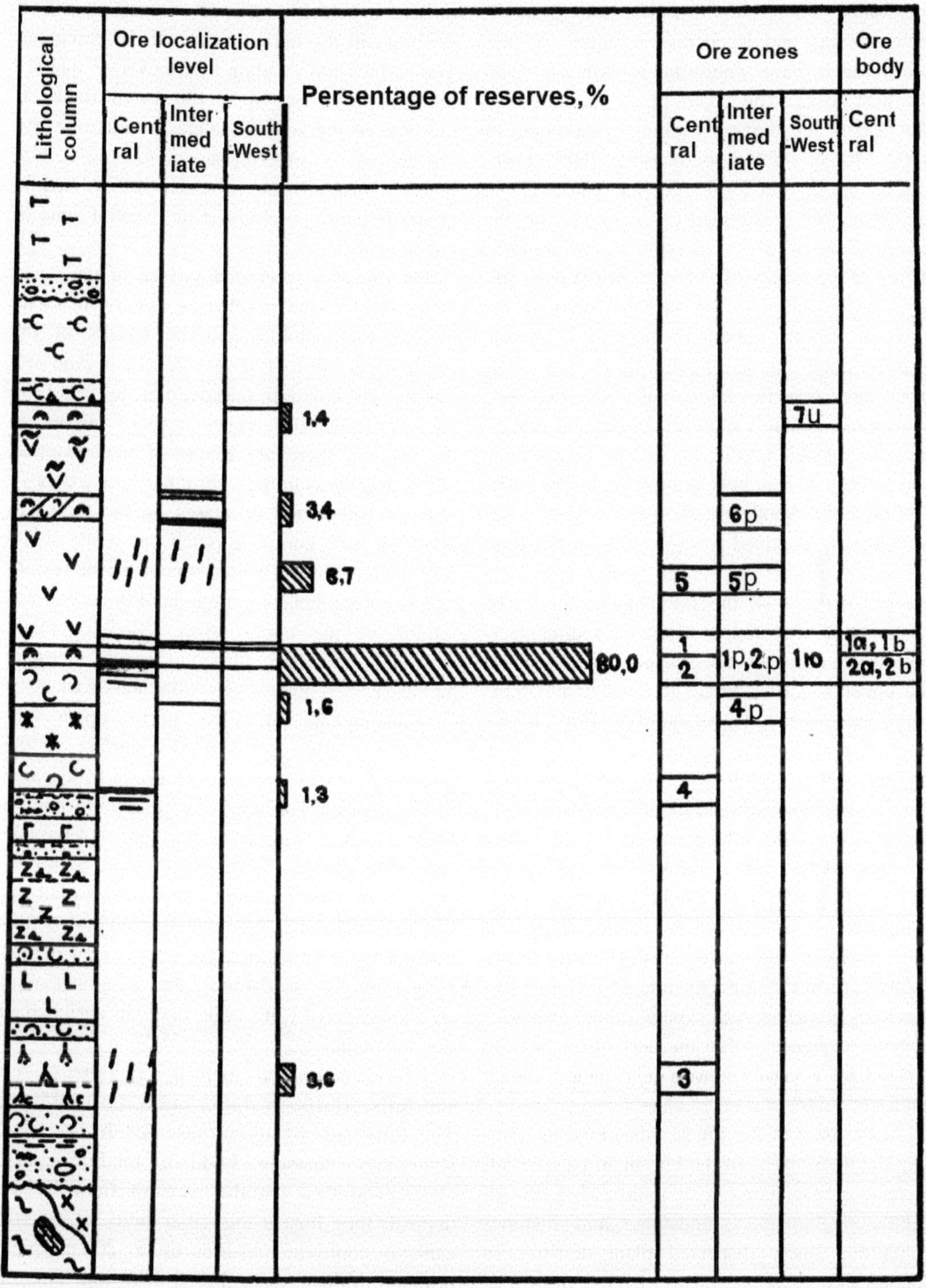

Fig. 4.12 Distribution of mineralization in the stratigraphic section of the Gurvanbulag deposit (according to Mironov 2003)

granitoids, gabbro-diorites, diorites, granodiorites, biotite and leucocratic granites are present, which have conformable contacts with the host rocks. The depth of the basement in the field is 500–800 m, reaching a maximum of 1000–1200 m on the southeastern flank. Fork-shaped uplifts and local trenches (paleovalleys) are established in the relief of the foundation, the combination of which determines the complex nature of the surface. The largest of the paleovalleys can be traced in the basement through the entire area of the deposit in the form of a wide (1.0–2.5 km) submeridional paleovalley. The paleovalley branches into smaller jelly-like incisions, predominantly of northwestern and sublatitudinal orientation, between which dome-shaped and swell-like basement uplifts are located. The Central and Southwestern ore-bearing zones are spatially confined to the slopes of the largest uplifts. Above the bottom of the depression are ore deposits of the Intermediate area. The composition of the upper Mesozoic sedimentary-volcanic deposits of the Bayantumen formation, the upper structural stage, is divided into three parts. The lower subformation of the Bayantumen Formation within the deposit is represented by intercalation of covers of quartz-feldspar porphyries, trachydacites and basaltic andesites with horizons of tuff-sedimentary rocks. The total thickness of the rocks of the subformation ranges from 150 to 450 m. The middle subformation is composed of 300–800 m thick felsic mantle volcanic rocks of various textures. The upper subformation includes a trachyandesite mantle a few hundreds of meters thick. The sedimentary-volcanogenic sequence within the deposit has a calm monoclinal occurrence with a southeast dip at angles of 5–20°.

At the base of the lower subformation, leveling the unevenness of the basement paleorelief, there is a basal horizon of interbedded mixed-pebble conglomerates, sandstones, and siltstones containing finely dispersed plant detritus. In single cases, there are lenses of felsic tuffs. The thickness of the horizon in the depressions of the basement relief reache 100–130 m, decreasing on the ledges to a few meters. Higher in the section, there is a relatively mature member of tuffites and dacitic tuffs, 50–70 m thick, which is absent in the uplifted areas of the basement. The tuffites are overlain by a cover of quartz–feldspar porphyries up to 100–140 m thick. At the base of the cover, fine-clastic tuffs of varying degrees of sintering are often present. Ore body 3, insignificant in scale, is localized in quartz-feldspar porphyry at the Central zone of the deposit.

On the weakly eroded surface of the cover of quartz–feldspar porphyries, aphyric massive, locally amygdaloidal andesite basalts of the lower cover, up to 120 m thick, occur. Andesite basalts are underlain and overlain by thin (15–20 m) horizons of terrigenous rocks. Higher up the section, there are covers of trachydacites, whose thickness in the central part of the deposit reaches 160 m and decreases on the flanks to 30–50 m with complete pinch out south of the Intermediate area. The deposits of the subformation are completed by amygdalolithic andesite-basalts of the upper cover, 20–30 m thick, separated from the underlying trachydacites by an unseasoned horizon of sedimentary rocks.

Volcanogenic formations of the upper subformation, lying on the eroded surface of the underlying rocks, are most common in the area of the deposit. The section of the subformation begins with a terrigenous horizon of relatively consistent thickness (20–30 m) and a complex internal structure. In the Central ore zone of the deposit, where it is the most studied and is ore-bearing, its composition includes mixed-pebble conglomerates, sandstones, and siltstones with an admixture of tuffaceous material enriched in organic matter.

Above the section, there is a member of felsite tuffs 70–100 m thick, rarely up to 150 m. The tuffs are mainly represented by massive, sometimes coarsely bedded, medium-coarse clastic varieties. On the western flank of the deposit, they host a lenticular body of subvolcanic oligophyric rhyolites up to 200 m thick, oriented in the submeridional direction. The top of the pack hosts the main layer-like ore body 2 (Fig. 4.13).

The rhyolite tuffs are overlain by a sequence of lithoid aphyric ignimbrites, which dominates

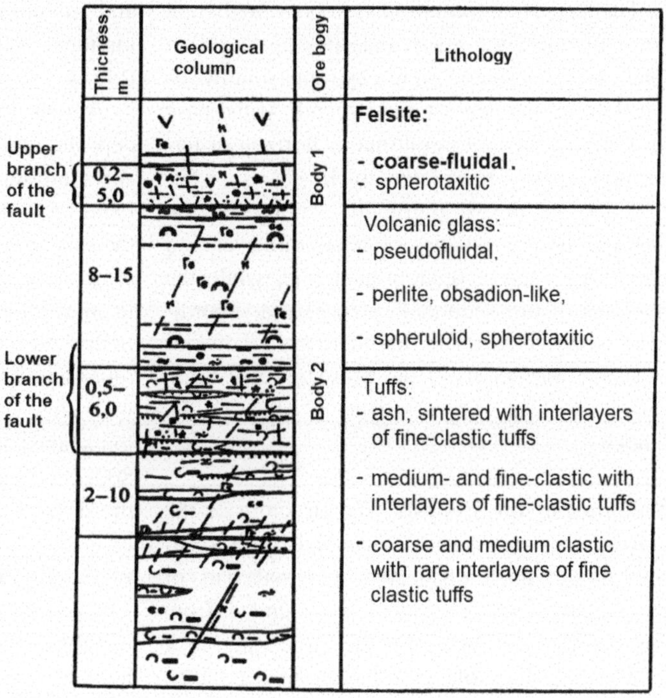

Fig. 4.13 Scheme of the structure of the ore-bearing zone of the Gurvanbulag fault (according to Mironov 2003)

in the section of the deposit. The stratum consists of two covers of felsite and a member of interbedded sintered ash tuffs and perlite separating them. The upper felsite cover is represented by coarse clastic varieties and is distributed over the entire area of the deposit. In the Central zone, its thickness is 400 m and successively increases eastward to 700 m. At the base of the upper cover lies a seasoned (8–15 m thick) member of volcanic glass that controls the position of the main ore-bearing level of the deposit. The lower felsite cover with a thickness of 0.2–5 m contains uranium mineralization of the third lithological-structural level. It is characterized by the development of spherotaxitic, glassy, microfelsite and vitroclastic varieties.

The felsite sequence completes the section of the Central zone; in the southern half of the deposit, it is overlain by younger volcanogenic formations. Its thickness decreases to 200–300 m. In the upper part of the ignimbrite sequence in the Intermediate area, two trough-shaped depressions linearly elongated to the northwest, possibly of erosion-tectonic origin, are established. Their dimensions in plan are 1300–1500 × 200–600 m at a depth of 80–100 m. The basins are filled with porous lavas of fluidal, ignimbrite-like felsite with thin (up to 5–7 m) lenticular interlayers of sintered tuffs, tuff breccias, and perlites. The development of small uranium veins and stockworks is associated with the zones of steeply dipping cracks in the axial parts of these structures. Coarse fluid felsite in the southern part of the deposit is overlain by a relatively mature (30–50 m thick) ore-bearing member of interbedded sintered ash and fine detrital tuffs with volcanic glass. The overlying cover of spheroid-fluid felsite is widespread in the southeastern part of the deposit, pinching out on its western flank. The thickness of the cover reaches 250 m. At the contact of the cover with the overlying rocks, a lensing horizon of volcanic glass 5–15 m thick is noted. Sanidine trachyrhyolites, which complete the middle subformation, form a thick (up to 600 m) cover in the southern part of the deposit.

The upper subformation is represented by a cover of microlitic trachyandesites up to 200 m thick; it is distributed on the extreme southwestern flank of the deposit and beyond. At the base of the cover lies a discontinuous horizon of terrigenous rocks a few meters thick.

Thus, in the stratigraphic section of the deposit, the cover facies of felsic effusives and ignimbrites play a predominant role, while the horizons of tuff-sedimentary rocks are subordinate. It should be noted that there are no dike or extrusive formations of the Mesozoic activation stage, which are characteristic of other DVTS blocks.

Tectonics of the Deposit. The structure of the Gurvanbulag deposit is determined by its position in the most dislocated marginal part of the Ulaan block. The block is bounded from the southwest and east by the large interblock Tsakhildag, Taliinolgoi and Mardaingol faults, from the northwest by the boundary of structural stages, and in the south it adjoins the Lower Cretaceous Tukhem depression. The main structural and tectonic elements of the deposit are systems of steeply dipping faults of the northwestern, northeastern, submeridional and sublatitudinal strikes, large gentle intraformational faults and zones of tectonic fracturing (Fig. 4.14).

Among the steeply dipping faults within the deposit, faults of the northwestern (310–330°) strike are most clearly manifested. The regional interblock structures of the first order include the subvertical Tsakhildag fault, the Mukhar tectonic zone, the Khureetnuur and Khugnu zones in the Intermediate section, and other large faults. As a rule, all faults in this direction are represented by tectonic sutures, which, together with the accompanying subparallel and echelon-shaped faults, form zones from 200–300 to 800–1200 m wide and 7–10 km long, dividing the deposit area through 1.2–1.8 km per series of linear blocks. Separate disturbances (Mukhar zone, etc.) have a deep character, controlling gabbro-diorite stocks, explosion pipes, polymetallic deposits in the basement rocks and its boundary. In the relief of the basement roof, they are expressed as long (1–4 km) trough-like depressions. According to the type of prevailing dislocations, the faults belong to right-sided fault-shifts with horizontal displacement amplitudes from 2 to 12 m and vertical ones up to 0.5–2.0 m. The northeastern (40–50°) trending

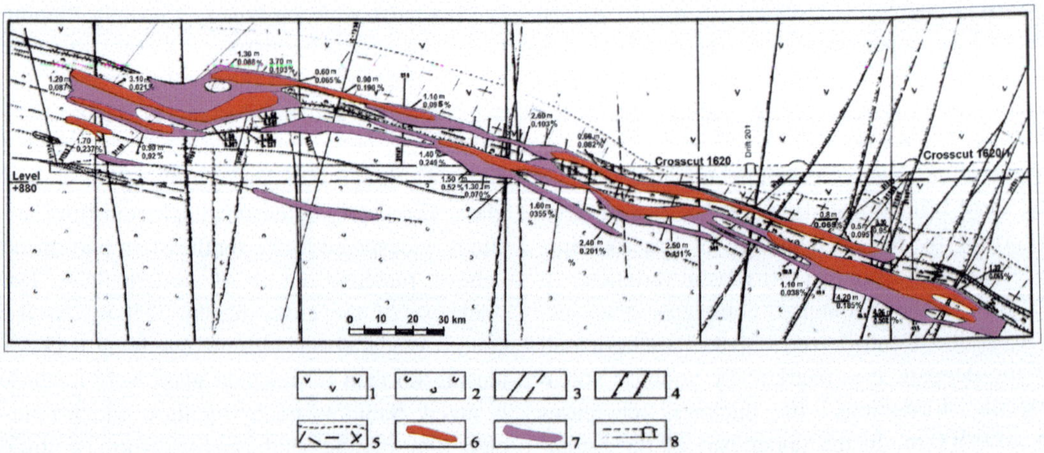

Fig. 4.14 Geological section of the Gurvanbulag deposit (according to Mironov 2003): 1–2—sedimentary-volcanogenic formations of the Bayantumen formation (J_3–K_1, bt): 1—felsite; 2—felsic tuffs; 3—steeply dipping tectonic faults; 4—gently dipping tectonic faults; 5—zones of increased fracturing; 6—ore bodies with economic uranium mineralization; 7—ore bodies with non-economic mineralization; 8—exploration and development mine workings

faults belong to the second order structures and are represented by faults traced through the entire area of the deposit for 8–12 km. Usually they are expressed by thin (up to 1–2 m) sutures with clay friction, which are accompanied by zones of intense fracturing or crushing from 5 to 30 m thick. At junctions with faults of a different orientation, especially with northwestern faults, the thickness of the zones increase to 1 m or more. In tuffaceous-sedimentary rocks, fractures are often represented by thin conductors or closely spaced fracturing.

Steeply dipping (75–90°) submeridional (350–10°) faults play an important role in the structure of the deposit. As a rule, these are zones of concentrated linear fracturing with a capacity of 2–8 m, increasing to 10–30 m at junctions with disturbances in other directions. Inside the zones of fractured rocks, central seams with friction clay 0.1–1.2 m thick develop, splitting in places into 3–4 closely spaced fault planes. In the immediate vicinity of the sutures, the rocks are intense crushed to form bands of fine clastic breccias (102 cm). These faults are accompanied by a series of closely spaced subparallel structures of higher orders, forming faulted sections 200–800 m wide and 4–8 km long.

Tectonic disturbances of a sublatitudinal (85–100°) strike are widespread and insignificant.

The mineral composition of the sutures and the accompanying hydrothermal alterations are the same for faults of different orientations. In addition to hematite and clay-hydromica minerals, fluorite, calcite, chamosite, and sulfides are present in small quantities. Seals of seams up to 0.3 m thick are lightened and hydromicaized.

The most characteristic feature of the deposit structure is the wide development of gently sloping faults. They are developed in the form of interstratal failures at the contacts of almost all horizons and covers of sedimentary-volcanogenic rocks, differing in strength properties, and often even within them. The largest of them are the Gurvanbulag, Khayakh and Khurtynbulag faults traced for tens of kilometers. The zones of these disturbances are associated with the formation of ore deposits of sheet-like morphology, which contain the main reserves of uranium.

The Gurvanbulag fault occurs at the base of the ignimbrite felsite cover. It has a northeast strike for 14 km and a gentle dip to the southeast. The fault is a zone of fragmentation and fracturing 20–40 km thick with intense hydrothermal alteration of rocks. In the central part of the zone, confined to the contact of felsites and volcanic glass, the rocks are transformed into friction clay a few meters thick. In the areas studied in detail, a wave-like morphology of the fault was established, caused by its bends along the dip and strike. Such bends are most clearly manifested in the areas where gentle faults meet steeply dipping faults. In these areas, in addition to the complication of the morphology of the gentle failure, there is a sharp increase in the thickness of the accompanying zones of increased fracturing.

The Khayakh fault, which is the second most important ore-bearing structure of the deposit, occurs in its southern part of the district and is confined to a pack of interbedded perlites and sintered tuffs between covers of coarse-fluidal and fluid felsite. The fault can be traced for 20 km in a sublatitudinal (80–100°) northeastern (30–45°) direction in the southwest with a general monoclinal dip to the south at angles of 5–15° according to the bedding elements of the host rock package.

The Khayakh fault is a zone of fracturing and uneven crushing up to 20–40 m thick, the central part of which is expressed by several (from 2 to 5) subparallel gentle faults along the contacts of the rocks composing the member. The internal structure of the fault, the nature of hydrothermal changes in the rocks and manifestations of mineralization are similar to the Gurvanbulag fault.

About the same size, the Khurtynbulag gentle fault, is oriented subparallel to the Khayakh fault and traced for 9 km.

It consists of two branches (upper and lower), connected at the surface in the southwestern part of the field. The upper branch runs along the contact of sanidine trachyrhyolites with the cover of fluidal felsite, while the lower branch runs along the contact of the latter with the underlying ignimbrig felsite. Both branches are represented by zones of uneven brecciation and crushing, 20–40 m thick, and are similar in their internal structure to the Gurvanbulag flat slip.

In the section of the sedimentary-volcanogenic sequence, layer-by-layer disturbances are widely developed and less significant in scale. They are established both at the boundary of large covers and inside the latter in areas of change in the internal structure. The field has a significant network of steeply dipping faults and fracture zones, but, unlike the Dornod field, these disturbances are less pronounced. The northeastern and submeridional faults have been studied the most. All of them have a steep (70–80°) dip to the east. In the rocks of the basement and the lower subformation, they are expressed by distinct tectonic sutures, accompanied by zones of increased fracturing. On the upper horizons, especially in felsite and trachyrhyolite, they are mapped as zones of dispersed fracturing. All steep discontinuities are low-amplitude (displacements up to a few tens of meters). In the structure of the deposit, in addition to well-manifested disturbances, a significant role is played by stockwork zones of fine fracturing, established at various stratigraphic levels. One of these zones was found in the cover of quartz-feldspar porphyries. It is represented by a system of small differently oriented cracks on an area of about 0.5 km^2.

Localization Conditions and Morphology of Ore Deposits. The Gurvanbulag deposit is represented by three spatially separated ore-bearing areas (Central, Intermediate, South-Western), which together form an ore-bearing zone 9 km long and 1.5–4 km wide, which is extended along the northwestern side of the Ulaan block.

Uranium mineralization with varying intensity is manifested in all parts of the geological section, with the exception of basement rocks, and is concentrated at three lithological-structural levels (Figs. 4.10, 4.11, 4.12 and 4.13). Its upper boundary is 15–40 m from the surface, and the established depth reaches 720 m. Mineralization is localized in ore-bearing fissure zones, with the main economical ore bodies confined to the gently sloping Gurvanbulag, Khayakh, and Khurtynbulag faults. In steeply dipping fractured zones, only small stockwork bodies are localized, which play a subordinate role.

A total of 12 ore bodies have been identified at the deposit. The main share of uranium reserves (80%) is concentrated in the Central site in sheet-like ore bodies 1 and 2, controlled by the Gurvanbulag gently dipping slip. The ore-bearing sites Intermediate and South-West are located 2.5 and 4.5 km to the south-west of the Central sites, respectively. Ore bodies within the plots are separated both in plan and in section by tens and even hundreds of meters.

Of the 12 ore bodies at the Gurvanbulag deposit, 9 have a sheet-like morphology and 3 have a stockwork-like one (Fig. 4.15). Separated ore bodies isolated and within ore sites have sizes from the first ten square meters to 0.5 km^2 in plan with insignificant thickness (Table 4.4).

Layered Ore Bodies. These include 9 out of 12 ore bodies identified at the deposit. The largest and most explored in detail is ore body 2, which contains 81% of the total category C reserves and 36% of category C_2 reserves. It is confined to the ore-bearing fissure zone of the Gurvanbulag fault and is localized in the lying side of its lower branch in the horizon of caked ash tuffs.

The body has a sheet-like shape and a gentle occurrence, repeating the position of the surface of the main seam. The upper boundary of the development of mineralization is usually the suture itself, the lower boundary is indistinct and approximately coincides with the base of the sintered tuff horizon. In plan view, the ore body extends in the northeast (40–45°) direction for 3.1 km with a width of 0.5–1.6 km. In the southeast direction, the body plunges in accordance with the dip of the ore-bearing zone and gradually pinches out at depths of about 600 m.

The morphology of the ore-bearing zone is determined by structural elements: the occurrence of the main displacement seam or the intersection nodes of steeply dipping faults with this seam. In such cases, the parameters of the deposit increase sharply, and along the

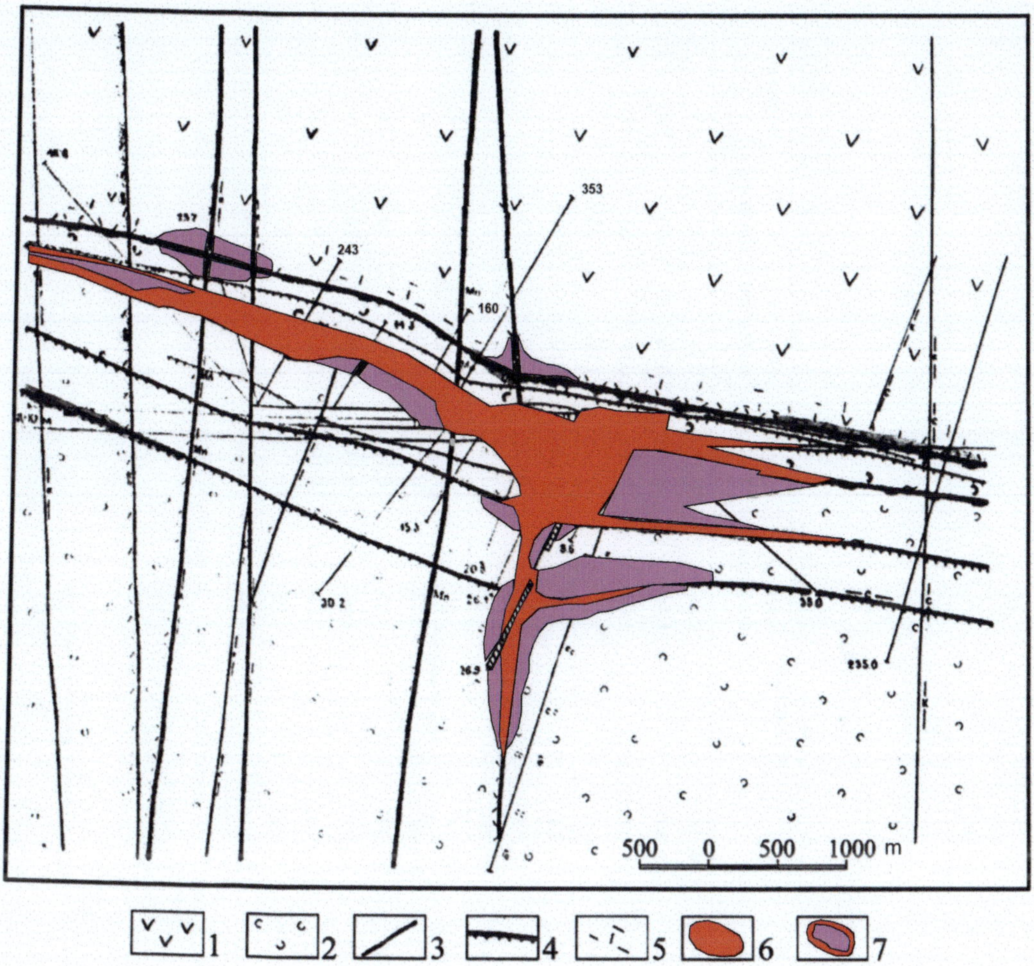

Fig. 4.15 Geological section along line 15+370 of the Gurvanbulag deposit (according to Mironov 2003): 1–2—sedimentary-volcanogenic formations of the Bayantumen formation (J_3–K_1, bt): 1—felsite; 2—felsic tuffs; 3—steeply dipping tectonic faults; 4—gently dipping tectonic faults; 5—zones of increased fracturing of effusive rocks; 6—ore bodies with economic uranium mineralization; 7—ore bodies with non-economic mineralization

junctions, rich bulges 5–30 m wide and up to 200 m long are formed. Steeply dipping faults outside the gently sloping ore-bearing zone are barren.

The quality of ore from deposit 2 is average. Intervals with a uranium grade of more than 0.3% (they make up 15% of ore sections) contain 47.5% of linear reserves. Rich ores are usually located in the central part of the deposit, and on its flanks average and low-grade ores with a uranium content of 0.060–0.150% are developed. Transitions between them are usually gradual.

Ore deposit 2 is divided into 17 independent layer-like ore bodies 2a, 2b, 2c, etc. The largest are bodies 2a and 2b, containing 46 and 24% of the deposit reserves, respectively; the rest contain from 0.3 to 6% of the total reserves of the deposit. Small ore bodies are thin beds with average and poor ores, localized along the main or feathering gentle seams. Out of balance ores border the ore bodies and fill the gaps between them in the plane of the deposit, forming a single wide halo extending beyond the area. Within this halo, there is a significant number of single intersections with conditional parameters,

Table 4.4 Characteristics of small ore bodies of deposit 2 (Mironov et al. 2012)

No. ore body	Share in deposit reserves (%)	Spatial position in lines	Depth occurrence from the surface (m)	Length (m)	Width (m)	Area (m^2)	Quantity intersections	Parameters			*Coef. variation (%)	
								Average thickness (m)	Average grade (%)	Ore coefficient	In thickness	In grades
2b	0.4	15+100–15+250	230–290	240	50	14,100	41	0.86	0.156	0.74	52	56
2g	3.0	14+200–15+100	150–280	720	80	61,700	28	1.30	0.193	0.74	100	80
2d	0.9	14+300–15+100	100–160	40	60	25,400	14	1.04	0.163	0.86	63	55
2e	0.4	16+350–17	180–220	100	40	4400	7	3.17	0.140	1.00	95	47
2j	0.6	17+50–17+100	210–300	170	60	10,300	21	3.30	0.123	0.62	59	102
2z	0.6	17+150–17+250	230–300	160	70	12,200	6	1.91	0.122	0.77	73	166
2i	4.1	17+200–18+300	260–410	510	150	79,900	34	5.31	0.127	0.31	75	71
2к	3.0	19–19+200	130–220	320	130	43,100	19	3.60	0.112	0.71	53	42
2l	1.7	18+350–19+100	210–290	260	70	19,800	10	3.18	0.136	0.80	59	50
2m	4.5	19–20	200–420	560	140	78,200	33	4.05	0.112	0.50	59	30
2n	5.8	14+150–15+100	420–500	470	170	86,100	5	4.06	0.110	0.60	74	66
2p	8	16+50–16+200	230–270	180	50	9500	33	3.09	0.199	0.83	83	75
2r	0.3	15+50–15+100	140–190	210	50	11,400	6	1.05	0.128	0.77	87	42
2s	0.8	15+50–15+150	360–420	210	100	18,100	9	1.84	0.113	0.67	54	107

*The coefficients of variation in thickness and grade for ore body 2a are 85 and 113%, respectively, for body 2b—81 and 81%

which, when detailed, can be distinguished into small ore bodies (Table 4.4).

Ore deposit 1, containing 22% of reserves of category C, is localized in spherotaxite felsite at the base of the cover of coarse-fluid felsite from the hanging side of the upper suture of the Gurvanbulag fault. The deposit has been traced for 1800 m. Its width is 400–700 m. The depth of the ore varies from 40 m on the northwestern flank to 440 m on the southeastern flank.

In terms of morphology, localized conditions and features of mineralization distribution of ore deposit 1 is basically similar to deposit 2 (Fig. 4.16). It also has a bedding consistent with the gentle fault, a flattened sheet-like shape, a heterogeneous internal structure and uneven development of mineralization.

The lower geological boundary of the distribution of conditioned mineralization is most often the main gentle seam; the upper boundary is indistinct and determined by the contact of spherotaxite and fluidal felsite. The maximum concentrations of uranium are observed in the lower near-contact part of spherotaxite felsite near the main seam and its subparallel branches, where vein-disseminated mineralization is distributed along gentle and steep cracks. Intervals with uranium contents of 1.815–1.937% over a thickness of 0.5–1.8 m are identified here.

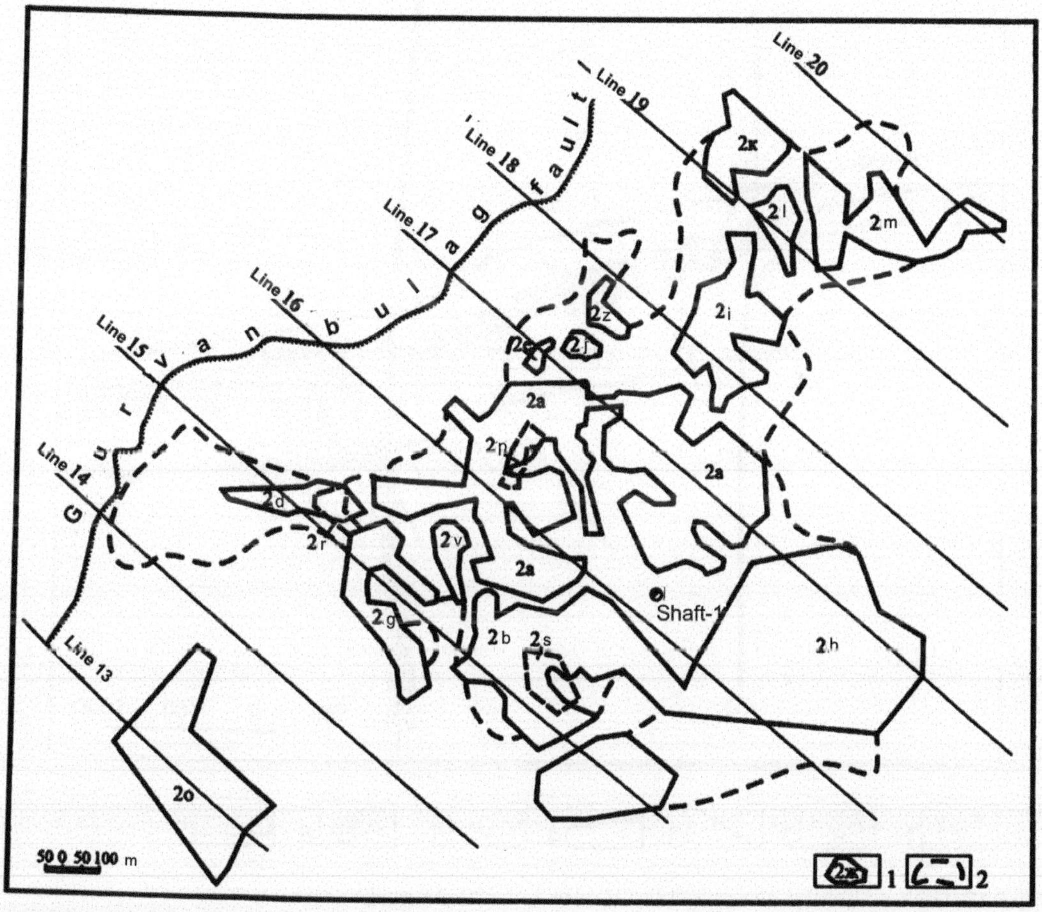

Fig. 4.16 Morphology of ore deposit 2 of the Gurvanbulag deposit: 1—ore bodies with economic mineralization of deposit 2 and their numbers; 2—contour of distribution of non-economic mineralization of deposit 2

Up and down the section, uranium concentrations decrease sharply. The ore deposits are average and poor. Intervals with uranium grades of 0.060–0.299% make up 92%; 60% of the uranium reserves of the deposit are concentrated in them. Pinching out of mineralization occurs gradually, usually with a transition to non-economic ores (Figs. 4.17 and 4.18).

Within the deposit, 14 spatially separated ore bodies are distinguished, of which the largest and most extensive are bodies 1a and 1n. The remaining 12 ore bodies are thin layers with average ores (average uranium grades are 0.118–0.204% for a thickness of 0.68–1.95 m), localized directly in the upper suture of the fault in areas where it intersects with steeply dipping faults (Table 4.5). Most of them have been explored in detail by mining, underground and surface wells, and their reserves are classified under category C.

Ore deposit 4 is identified in the central part of the site and stratigraphically confined to the terrigenous horizon at the base of the middle subformation (lithological-structural level 4). The host rocks are represented by alternating pebbly conglomerates, sandstones, and siltstones

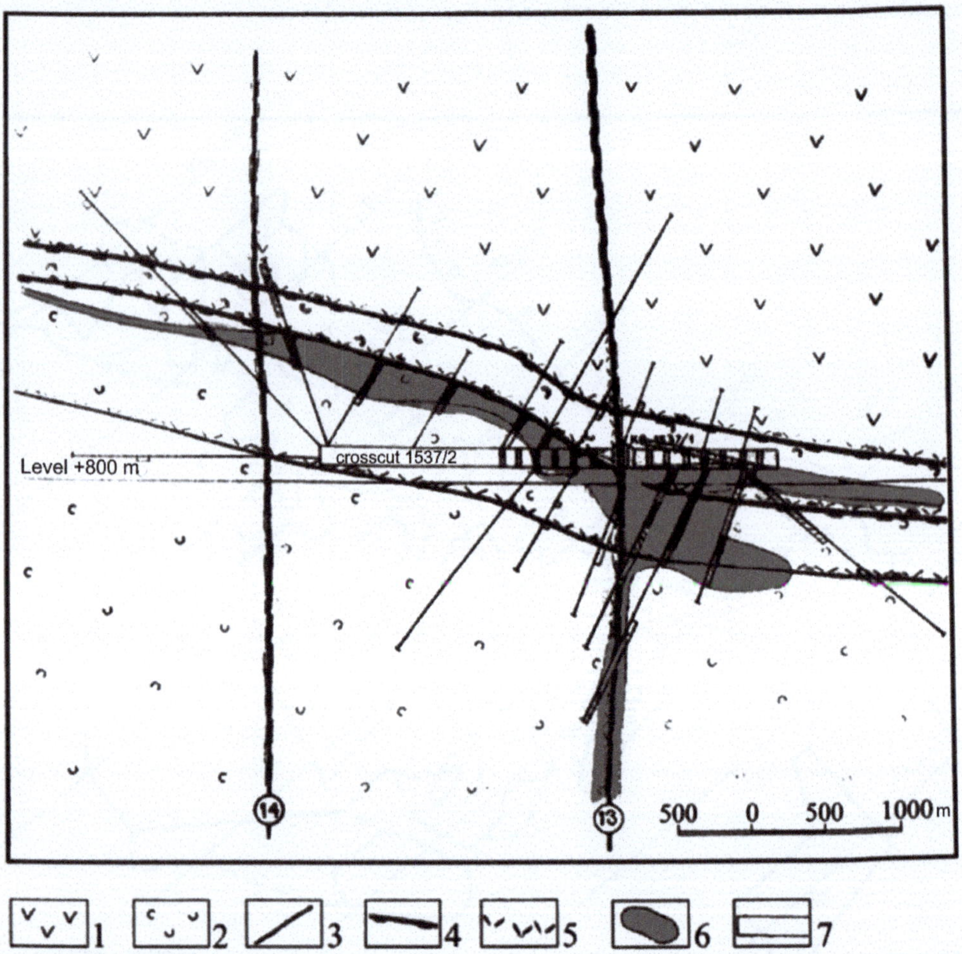

Fig. 4.17 Geological section along line 15 + 373 of the Gurvanbulag deposit (according to Mironov 2003): 1–2, sedimentary-volcanogenic formations of the Bayantumen formation: 1—felsite; 2—felsic tuffs; 3—steeply dipping faults; 4—gently dipping tectonic faults; 5—zones of increased fracturing; 6—ore bodies with economic mineralization; 7—underground mine workings

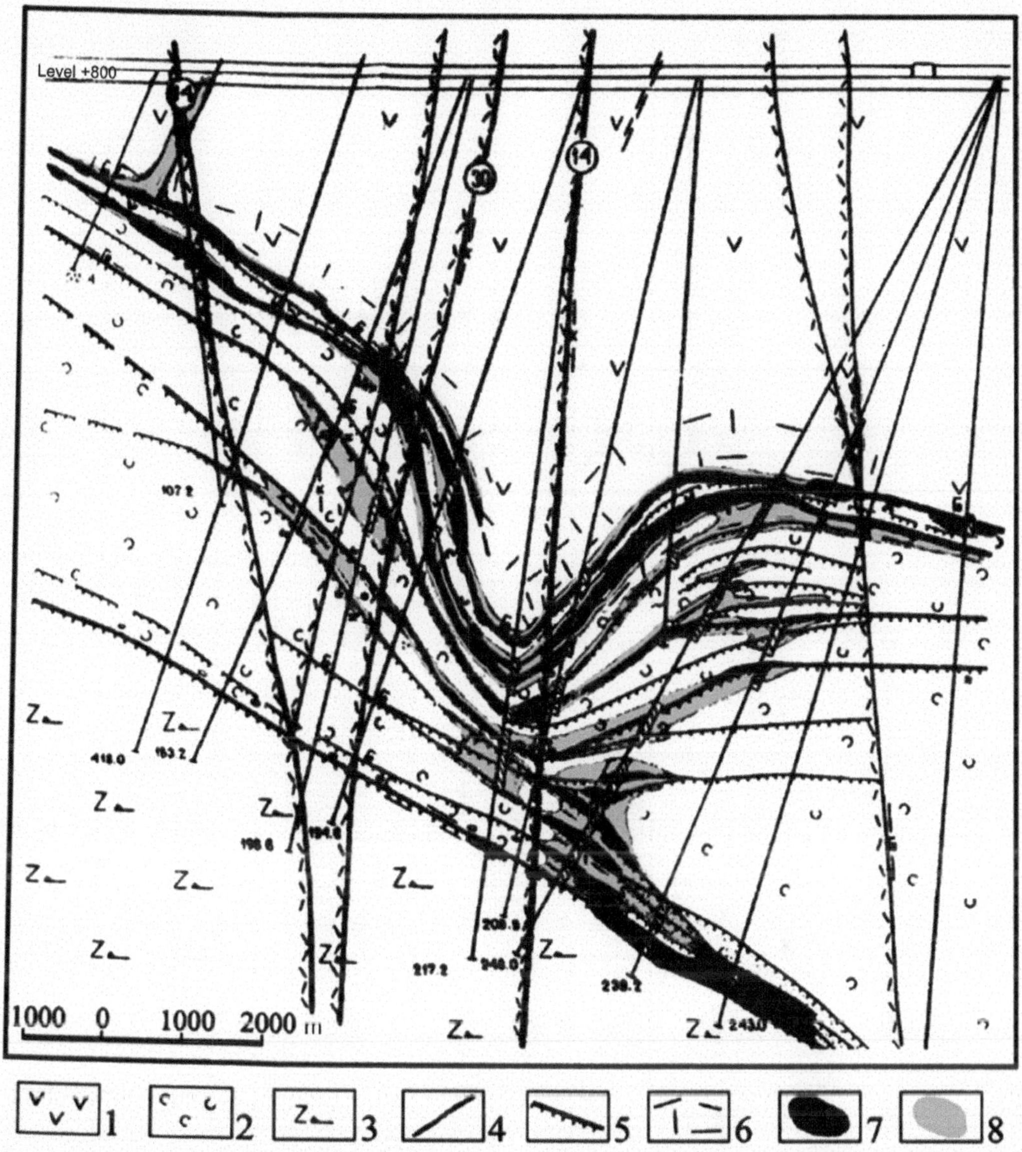

Fig. 4.18 Geological section along line 15+100 of the Gurvanbulag deposit (according to Mironov 2003): 1–3—sedimentary-volcanogenic formations of the Bayantumen formation: 1—felsite; 2—felsic tuff; 3—trachydacite; 4—steeply dipping tectonic faults; 5—gently dipping tectonic faults; 6—zones of increased fracturing; 7—ore bodies with economic uranium mineralization; 8—ore bodies with non-economic uranium mineralization

with an admixture of tuffaceous material and are enriched in finely dispersed organic matter. The deposit is controlled by a low-angle fault, widespread at the boundary of the lower and middle subformations, dipping to the southeast at angles of 10–20° and expressed by a series of small low-angle faults. It holds 1.3% of the field's reserves. The depth of occurrence of ores is 400–500 m from the surface.

The deposit is elongated in the meridional direction along fault 14 by 320 m with a width of 20–55 m and consists of a number of ore layers with a variable thickness of 11.75–30.5 m and a length of 50–200 m (Table 4.6).

Table 4.5 Characteristics of ore bodies of deposit 1 (Mironov et al. 2012)

Ore body No.	Share in deposit reserves (%)	Spatial line position	Size				Number of intersections	Parameters			[a]Coefficient of variation (%)	
			Depth from the surface (m)	Length (m)	Width (m)	Area (m^2)		Average thickness (m)	Average grade (%)	Ore coefficient	In thickness	In grade
1b	5.7	15+50–15+250	160,210	210	110	22,900	23	1.24	0.204	0.70	69	58
1v	1.3	14+300–15	150–210	150	60	8600	6	0.98	0.142	1.00	36	38
1g	1.4	14+100–14+200	160–260	310	60	20,600	8	0.68	0.125	0.62	31	43
1d	4.3	14+350–15+100	210–310	360	70	28,200	21	1.14	0.134	0.76	52	73
1e	0.9	16+25–16+100	210–240	120	60	7000	22	1.18	0.127	0.64	78	38
1j	2.6	16+150–16+300	210–270	180	80	14,600	36	1.39	0.138	0.69	63	56
1z	1.6	16+300–17	210–260	160	60	10,100	15	1.32	0.121	0.73	39	71
1i	3.3	17+50–17+200	180–270	200	90	18,400	17	1.46	0.119	0.76	75	38
1к	3.9	16+350–17+50	270–340	230	60	16,200	36	1.60	0.179	0.64	98	47
1l	2.7	17–17+50	350–400	250	70	17,900	16	1.05	0.133	0.77	73	53
1m	1.0	17+100–17+250	350–400	200	60	12,100	6	0.70	0.176	0.50	31	17
1n	5.3	18+50–18+300	100–180	300	70	21,800	8	1.95	0.118	0.76	82	50

[a] Capacity and grade variation coefficients for ore body 1a are 53 and 104%, respectively, and for ore body 1n, 67 and 701%

Table 4.6 Characteristics of the layer-like ore deposits of the Intermediate and South-Western ore-bearing area (Mironov et al. 2012)

No. of ore zone	Share in reserves in zones (%)	Spatial position in lines	Depth from the surface (m)	Sizes			Number of intersections	Parameters		
				Length (m)	Width (m)	Average thickness (m)		Average thickness (m)	Average grade (%)	Ore coefficient
Intermediate ore-bearing area										
1p	15	51–51 + 300	380–470	1500	5	126,500	5	6.64	0.101	0.2
2p	14	53–54	610–650	500	5	158,500	5	8.63	0.104	0.1
4p	9	53 + 200–54 + 200	680–750	450	6	174,750	6	1.17	0.104	0.5
6p	19	51 + 200–52 + 200	210–310	1000	8	154,000	8	9.16	0.118	0.1
South-Western ore-bearing area										
1u	70	57–57 + 200	370–440	750	8	212,250	8	1.19	0.164	0.5
7u	30	56–57	230–270	600	4	84,000	4	0.92	0.153	0.7

The ores are average and poor, the distribution of uranium in the deposit is uniform. The uranium grade varies from 0.060 to 0.170% and only in a single case reaches 0.489%. The average grade for the deposit is 0.125%; ore bearing coefficient is 0.26. Uranium mineralization is mainly disseminated, nested-disseminated in nature.

Stockwork-Like Deposits. Deposits of this type contain 12% of the total reserves of the field and are located in the Central and Intermediate areas (Fig. 4.19). These include deposits 3, 5 and 5p, localized at the 2nd and 6th lithological-structural levels. The main ore-controlling elements of these deposits are steeply dipping submeridional faults and the fractured zone of the Khureetnuur fault (deposit 5p). Ore deposit 3 contains 4% of the deposit's reserves. Localized in the cover of quartz-feldspathic porphyries in the Central area. Structurally, it is confined to a zone of steeply dipping fracturing along a series of closely spaced submeridional faults and lies at depths of 280–750 m from the surface. The area of the deposit is 880 × 170 m, the vertical span of mineralization reaches 100–120 m. Uranium mineralization develops directly along tectonic sutures and cracks and gradually fades away with distance from them. The thickness of ore intersections ranges from 0.7 to 6.4 m with a uranium grade of 0.061–0.179%. Average grade 0.117%.

Ore deposit 5 contains 1% of the deposit reserves. It is located in the central part of the site in the cover of coarse fluid felsite, where it is confined to a zone of steeply dipping submeridional tectonic sutures at a depth of 15–150 m from the surface. It has a complex stockwork-like structure due to the close subparallel veins and lenses a few hundred meters long, separated by blocks of out-of-balance ores and waste rocks. The deposit is extended along the ore-controlling seams for 480 m with a width of 20–50 m; the vertical range of mineralization is 60–110 m.

The thickness of the ore intervals ranges from 0.6 to 12.6 m, uranium grade—from 0.061 to

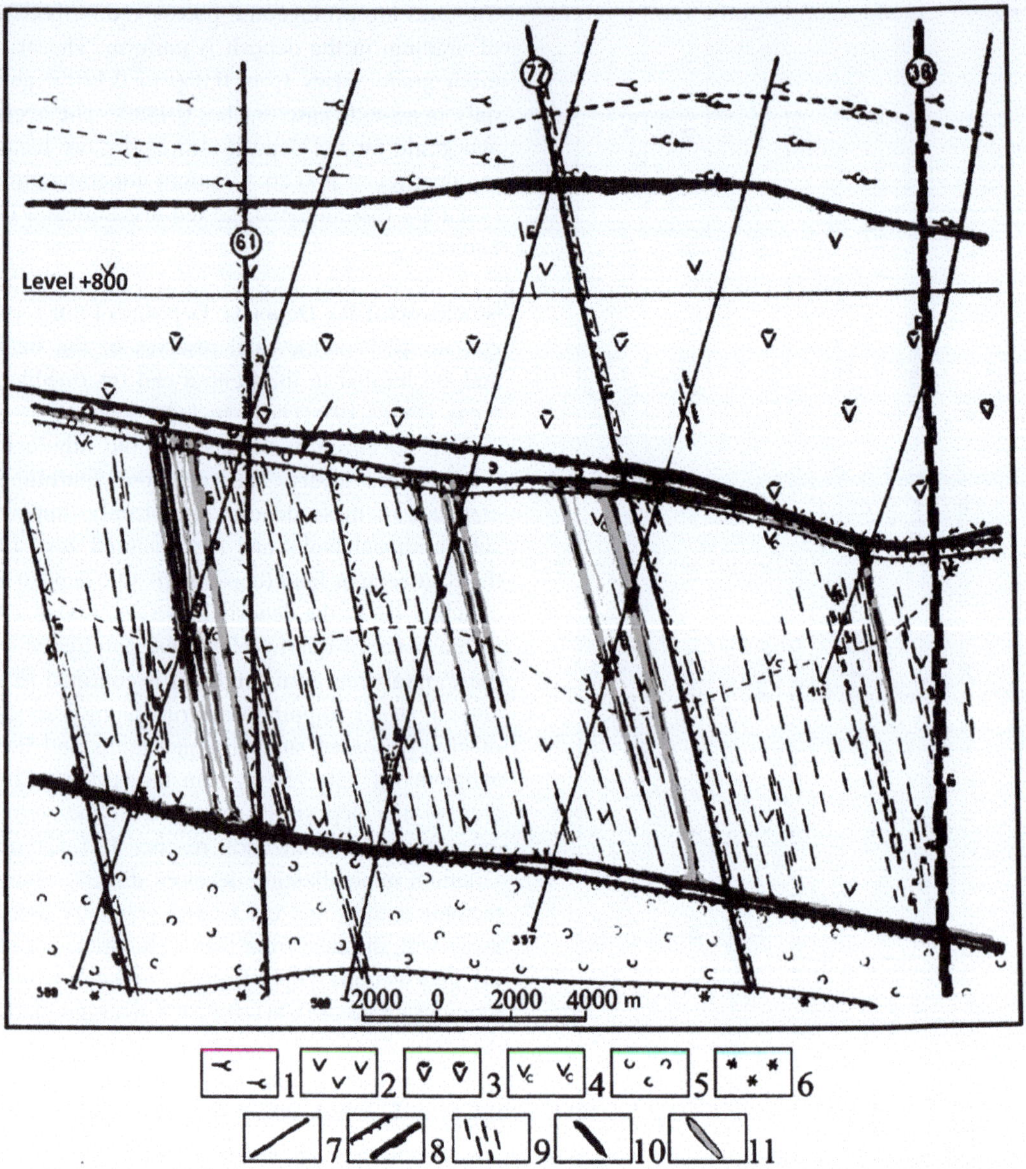

Fig. 4.19 Geological section along line 52+200 of the Gurvanbulag deposit (according to Mironov 2003): 1–6—sedimentary-volcanogenic formations of the Bayantumen formation: 1—sanidine trachyrhyolites; 2—felsite; 3—coarse fluid felsite; 4—felsite tuff; 5—felsic tuff; 6—oligophyric rhyolite; 7—steeply dipping tectonic faults; 8—gently dipping tectonic faults; 9—zones of increased fracturing in effusive rocks; 10—ore bodies with economic uranium mineralization; 11—ore bodies with non-economic uranium mineralization

0.216% (average grade 0.114%) Uranium mineralization is of a vein-disseminated nature.

Ore deposit 5p (43% of the uranium reserves of the Intermediate area) is located in the upper part of the cover of coarse fluid felsite and is controlled by the fracture zone of the Khureetnuur fault (the second lithological-structural level), stretching along it for 1200 m at a width of 50–370 m. The range of mineralization varies from 40 up to 120 m; The length of the ore bodies is estimated at a few hundred meters. The thickness of the ore bodies ranges from

0.70 to 80 m, uranium grade ranges from 0.068 to 0.330%. The ore content coefficient for the deposit is 0.026.

Mineral Composition of Ores and Alterations. According to the mineral composition and type of hydrothermal-metasomatic alterations in the host rocks, the ores of the Gurvanbulag deposit (Fig. 4.20) belong to the pitchblende-coffinitic type in argillicites. Rock alterations and the formation of uranium mineralization itself occurred in many stages.

The earliest alterations associated with the pre-Late Mesozoic period of development include the processes of granitization, K-feldspathization, albitization, and greisenization of substrate rocks. The halos of these alterations are extensive and go far beyond the field, concentrating along large tectonic faults of ancient origin. The first stage of Late Mesozoic hydrothermal-metasomatic alterations of sedimentary-volcanogenic rocks of the Bayantumen Formation is paragenetically closely related to post-volcanic, primarily autometamorphic and fumarolic-solfatarian processes. Two stages of this process are outlined.

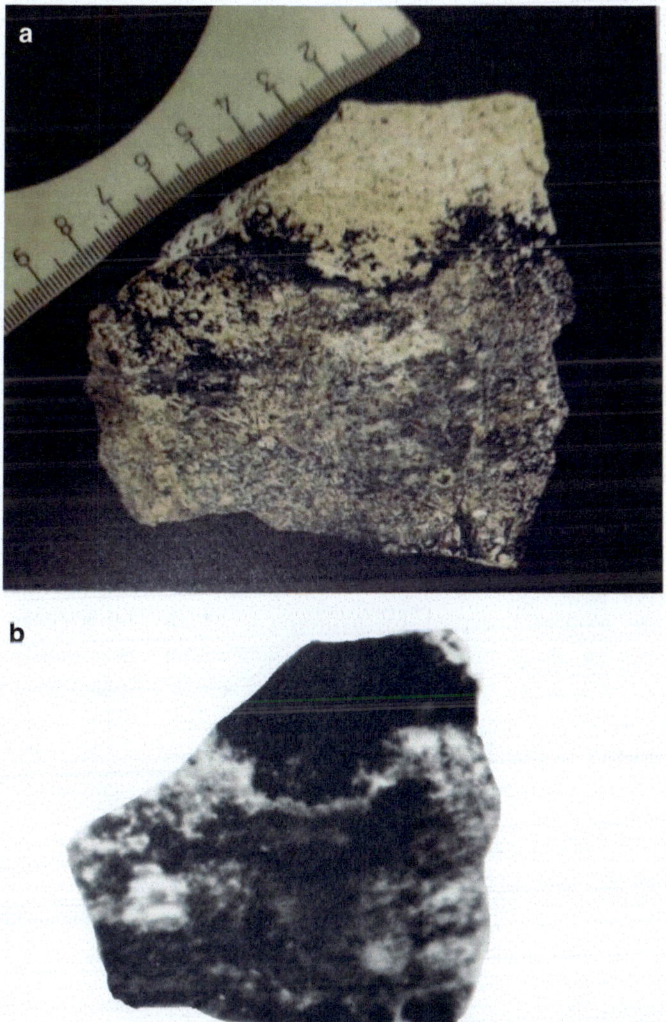

Fig. 4.20 Nest-disseminated pitchblende-coffinitite ore in manganese and argillized tuffs of acidic composition. Gurvanbulag deposit, ore zone 2: **a** photo of the sample (natural color); **b** radiography (exp. 3 days) (Mironov et al. 2012)

The early stage is high-medium temperature, expressed in areal K-feldspathization, beresitization, propilitization and silicification of host rocks. More locally, this stage manifested quartz-muscovite, quartz-tourmaline, quartz-albite alterations. The first two mineral associations are identified in the eastern part of the deposit in the rocks of the lower subformation in the central part of the section among mixed-clastic tuffs and volcanic glass. The intensity of these changes is small. Newly formed minerals are present in the range of 5–10%.

The second stage of autometamorphic alterations of rocks of the sedimentary-volcanogenic complex is caused by the processes of low-temperature acid leaching (hydrolucidization). This process manifested itself widely and intensively in the field. Almost throughout the entire section of the host rocks, the rock-forming minerals were, to one degree or another, metasomatically replaced by hydromicas, mixed-layer minerals, carbonate, chlorite, and quartz, forming powerful extended halos of hydromica-carbonate alterations. Often these halos extend beyond the field area, being located along the main steeply dipping faults.

The ore-forming process at the Gurvanbulag deposit proceeded in three stages: pre-ore, ore and post-ore (Table 4.7).

In terms of the composition of mineral associations, the pre-ore stage differs relatively little from the low-temperature hydrothermal alterations of the pre-ore stage. The leading place here is also occupied by hydromica, which develops along small cracks and voids within the ore-bearing zones. They are represented in mixed-layered montmorillonite-hydromicaceous varieties containing 15–20% clayey-hydromicaceous packages. Quartz and carbonates are represented in a smaller volume than at the previous stage, hematite is noted. These changes primarily affected the zone of the Gurvanbulag fault and a member of mixed-clastic tuffs. Volcanic glass is hydromica depending on the degree of their recrystallization. The main ore-bearing rocks—ash, medium-, fine-clastic tuffs and spherotaxitic felsite are weakly affected by changes in the pre-ore stage.

The ore stage is represented by three mineral associations successively replacing one another. The earliest association of minerals is quartz-sulfide.

It is composed of coarse-grained quartz with variable amounts of pyrite, marcasite, pyrrhotite, arsenopyrite, and siderite. Sometimes there are small allocations of hematite and jordizite. Ankerite veinlets with sphalerite, chalcopyrite, galena, and native silver are superimposed on these minerals.

The uranium ore mineral association is represented by pitchblende, radioactive leucoxene, jordisite, chlorite, kaolinite, and montmorillonite. The typhomorphic mineral here is pitchblende, which forms two generations. The first generation of pitchblende produces systems of abundant pores and partially replaces the host rocks. The pitchblende of the second generation is distinguished in the form of veinlets, individual spherulites or their intergrowths, as well as kidney-shaped crusts enveloping fragments of hydromica rocks. Iordisite is present in the form of small nests of irregular shape. Its scales often cover or cross pitchblende-1. Chamosite, montmorillonite and hydromica, present in small quantities, surround pitchblende grains and sometimes grow into it.

The coffinite-quartz-fluorite association is the main one at the deposit. The leading mineral here

Table 4.7 Mineral composition of involved in the uranium ore process (Mironov et al. 2012)

More common minerals	Common minerals	Less common minerals	Rarely occurring minerals
Quartz, montmorillonite, kaolinite, hydromica, chamosite, fluorite, coffinitite, calcite	Pyrite, marcasite, jordisite, galena, pitchblende, zeolites, potassium feldspar, albite, analcime, leucoxene, ankerite, siderite, manganese hydroxides	Sphalerite, pyrrhotite, arsenopyrite, chalcopyrite, arsenic, bertrandite, rutile, barite, dolomite, graphite, dolomite, hematite, muscovite, tourmaline	Cassiterite, pseudobrookite, sphalerite, native silver

is fluorite. Following it, pyrite, sphalerite, chalcopyrite, fahlore, galena and coffinite were deposited. The latter makes up 55–60% of the total amount of uranium minerals. It pseudomorphically replaces pitchblende-1 and pitchblende-2, preserving their textural and structural features, and is often localized around sulfide grains. Like pitchblende, coffinite has a heterogeneous composition; sometimes powdery formations such as uranium blacks are noted (Fig. 4.33).

Fluoritization is manifested in almost all lithological varieties of rocks. In felsite, fluorite forms a rare small dissemination; in tuffs, trachydacites, and quartz-feldspar porphyries, it forms the veinlets and cement of tectonic breccias.

A specific feature of the Gurvanbulag deposit is the extremely intense manifestation of post-ore formations, especially along the Gurvanbulag fault and its feathering ones. They are represented by rectorite, smectite, clinoptilolite, mordenite, analcime, kaolinite, fluorite, and calcite.

Supergene changes in ore deposits are not clearly expressed, since the bulk of primary ores do not reach the surface. The oxidation zone of the hydroxide-silicate type develops only along large tectonic disturbances, reaching a depth of 250–300 m. Uranium minerals in the oxidation zone are represented by uranyl hydroxides and silicates (uranophane, beta-uranotyl, etc.). The most intense oxidation processes occur in the central part of ore deposit 2. In addition to uranium minerals, hydroxides of iron and manganese containing small amounts of uranium are common in the oxidation zones.

According to geological data, the age of hydrothermal uranium mineralization is Early Cretaceous. The absolute age of uranium ores, established in the IGEM laboratory using the lead isotope method, is 136–153 million years (J_3–K_1).

The ores of all deposits of the Gurvanbulag deposit generally have a homogeneous pitchblende-coffinitic mineral composition. The bulk of the ore material was deposited in the form of thin impregnations and nests as a result of metasomatic replacement of ore-hosting rocks, partly due to the filling of cavities. The main uranium minerals are coffinite, pitchblende and uranophane, developed independently or in association.

Depending on the quantitative predominance of minerals, three types of ores are distinguished: coffinite, pitchblende, and uranophane. The most widespread is the coffinite variety, which composes the main part of deposits 1, 2, almost the entire deposits 3, 4, 5 and all small deposits of the Intermediate and South-Western sections. This type of ores is characterized mainly by disseminated, dispersed-disseminated, spotted-disseminated, less often banded-disseminated, vein-disseminated, earthy textures and collomorphic, granular structures.

Pitchblende and uranophane varieties are distributed mainly in the central and northeastern parts of deposit 2. The textures of these ores are veined, banded-disseminated, nested-disseminated, veined-disseminated and spotted. In steeply dipping fracture zones and especially where they intersect with gently sloping joints in rich ores, breccia textures develop. The structures of the ores are looped, reticulated, spherulitic, entangled fibrous and lamellar.

The ores are mainly composed of an aggregate of fine-grained, to a lesser extent finely dispersed uranium minerals with sinuous rounded, complexly filled boundaries. Unit sizes range from 0.001 to 2 mm.

Chemical Composition of Ores. The chemical composition of ores was studied based on analyzes of group samples characterizing the balance or out of balance ore interval (Table 4.8).

The data obtained clearly determine the classification of the ores of the Gurvanbulag deposit as a single aluminosilicate technological type with a low content of acidic components.

In terms of chemical composition, the uranium ores of the Gurvanbulag deposit are characterized by an insignificant content of carbonates and fluorite, a high degree of oxidation of iron with a low total content. The calcite content in ore intersections does not exceed 1%; iron, the content of which reaches 4.42%, is mainly present in the form of hematite; the maximum content of ferrous iron, reaching 1.92% in some intersections, are confined to areas of pyritization and

Table 4.8 Chemical composition of ores of Gurvanbulag deposit (Mironov et al. 2012)

SiO_2	74.24	CaO	0.39	H_2O	0.86
Al_2O_3	11.90	MgO	0.39	$CaCO_3$	0.10
Fe_2O_3	2.39	MnO	0.08	CaF_2	0.16
TiO_2	0.18	K_2O	4.36	S total	0.10
P_2O_5	0.10	Na_2O	1.80		

chloritization. Molybdenum, lead, zinc, zirconium, copper, and silver present in uranium ores in high concentrations have no economic value.

The distribution of molybdenum in ores is uneven and has a weak correlation. Increased concentrations (up to 0.033%) in the ores of deposit 1 are sporadically observed only in the western part of the Central site. For deposit 2, the highest molybdenum grades (from 0.005 to 0.100%, maximum—0.237%) are found within the eastern part of ore body 2a. The average molybdenum grade here is 0.017%.

Lead and zinc occur in the form of galena and sphalerite (cleiophane). Their quantity in standard ores is slightly higher than in substandard ores. Thus, the average grade of lead in standard ores is 0.091%, in substandard ores 0.03–1%; zinc 0.041 and 0.026%, respectively.

Zirconium, the grade of which in ores varies from 0.012 to 0.066%, is present in accessory zircon and, apparently, in small quantities in coffinite.

The behavior of copper, silver and other elements in uranium ores, due to their low concentrations, has not been studied.

Geochemical Characteristics of Rocks and Ores. The geochemical features of the host rocks and ores of the Gurvanbulag deposit were studied on representative material (51 thousand samples).

Analysis of the average grades of elements allows us to determine the specialization of ore-bearing rocks (%): U—0.001–0.0014, Th—0.0035–0.0049, Nb—0.0064–0.0119, Sn—0.0011–0.0022, Pb—0.0055–0.0071 have a high correlation with uranium (Kк 0.4–0.8); Bi—0.00007–0.00016, Ga—0.0036–0.0101, Be—0.0014, W—0.0006–0.0007—weak bond with uranium (Kk 0.1–0.4). This group is characterized by high background concentrations (Kк>3). The rocks are enriched to a lesser extent (Kc 1.5–3) with Mo—0.0001–0.0003, Zn—0.0117–0.015, Ag—0.046–0.083 g/t; Tl—0.00012–0.00028, Zr—0.031–0.035, Hf—0.0009–0.0011, Y—0.005–0.0081, Yb—0.0006–0.0009, and Li—0.0034–0.0088%.

A feature in the rocks of the Central area is the increased background content of Ta—0.0023 and Gd—0.0013%. A feature in the rocks of the South-Western area is the increased background content of Th—0.0077, Fe—7.8, Na—3.6 and K—5.5%. The distribution of elements is homogeneous, the coefficients of variation range from 30 to 60%. It should be noted that the ore-bearing rocks of the Central area are characterized by a wider range and higher concentration of elements than other ore-bearing areas of the deposit.

In scattering halos, ores, and uranium-concentrating minerals, all elements can be divided into three groups according to the level of accumulation.

The first includes uranium, lead, arsenic, and molybdenum, whose concentrations in ores and minerals are maximum (100–1000 clarks or more), and in scattering halos 3–5 times higher than the local geochemical background. Correlations are high (0.4–1.0), the distribution is extremely heterogeneous (variation coefficients exceed 150%). These elements can be attributed to direct indicators of mineralization.

The second group includes elements with a low level of accumulation (less than 2 backgrounds, for some ore-bearing areas up to 5 backgrounds). Bismuth, cobalt, manganese, silver, zinc, copper have a positive bond with uranium (Kc 0.2–0.5) and a very inhomogeneous distribution (Kv 100–200%). The maximum contents of these elements are recorded in rich ores, with the exception of zinc (maximum in non-economic ores and halos).

Of interest are tin, gallium, gadolinium, cerium, lanthanum, magnesium, iron, which are concentrated in primary scattering halos and non-economic ores (accumulation level 1.5–2, less often 5 backgrounds). In rich ores, these elements are characterized by minimal grades (0.4–0.8 background). Accordingly, the connection of these elements with uranium is negative. The distribution is heterogeneous (Q 50–100%). All elements of the second group can be used as indirect indicators of mineralization.

The third group includes tantalum, niobium, hafnium, beryllium, tungsten, lithium, thorium, ytterbium, yttrium, thallium, nickel, barium, the concentrations of which in ores and halos range from 1 to 1.5, less often up to 2 backgrounds. The distribution is uniform, less often homogeneous (K_V 40–60, less often 100%). The Kv values are low (up to ±0.2), the halos of these elements are not consistent with ore bodies. For zirconium, titanium, vanadium, scandium, chromium, germanium, strontium, phosphorus, aluminum, silicon, potassium, sodium and calcium, the contents are background (0.7–1.3), uniform distribution (Q 30–60%), connection with no uranium.

Tellurium, boron, indium, antimony, and cadmium were found in single samples; gold, platinum are absent (their grade in ores is below the sensitivity threshold of the analysis).

Along with the general patterns of distribution of elements, there are features that are characteristic of each ore-bearing area.

For the Southwestern ore-bearing area, an association of the main indicator elements U, As—0.017, Zn—0.027, Cu—0.0018, Mo—0.0007%, having a high level of accumulation (more than 2 backgrounds) and a positive relationship with uranium, was identified. Of the minor elements, they are marked by increased grades of Co—0.0002, Bi—0.00032, Mg—0.09, Th—0.0061%, Ag—0.061 g/t and Ca—0.4%.

In the Intermediate ore-bearing area, the association of the main satellite elements includes (%): U, Pb—0.015, As—0.0084, and of the minor ones—Sn—0.0017, Y—0.01, Tl—0.00024, Ba—0.0049 and W—0.0009.

The central area has been geochemically studied in the most detail. The leading elements in the ores are Pb—0.112–0.326, As—0.0052–0.037, Mo—0.0044–0.0092%. In elevated concentrations Ta—0.0049–0.0141, Nb—0.021–0.042, Hf—0.0016–0.0035, Yb—0.0018–0.0023, Li—0.0082–0.01127, Mn—0.074–0.191, Gd—0.0021–0.007, La—0.0045–0.0134, No—0.1004–0.0006, Ce—0.0045–0.0137, Fe—1.1–1.8%.

An analysis of the average grades of elements by ore classes shows that the maximum concentrations of indicators are recorded in rich ores. Moreover, their grade is on average 2–6 times higher than in non-economic ores. The values of the coefficients of variation and correlation coefficients are also higher (1.5–3 times).

In a sample of 547 samples, an analysis of the occurrence of highly anomalous samples was carried out.

It follows from the data presented that the occurrence of abnormal samples in halos is less than 1, less often 2–8%. At the same time, the majority (60–80%) of anomalous samples are observed in economic ores.

Primary dispersion halos are much larger than ore bodies, merge with each other, forming a common ore deposit halo. The configuration of primary halos is determined by the geological and structural conditions of the formation of the deposit, primarily by the morphology of ore deposits, ore-bearing rocks and structures. The aureoles of all satellite elements are elongated along the ore deposits and are controlled by gently dipping faults. In screening rocks (felsite, trachydacite), the occurrence and parameters sharply decrease.

The most extended and contrasting (more than 2 backgrounds) are uranium scattering halos. Their thickness ranges from a few to a few tens of meters on the flanks of ore deposits and reaches 50–100 m in the central part. Within the deposit along the strike, the halos can be traced along the entire length of the Gurvanbulag fault and move away from the ore deposits by more than 500 m.

In the superore strata above the richest ore areas, low-contrast scattering halos (1.5–2 backgrounds) are recorded, associated with steeply dipping faults. The power of the primary uranium scattering halos is a few meters with a length of tens of meters. They are observed in the interval of 150–200 m above the deposit.

Contrasting uranium scattering halos exceed the size of the ore deposit by 2–3 times. Low-contrast primary scattering halos (less than 1.5–2 backgrounds) are developed throughout the field. Their distribution area for the Gurvanbulag productive level exceeds 70 km^2.

The scattering halos of each satellite element individually are inferior in all parameters to the halos of uranium (Table 4.9). Multiplicative halos are more effective.

Their parameters are comparable, and even exceed the halos of one element by 5 times or more in contrast. The maximum intensity halos of satellite elements: Pb—0.1–0.5 to 1, As—0.01–0.05 to 0.5, Mo—0.01–0.05 to 0.5, Zn—0.1–0.5%, Ag—0.1–1–20 g/t, Cu—0.005–0.01 to 0.2% tend to the central part of the ore-bearing area, to the intersection of latitudinal and meridional faults. Scattering halos are recorded mainly in the near-contact part of tuffs with felsite and in the supra-ore sequence in steeply dipping cracks above rich ores. The length of scattering halos with the richest contents of satellite elements varies from a few tens to 150 m, with a thickness from a few to 10–20 m in the richest ore area. It should be noted that the dispersion halos of the main satellite elements (lead, arsenic, molybdenum) are developed mainly within the ore bodies, and the minor ones (zinc, silver, copper)—in the host rocks. This pattern is emphasized both by the location of contrasting multiplicative halos (the maximum intensity multiplicative halos of indicator elements are mapped in areas with rich ores), and by the results of correlation analysis, where in the selected association the elements are closely related to each other (Kc 0.5–1.0).

A similar connection is noted with indirect satellite elements, some of which accumulate, while others (zinc, tin, gallium, cerium, gadolinium, lanthanum) are removed from ores. Accordingly, the correlation coefficients have different signs.

The results of geochemical studies showed the absence of industrial concentrations of other minerals other than uranium at the deposit.

4.4 Mardaingol Deposit

The Mardaingol uranium deposit is located in the northern part of the Dornod volcanotectonic structure (DVTS), within the Mardaingol tectonic block (Jamsrandorj et al. 2017b), 6 km northwest of the Dornod deposit. Discovered in 1979 by geologists of the Mongolian Geological Survey Expedition of the USSR Ministry of Geology. It was explored until 1989. Structurally, the field is confined to the junction of the main Dagai zone with a northwestern strike with the Western zone of submeridional faults.

Geological Structure of the Deposit. The geological section of the deposit is represented by formations of two structural levels. The lower structural stage formed in the Pre-Late Mesozoic period is dominated by granitoids of the Mod-Khudag palingene-metasomatic complex represented by Early Paleozoic gneissic biotite granites and granodiorites. The granitoids are injected with Late Paleozoic leucocratic and biotite granites, dikes of aplites, pegmatites, and microdiorites. Everywhere the rocks of the substrate are granitized due to the manifestation of silicic-alkaline metasomatism.

Table 4.9 Occurrence of highly anomalous samples in ores of the Gurvanbulag deposit (Mironov et al. 2012)

Groups elements	Grade of elements (%)	Elements (number of abnormal samples in the sample)
1	More 1.0	Lead (13), manganese (24)
2	0.1–1.0	Thorium (45), arsenic (21), niobium (19), zirconium (18), molybdenum (21), strontium (9), zinc (7), titanium (4), barium (3), tantalum (2), yttrium (2), cerium (1), tungsten (1), lithium (1)
3	0.01–0.1	Gallium (40), hafnium (180), beryllium (8), chromium (7), copper (6), ytterbium (4), bismuth (2), tin (1), cobalt (1)
4	Less than 0.01	Vanadium (23), boron (19), cadmium (12), thallium (8), antimony (8), tellurium (6), indium (4), scandium (4)

Geophysical work and exploratory drilling in the relief of the foundation, the depth of which reaches 500–800 m, mapped two swell-like uplifts in a submeridional and northwestern direction. Between them there is a wide paleo-valley, complicated by uplifts and troughs of the second and higher orders.

The upper structural floor is represented by a thickness of sedimentary-volcanogenic rocks of the Late Mesozoic Bayantumen formation with a total thickness of 700–800 m. At the site of the deposit, the lower and middle subformations developed, the rocks of which lie on the basement with inclination angles of 10–20° to the northwest.

The lower subformation is characterized by alternating units of terrigenous sedimentary rocks with covers of basaltic andesites, trachydacites and trachyrhyolites (Fig. 4.21). The section of the subformation is presented in the following form (from bottom to top):

- horizon of basal boulder-block conglomerates, leveling the sharply dissected surface of the basement—220 m;
- cover of massive basaltic andesites—150–200 m;
- a pack of frequently interbedded terrigenous-sedimentary rocks—gravelites, sandstones, siltstones with interlayers and lenses of tuffites and tuffs of mixed composition—20–40 m;
- a unit of volcanogenic-sedimentary rocks, consisting of frequent interlayering of thin covers of basaltic andesites, trachytes, trachydacites, trachyrhyolites with thin layers of tuffaceous-sedimentary rocks; the deposit area is developed fragmentarily up to 150 m;
- cover of olivine basaltic andesites; in the base part it is represented by spherical lava breccias, turning into massive varieties toward the top of the cover—80 m;
- the middle subformation is represented by polyfacies felsic volcanics and their tuffs, the total thickness is 300–450 m. Section of the subformation (from bottom to top);
- member of interbedded siltstones, sandstones, gravelstones with conglomerate lenses and tuffite horizons; there are numerous remains of unified plant detritus—60–70 m;
- horizon of mixed-clastic, mostly agglomerate tuffs—50 m;
- cover of oligophyric rhyolites, to the south-east of the deposit turning into extrusion—up to 350 m;
- horizon of fine clastic tuffs with interlayers of tuffaceous sandstones containing unified detritus in felsite-porphyry covers up to 300 m.

Along with stratified volcanics, subvolcanic and vent felsic facies and mafic dikes, spatially gravitating toward near-meridional faults, are widely developed, especially in the central and western parts of the deposit.

Tectonics. The structural position of the deposit is determined by the intersection of steeply dipping faults of the northwestern (Dagai zone) and near-meridional (Western zone) directions. These fracture systems play a leading role in the placement of ore mineralization. The main ore-controlling fault here is the Dagai zone, which crosses the Mardaingol tectonic block in the northwest (320–330°) direction and is traced for 14 km with a maximum zone width of about 1.5 km. The internal structure of the Dagai zone is very complex (Fig. 4.22). It consists of three main faults a central one (2) and two extreme faults that limit the zone from the northeast (1) and southwest (3). The axial fault is characterized by bends along strike and a dense network of small feathering cracks; fault 1 controls mineralization on its flank; southwestern fault 3 is accompanied by diagonal cleavages of near meridional direction. On the whole, the Dagai zone is a combination of northwestern subparallel echelon-like stepped faults and tectonic sutures separated by wide (up to 100 m) bands of intensely crushed and fractured rocks. Numerous feathering ruptures of the northeastern and submeridional directions are also noted here. The northwestern faults are characterized by dip angles of 65–85° with an inclination to the southwest or northeast.

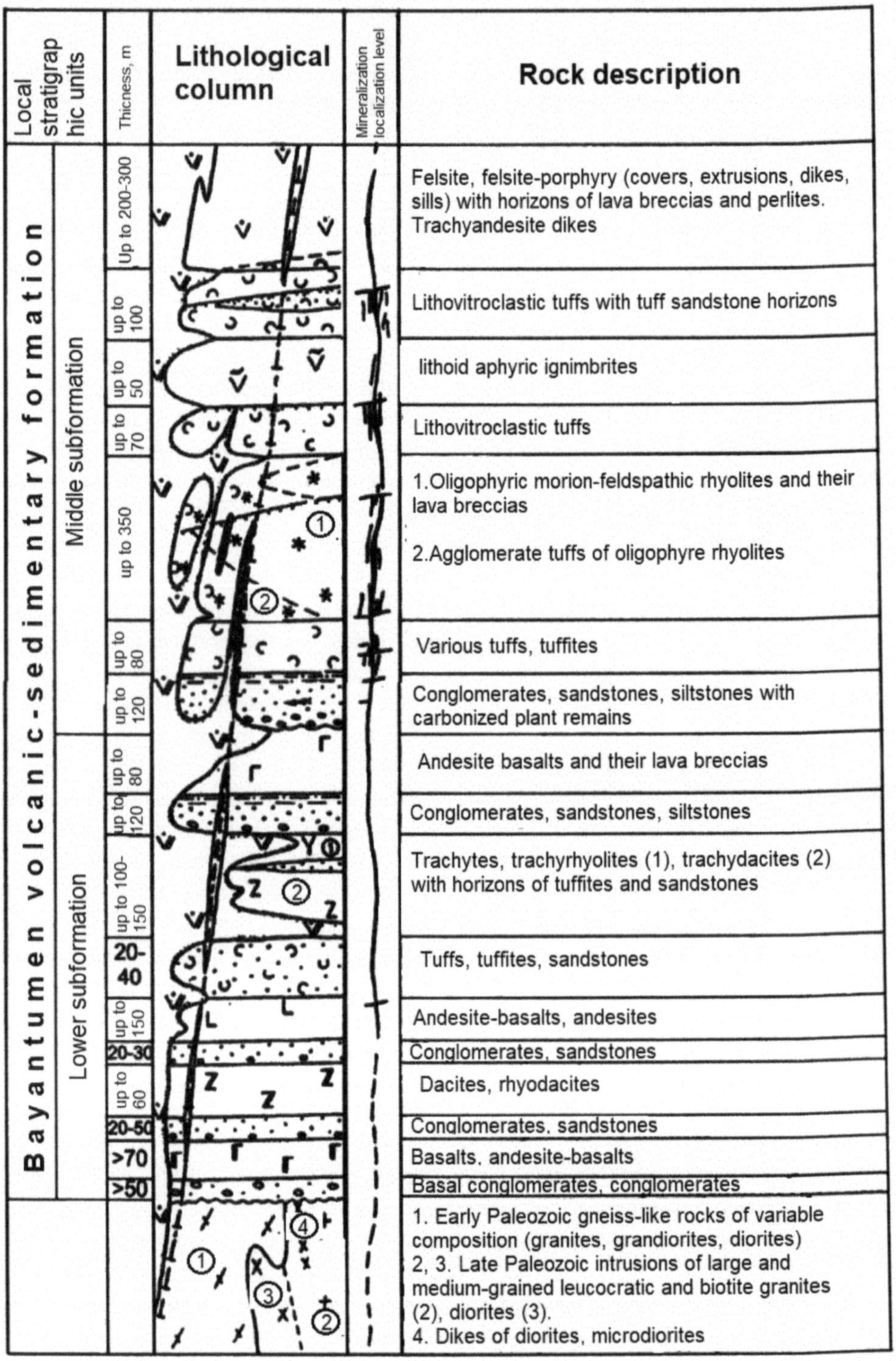

Fig. 4.21 Stratigraphic column of the Mardaingol deposit (according to Mironov 2003)

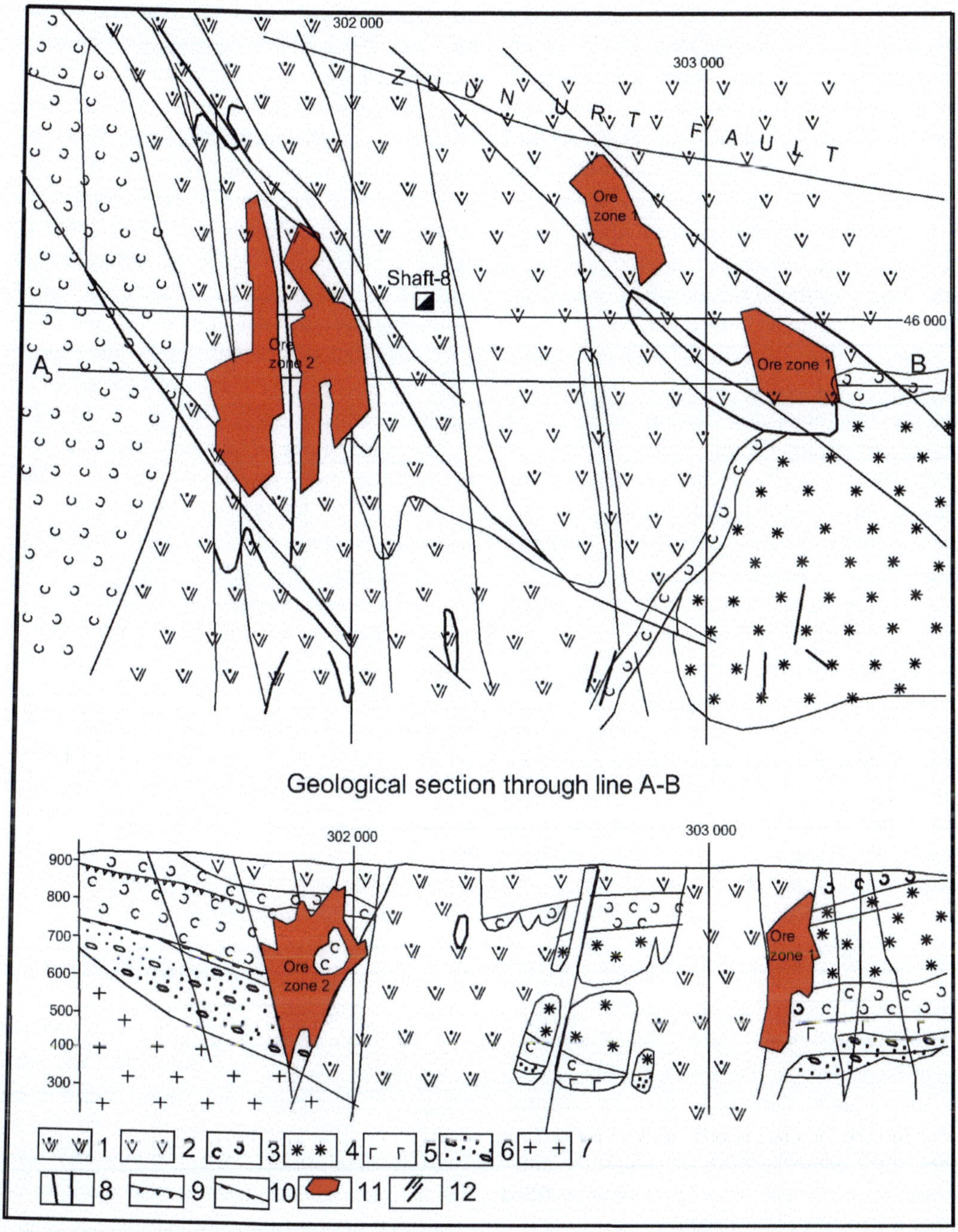

Fig. 4.22 Geological plan and section of the Mardaingol deposit (according to Mironov 2003): 1–6—sedimentary-volcanogenic formations of the Bayantumen formation: 1—dikes, sills, extrusions of felsite porphyries; 2—felsite; 3—tuffs, tuffites, tuff sandstones; 4—oligophyric rhyolites and their tuffs; 5—basaltic andesites; 6—conglomerates, sandstones; 7—undivided basement rocks; 8—circular tectonic faults; 9—gently dipping faults; 10—gentle falls; 11—combined contours of ore-bearing zones; 12—ore bodies

An important role in the structure of the deposit belongs to tectonic disturbances of the near-meridional direction, which have a significant impact on the localization of mineralization. The largest of them (4, 5, 7) are located at a distance of 200–400 m from each other and stretch for 3–4 km. Numerous subparallel tectonic sutures develop between them, forming non-extended weakened strips up to 170 m wide. Faults in this direction often control the placement of trachyandesite dikes.

Faults in other directions, primarily northeastern and latitudinal, are located subparallel near the Talintolgoi and Zuunurt faults. All of them are practically barren.

Steeply dipping faults are accompanied by halos of clayey-hydromica alterations in the host rocks, veinlets of calcite and fluorite, and rare dissemination of sulfides.

Gentle interstratal faults along the contacts of stratified deposits and within horizons of sedimentary rocks and tuffs have become widespread in the area of the deposits. The largest gentle fault has been studied on the western flank of the deposit, where it screens the development of mineralization along the rise. At the same time, sheet-like ore bodies are present in local areas. The failure is represented by fractured crushed rocks up to 1 m thick with two main tectonic sutures, accompanied by flaring and friction clay.

Small, gently sloping and steeply dipping fracturing is manifested in a variety of ways, developing near major faults, subvolcanic bodies, at rock contacts that differ in physical and mechanical properties, and especially at intersection points of multidirectional faults. Fracture zones are a dense network of complexly interconnected cracks. The latter are sometimes filled with discontinuous friction clay.

Conditions for Localization of Mineralization and Morphology of Ore Bodies. Uranium mineralization developed within the Mardaingol deposit is recorded over an area of about 1.1 km^2. Most of it is concentrated in two (1 and 2) ore-bearing zones, respectively, on the eastern and western flanks of this area. The ore-bearing zones are separated from each other by a strip of barren rocks 750 m wide. In general, the geological position of these ore-bearing zones is determined by their location at the center of the intersection of the main northwestern Dagai zone of steeply dipping faults with a tectonically weakened zone of meridional strike. Spatially, ore-bearing zones gravitate toward exocontacts of felsite-porphyry extrusion (Fig. 4.23). Uranium mineralization is distributed over almost the entire section of sedimentary-volcanogenic rocks, but the richest mineralization is confined to the lithological-structural level of oligophyric rhyolites with tuffs similar in composition.

The upper limit of uranium mineralization is 90–130 m from the surface, the lower one is at a depth of 500–650 m. The maximum ore saturation is observed in the depth interval of 90–420 m from the surface. With an overall significant length of ore-bearing faults, mineralization develops in local areas: in the nodes of conjugation of ore-bearing faults with ruptures in other directions, in places where feathering fracturing is concentrated, in areas of their bends in plan and section. Economic accumulations of uranium are represented by small lenses or nests developed inside stockwork-like and vein-like, less often sheet-like bodies of out of balance uranium mineralization. In most cases, ore bodies with standard mineralization do not have a clear morphology, being localized in areas of development of porous multiclastic tuffs and tuff sandstones enriched in organic remains.

Ore-bearing zone 1 contains 33% of all the uranium reserves. It is located in the eastern part of the deposit and can be traced along a strike of 1700 m with a width of 80–180 m. Two ore deposits of stockwork-like and vein-like morphology have been identified within the zone. The northwestern ore deposit was traced by exploration to 340 m and has a width of 80–140 m.

The parameters of the southwestern deposit are 220 × 180 m. The vertical range of ore saturation in the 1st zone is from 80 to 280 m. Ore deposits consist mainly of non-economic mineralization. Standard ores within ore deposits are represented by small lenses and pockets

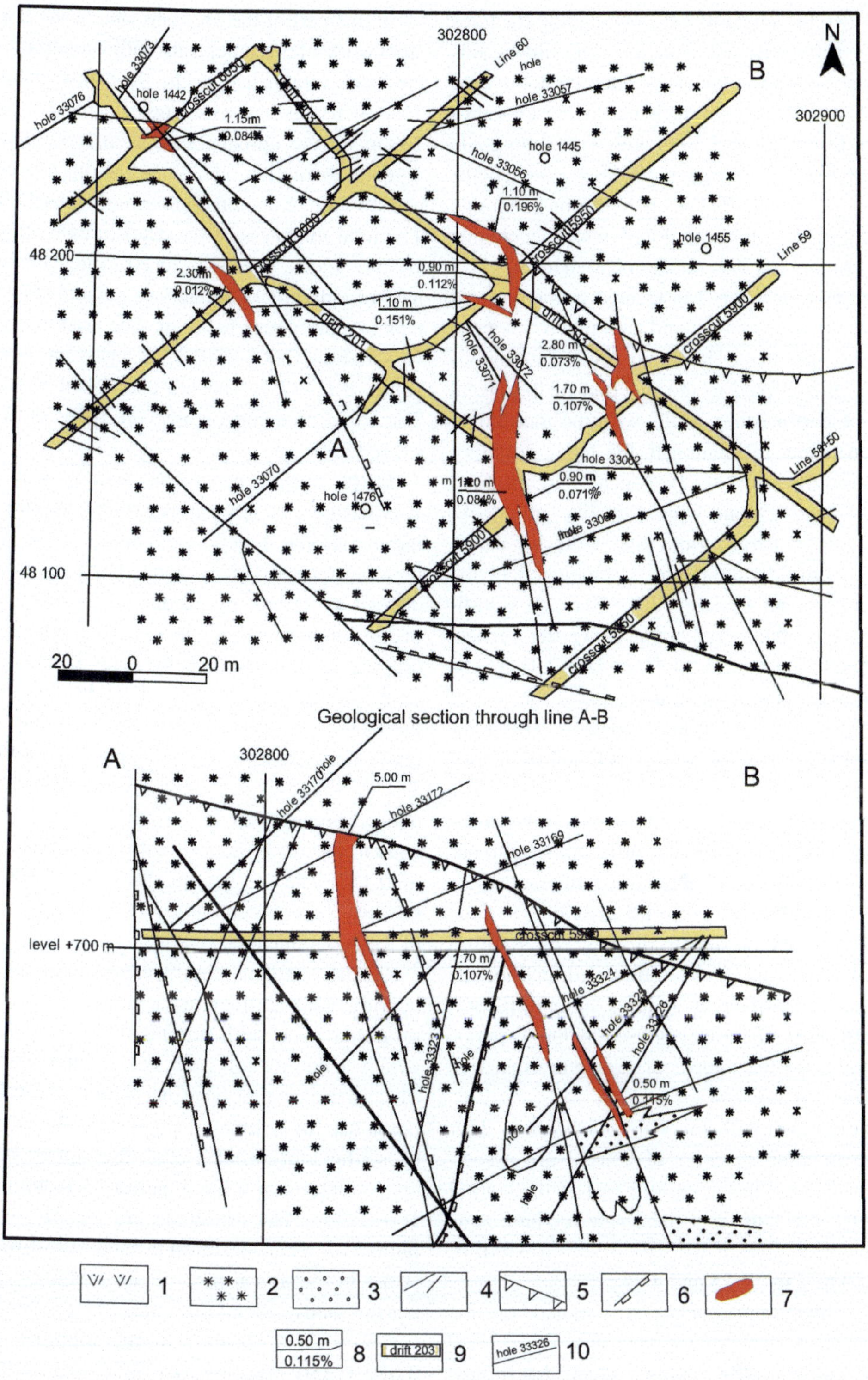

Fig. 4.23 Geological plan (**a**) and section (**b**) of the ore-bearing zone 1 of the Mardaingol deposit: 1–3, sedimentary-volcanogenic formations of the Bayantumen formation: 1—dikes and extrusions of rhyolites; 2—tuffs of oligophyric rhyolites; 3—sandstones, 4—steeply dipping tectonic faults; 5—gentle falls; 6—fracture zones along tectonic faults; 7—ore bodies; 8—parameters of ore bodies (numerator—thickness, m; denominator—grade, %); 9—underground mine workings; 10—exploration lines

controlled by subparallel steeply dipping fractures with a submeridional direction. The dimensions of these bodies in plan and section do not exceed 10–15 m. Usually, the thickness of ore crossings ranges from 0.85 to 5.30 m with a uranium grade of 0.064–0.437, averaging 0.098%. Ore coefficient is 0.012–0.030. The share of non-economic mineralization, in the halo of which economic ores develop, is 109% in relation to them.

Ore-bearing zone 2 is located in the western part of the deposit area and contains 67% of the uranium reserves. The main mineralization of zone 2 is confined to the junction of the Dagai zone of northwestern faults with the meridional fracture zone. In relation to the felsic porphyry extrusion, it covers its exocontact part. In plan, the ore-bearing zone with sheet-like and stockwork-vein mineralization, the width of which varies from 140 to 450 m, is elongated in the near-meridional direction by 800 m. A significant range of mineralization along the vertical is characteristic: from 80–90 to 180–230 m. The lower boundary of mineralization is the top of the upper cover of basaltic andesites. The highest productivity in the section is found in tuffs of oligophyric rhyolites injected with subvolcanic felsite-porphyry bodies. Layer-like mineralization is localized mainly in tuffaceous-sedimentary rocks with coalified plant organic matter.

The distribution of mineralization in zone 2 is controlled by steeply dipping faults with a northwestern and near-latitudinal directions, as well as layer-by-layer failures among favorable horizons of tuffaceous-sedimentary rocks. According to the degree of ore saturation, five ore zones are present.

The largest of them, with which about 37% of the uranium reserves of zone 2 are associated, is located in the eastern part of the central northwestern fault 2 of the Dagai tectonic zone and extends along it for 580 m with a width of 60–120 m (Figs. 4.23 and 4.24).

The maximum thickness is noted on the southeastern flank of the site, at a sharp bend in the ore-controlling suture, which determined the widespread development of tectonic disturbances of higher order.

In this part of the ore zone, one of the largest stockwork-like ore deposits with economic ores has been established in the deposit, containing 15% of its reserves. Its length in plan is about 100 m. In the uprising section, it is limited by a gentle disruption, and along the dip it is adjacent to fault 2, its length does not exceed 60 m. The total thickness of the deposit, consisting of 1–3 ore lenses, does not exceed 25 m. The thickness of individual ore intervals range from a few tens of centimeters to 7.6 m. The uranium grade varies from 0.066 to 0.367% with an average value of 0.187%. The ore coefficient is 0.218. The share of non-economic ore halos does not exceed 26%.

In the western part of the main central fault 2, a narrow ore deposit is present, oriented in the meridional direction, containing 21% of the uranium reserves of the zone. The length of the deposit is 450 m, the width is 25–60 m. Ore-bearing zones within the site are both steeply dipping tectonic sutures limiting it, and feathering subparallel fissures. The dimensions of individual vein bodies with non-economic mineralization reach 40–60 m in plan and section. The thickness of economic ore intervals among non-economic ores is 5.20 m with a uranium content of 0.077–0.252% with an average grade of 0.117. The ore-bearing factor is 0.028. The quantity of non-economic ore makes up 124% of the economic ore.

In the extreme southwestern part of ore-bearing zone 2 there is a wedge-shaped ore-saturated zone with a total area of 0.02 km^2, which accounts for 20% of its total reserves. The vertical extent of mineralization, the lower boundary of which is close to the horizon of sedimentary rocks, and the upper boundary is determined by a gentle detachment, is 160–240 m. The ore hosts are steeply dipping faults of near-meridional and northwestern strike in the apical part of the felsite porphyry tearing body. Stem thickness of ore intervals varies from 1.40 to 6.40 m; uranium grade 0.070–0.356, average 0.112%; ore-coefficient 0.046; the share of non-economic mineralization reaches 121%.

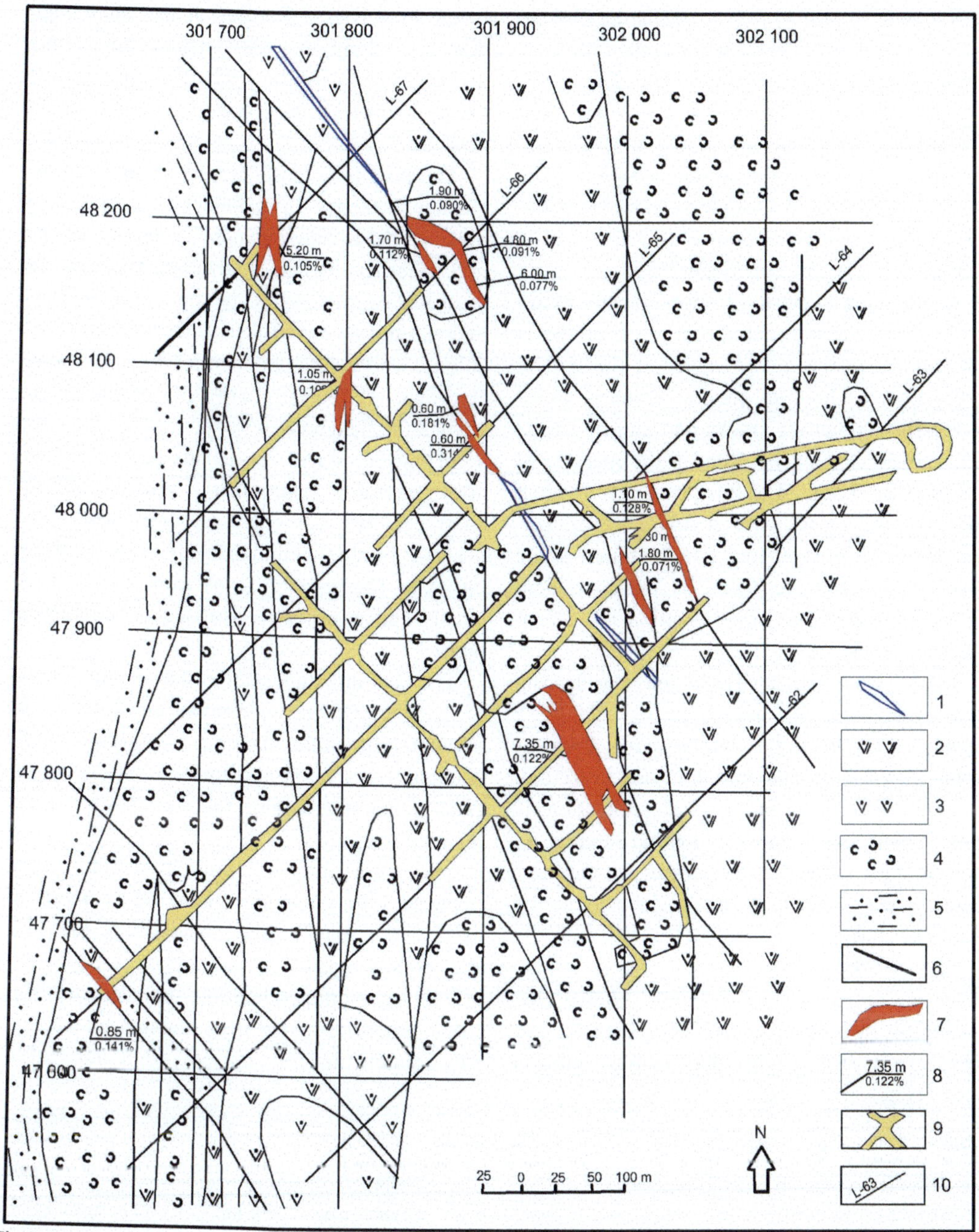

Fig. 4.24 Schematic geological plan of ore-bearing zone 2 (horizon + 700) of the Mardaingol deposit: 1–5—sedimentary-volcanogenic formations of the Bayantumen formation: 1—trachydacite and basaltic andesite dikes; 2—rhyolite extrusions and dikes; 3—coarse-fluid felsites; 4—rhyolite tuffs; 5—sandstones, siltstones; 6—tectonic disturbances; 7—ore bodies; 8—parameters of ore bodies (numerator—thickness, m; denominator—grade, %); 9—underground mine workings; 10—exploration lines

Layered Ore Deposits. A significant portion of uranium mineralization is concentrated in ore deposits with tabular morphology. They account for about 9% of the uranium reserves of ore-bearing zone 2. One of these deposits is located on the western flank in the upper part of the section and is controlled by a gentle detachment in the horizon of tuff sandstones in the tuff mass. Uranium mineralization is traced in the meridional direction along fault 4A for 270 m with a width of 100–180 m and a dip of 90–140 m. The total area of mineralization is 0.04 km^2. The thickness of standard ore intervals here is 0.34–2.11 m; uranium grade from 0.065 to 0.249%, average 0.117%; ore-coefficient 0.252.

Another ore deposit of sheet-like morphology is located in the central part of the ore-bearing zone 2, in the bottom layer of the middle subformation. Uranium mineralization is multi-tiered, fills numerous cracks and can be traced in a narrow strip of 30–50 m in the meridional direction for 310 m. The thickness of individual flat-lying ore lenses with standard mineralization varies from 0.60 to 5.5 m; uranium grade 0.084–0.204, average 0.132%; ore-coefficient 0.116; non-economic mineralization is 42% of economic ores.

Along with the characterized ore-bearing areas at the deposit and its nearest flanks, a number of isolated ore bodies with non-economic and, less often, economic ores were discovered. The scale of this mineralization is insignificant and is not of industrial interest. In the summary Table 4.10 shows data on the morphology and parameters of ore deposits at the Mardaingol deposit.

Mineral Composition of Ores and Alterations. According to the material composition and type of hydrothermal-metasomatic changes in the host rocks, the ores of the Mardaingol deposit belong to the pitchblende-coffinite mineral type in argillites. There are three stages of the hydrothermal process: pre-ore, ore and post-ore.

The pre-ore stage is closely related to post-volcanic changes in sedimentary-volcanogenic rocks: K-feldspathization, pyritization, seritization, silicification and albitization. However, the most intense process of hydromica, represented by hydromicas, mixed-layer minerals, carbonate, chlorite, and quartz, are manifested at this stage. Hydromica-carbonate aureoles often extend beyond the field area.

In the ore stage, ore-accompanying ferruginous chlorite, hydromicas, montmorillonite, quartz, fluorite, carbonate, kaolinite, and hematite are formed simultaneously with uranium minerals. This mineral association is distributed locally and is primarily isolated to ore-bearing faults.

The post-ore stage at the Mardaingol deposit is characterized by widespread development of fracture-vein mineralization, represented predominantly by montmorillonite-kaolinite aggregates developing along the planes of various oriented fractures.

The formation of uranium mineralization occurred through metasomatic replacement of rocks and is partly due to the filling of voids and slightly open cracks. The most characteristic uranium mineralization includes disseminated, nested and vein disseminated textures of ores; cement and breccia are often noted. For

Table 4.10 Ore deposit parameters of Mardaingol deposit (Mironov et al. 2012)

Ore deposits	Size (length) (m)		
	Along strike	By fall	Thickness
Ore zone 1	1700	80–280	80–180
Northwestern ore deposit with stockwork-vein mineralization	340	80–120	80–140
Southwestern ore deposit with stockwork-vein mineralization	220	80–280	180
Ore zone 2	800	160–240	60–120
Eastern ore deposit with stockwork-vein mineralization	580	160–240	60–120
Western ore deposit with stockwork-vein mineralization	450	80–120	26–60
Southwestern ore deposit with stockwork-vein mineralization	320	160–240	15–50
Western ore deposit with sheet-like mineralization	270	100–180	10–20
Central ore deposit with sheet-like mineralization	310	30–50	10–30

cloak-shaped ore bodies, the textures are usually banded and linearly disseminated. The structures of the ores are granular, colloform-spherulite, loop-shaped and reticulated. The ores are mainly composed of aggregates of fine, fine-grained and finely dispersed uranium minerals with irregular, sinuous boundaries. The size of the aggregates range from 0.001 to 0.5 mm, less common are veinlets with a thickness of 0.01 to 1–2 mm, with isolated veins up to 0.5 cm.

The main uranium minerals are coffinite and pitchblende; uranophane and betauranotyl are found in small amounts (up to 1%); and in isolated cases, curite.

Coffinite forms spherulites of a radial-shell-concentric structure, kidney-shaped aggregates, partially or completely replace pitchblende deposits, and often envelop grains of sulfides and leucoxenized titanomagnetite. In the voids, on the planes of microcracks and fissures, powdery varieties of finely dispersed coffinite are noted, which are closely associated with post-ore mineralization.

Pitchblende in ores make up no more than 1–2% and occur in the form of relics not replaced by coffinite, often forming kidney-shaped crusts up to 0.01 mm thick (with a few up to 0.5 cm), sometimes growing on sulfides. In rich ores, isometric and irregular metasomatic segregations are observed with the formation of looped and reticulate structures (Fig. 4.25).

Uranophane, beta-uranotyl, and in isolated cases curite and welsetsdorfite are extremely rare and form thin veins, crusts, and small accumulations in oxidized ores, which are sometimes observed along steeply dipping faults to a depth of 400 m.

Sulfides are often present in ores, but the total content does not exceed 1%. Quantitatively, pyrite, galena, and marcasite predominate; arsenopyrite, sphalerite, and occasionally chalcopyrite and molybdenite are less common. The size of sulfide deposits range from 0.005 to 0.01 cm (pyrite and galena—up to 2 mm). Molybdenite is observed sporadically, developing along cracks and brecciation zones in the form of thread-like veins, small (up to 0.01 mm) flakes and nests.

Along with uranium minerals and sulfides, hydromicas, montmorillonite, kaolinite, and chlorite are developed in the ores, the amount of which ranges from 10 to 35%. Carbonates, represented by calcite and ankerite, make up no more than 1%. Fluorite occurs in the form of small impregnations and veinlets, and is less commonly observed in breccia cement. Its amount is insignificant (0.4–0.6%).

Post-ore, predominantly fissure-vein mineralization is observed quite widely in ores and is most often represented by kaolinite, less often quartz, fluorite, montmorillonite, calcite, chlorite and hydromica. In the oxidation zone, which occurs in the deposit along steeply dipping faults to a depth of 400 m, brown iron hydroxides and, less commonly, sooty deposits of manganese hydroxides are noted in the ores.

The age of hydrothermal uranium mineralization, according to geological data, is Early Cretaceous. The absolute age of uranium ores, established in the IGEM laboratory using the lead isotope method, is 136–152 million years (J_3–K_1).

Chemical Composition of Ores. The ores are of the aluminosilicate type with a low content of carbonates and sulfides (Table 4.11). The ores are monometallic; elevated concentrations of lead (up to 0.03%) and molybdenum (0.005–0.14%) sometimes noted in them occur sporadically and are not of economic significance. In general, the ores of the deposit are characterized by insignificant contents of harmful impurities that negatively affect the technology of their processing.

Geochemical Characteristics of Rocks and Ores. It is based on the results of spectral analyzes of rock chip, core and furrow samples. Separately, for ore-bearing zones 1 and 2, the background and anomalous content of chemical elements in host rocks and by ore classes were calculated. Correlation between uranium and accompanying elements were established, and their accumulation levels were determined.

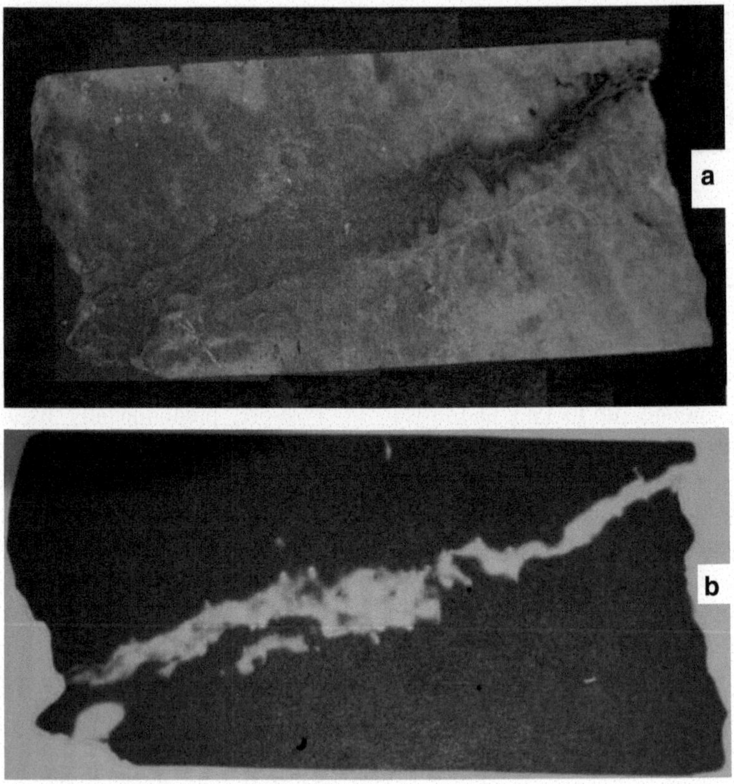

Fig. 4.25 Veins of pitchblende in felsite porphyries. Mardaingol uranium deposit. Ore zone 1: **a** photo of the sample (natural); **b** radiography (exp. 3 days)

The rocks and ores of the deposit contain a wide range of satellite elements characterized by varying degrees of concentration and correlations with uranium (Table 4.12).

As can be seen from Table 4.13, the main group of elements characterized by a clear tendency to accumulate in ores and halos, along with uranium, are molybdenum, lead, silver, and arsenic. The distribution of elements with this association, which are most closely related to the ore process, are distinguished in most cases by stable positive correlations with uranium. Analysis of the levels of accumulation of elements by ore class indicates that the maximum concentrations of most satellite elements are recorded in rich ores. At the same time, molybdenum is characterized by the maximum accumulation level (> 30 GF). At the same time, the highest accumulation of molybdenum and arsenic is observed in ores localized in rhyolite tuffs.

Negative correlations of uranium are observed with potassium, calcium, magnesium, cerium, zinc, gallium, boron and tin, with magnesium being the most significantly removed.

Ore-bearing zones 1 and 2 are generally of the same type in terms of the distribution of satellite elements with relatively higher levels of accumulation of uranium and elements of the ore association correlated with it in zone 2.

Primary uranium halos are significantly larger in size than individual ore bodies, collectively forming a single halo of anomalous radioactivity, which usually does not extend beyond the ore-bearing zones. In the section, the halo of anomalous radioactivity gravitates toward the upper boundary of productive sedimentary-volcanogenic formations and breaks away from it no further than the first tens of meters. Inside the ore-bearing zones, the most contrasting (> 10 GF) uranium halos, the width of which is an order of

Table 4.11 Chemical composition of ores of Mardaingol deposit (Mironov et al. 2012)

Number samples	Grade (wt.%)															
	SiO_2	Al_2O_3	Fe_2O_3	FeO	TiO_2	P_2O_3	CaO	MgO	MnO	K_2O	Na_2O	nnn	H_2O	CO_2	SO_{tot}	U
GTP-36	62.29	13.33	1.71	1.76	0.61	<0.01	1.52	0.70	0.05	3.70	1.23	5.44	0.92	0.85	?0.25	0.12
TP-1-M	75.49	13.03	1.38	0.47	0.09	<0.1	0.50	<0.5	<0.05	3.38	3.70	1.50	0.24	–	–	0.022
TP-2-M	74.21	12.64	1.84	0.54	0.09	<0.01	1.40	<0.5	0.08	3.62	3.14	2.45	0.43		–	0.014
TP-Z-M	74.34	12.53	1.45	0.31	0.10	<0.01	0.84	<0.5	0.15	5.00	2.00	2.31	0.59		–	0.189
TP-4-M	72.87	12.95	2.62	0.40	0.09	<0.01	1.12	<0.5	0.08	4.30	3.56	1.86	0.30		–	0.052
TP-5-M	75.16	12.65	1.03	<0.25	0.09	<0.01	0.84	<0.5	0.06	4.80	2.77	1.74	0.46		–	0.302
TP-6-M	75.32	13.21	1.72	0.34	0.08	<0.01	0.56	<0.60	<0.05	4.30	0.73	3.06	0.58	–	–	0.058
TP-7-M	76.49	11.72	1.62	<0.25	0.10	<0.01	0.56	<0.5	0.06	4.10	2.55	2.16	0.44			0.042

magnitude greater than the size of the ore bodies, are spatially confined to ore-bearing faults and permeable horizons of ore-bearing rocks.

The halos of satellite elements individually are inferior in all respects to the halos of uranium. In this regard, only the multiplicative halos of the main indicator elements of uranium, represented by molybdenum, lead, and silver, are of interest.

Their sizes are comparable, and their contrast is 10–100 times or greater than that of the uranium halos. At a low (3–10 GF) level of anomaly, multiplicative halos, unlike uranium, are recorded on the surface; the most contrasting of them extend along the main ore-controlling faults.

Vertical geochemical zoning at the deposit is expressed in the displacement of the geometric centers of the halos of indicator elements relative to the ore bodies. This is clearly fixed by the change in the linear and areal productivity of halos at different hypsometric levels. The generalized series of vertical zonation, calculated from the analysis of samples from the core of wells, in the direction from the supra-ore part to the bottom of the section has the form: (arsenic, cobalt, vanadium, nickel)—silver—germanium—lead, zinc, uranium—molybdenum—thallium—bismuth—beryllium (nickel, arsenic, vanadium)—copper. The most characteristic features of vertical geochemical zoning are:

- close proximity in the zoning series of uranium, silver, lead, zinc and molybdenum;
- there is some displacement of the silver halos relative to the ore bodies in the superore part, and of the molybdenum in the sub ore part;
- a significant distance of copper in the zoning series from uranium to the subore part was noted.

The reserves of the Mardaingol deposit, approved by the State Reserves Committee of the USSR and registered into the account of the Mineral Reserves of Mongolia based on the results of exploration, were additionally explored and recalculated. Detailed mining and drilling exploration showed that within the

Table 4.12 The level of accumulation of elements-day lighters in aureoles and ores (Mironov et al. 2012)

Magnitude savings elements in geochemical backgrounds	Grades of uranium content in halos and ores (%)			
	0.01	0.01–0.05	0.05–0.3	>0.3
More 10	–	Molybdenum (0.009)	Molybdenum (0.0038) Silver (0.22 g/t)	Molybdenum (0.023) Lead (0.052)
2–10	Molybdenum (0.0012) Silver (0.07 g/t)	Lead (0.026) Silver (0.22 g/t) Bismuth (0.0002)	Lead (0.046) Manganese (0.14) Arsenic (0.0035) Barium (0.01)	Arsenic (0.0010) Silver (0.22 g/t) Cobalt (0.0005) Manganese (0.097) Lithium (0.006)
1.3–2	Bismuth (0.0001) Tin (0.0014) Lead (0.01) Lithium (0.004) Thallium (0.0001) Gallium (0.005)	Boron (0.022) Chromium (0.002) Arsenic (0.0015) Thallium (0.0002) Copper (0.0014) Lithium (0.004) Hafnium (0.002) Germanium (0.0001)	Boron (0.022) Chromium (0.002) Arsenic (0.0015) Thallium (0.0002) Copper (0.0014) Lithium (0.004) Hafnium (0.002) Germanium (0.0001)	Thallium (0.0002) Barium (0.007) Bismuth (0.0001) Chromium (0.0015) Niobium (0.012) Germanium (0.0001) Ytterbium (0.0012) Thorium (0.0038)

Note The level of accumulation is calculated relative to the local background of host rocks, the average grade of elements is given in percent

Table 4.13 Morphology and parameters of ore bodies (zones) of Mardaingol deposit (Mironov et al. 2012)

Ore bodies (zones)	Dimensions (length) (m)			
	Along strike	By dip	Thickness	Azimuth and incidence angle
Ore-bearing zone 1: eastern sheet-like body	1100–600	100–480	1.4	Subhorizontal occurrence
2: western ribbon-shaped body	1000	140–200	0.3–3.4	Subhorizontal occurrence
Ore-bearing zone II: central sheet-like body	400	100–200	1.4	Dip azimuth 50–70°, angle 10–12°
North-western sheet-like body	350	200	0.5–5.2	Dip azimuth 40–60°, angle 8–12°
Ore-bearing zone III	400	50–150	0.3–9.6	Subhorizontal occurrence

ore-bearing zones there are localized ore bodies of sheet-like, vein-like and stockwork-like type deposits. They are small in size with minimal reserves and have average to poor uranium grades. Due to the insignificant sizes of ore bodies and low ore saturation of ore-bearing zones mining would necessitate spatial delineation of each ore body; such delineation would require significant volumes of mining and was considered uneconomic. Considering the small size and small reserves of the ore bodies and low uranium grade, the deposit's reserves are qualified as reserves of category C.

4.5 Nemer Uranium Deposit

The Nemer deposit is located in the northwestern part of the Mardaingol block of the DVTS, which is a local volcano-tectonic structure close to the caldera type, and is controlled by the large Mardaingol and Dagai fault zones with a northwestern strike, as well as the Dzunurt sublatitudinal deep fault (Fig. 4.26). The deposit was discovered in 1987 by geologists of the Mongolian Geological Expedition of the USSR Ministry of Geology during geological prospecting work.

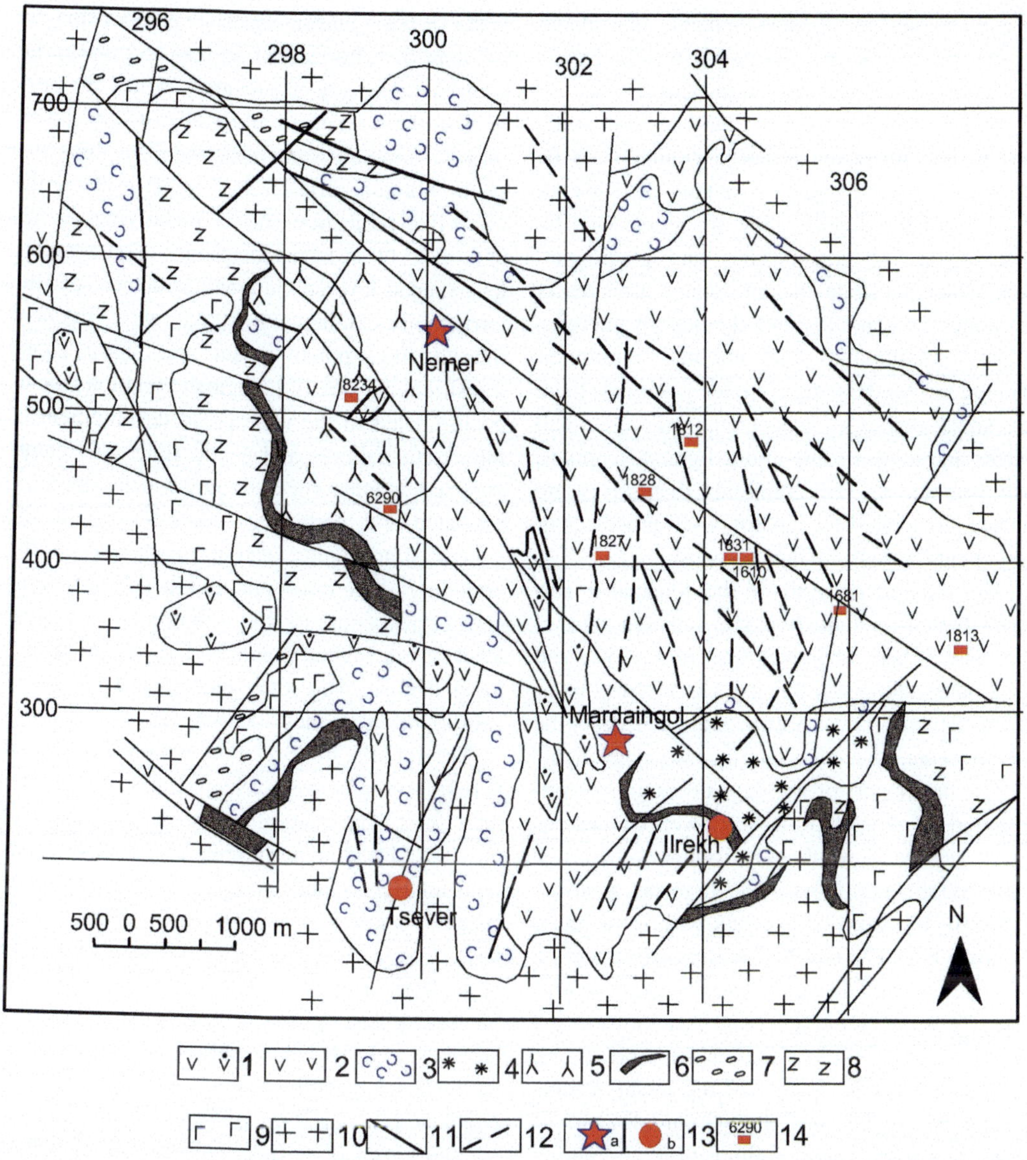

Fig. 4.26 Schematic geological map of the Mardaingol tectonic block with the Mardaingol and Nemer uranium deposits (according to Mironov 2003): 1–9—sedimentary-volcanogenic formations of the Dornod complex: 1—dikes and extrusions of felsite; 2—felsite; 3—mixed-clastic tuff; 4—oligophyric rhyolite; 5—quartz porphyry; 6—sandstone, siltstone; 7—conglomerate; 8—dacite; 9—basaltic andesites; 10—basement granitoids; 11—tectonic disturbances and their numbers; 12—tectonic disturbances identified according to structural geophysics data; 13—deposits (a) and ore occurrences (b) of uranium; 14—holes with economic uranium mineralization and their numbers

Geology of the Deposit. The peculiarities of the geological structure of the Nemer deposit are due to the heterogeneous section of the host rocks, the widespread development of volcanic bodies and felsic porphyry sills, the presence of fault tectonics and a variety of superimposed hydrothermal-metasomatic processes.

Like other uranium deposits of the Dornod ore cluster, the geological section of the Nemer deposit consists of two parts—the lower pre-late

Mesozoic period of development, considered as the foundation, and the upper, formed by the processes of late Mesozoic destruction.

A characteristic feature of the basement structure is the wide development of granitoids of the Modkhudag palingene-metasomatic complex. In the area of the deposit, they are represented by biotite granites, granodiorites and plagiogranites. Older substrate metamorphites are present as numerous xenoliths and skialites of gneisses, schists, amphibolites, and marbled limestones. Among the Modkhudag complex granitoids there are small bodies of leucocratic granites, small stocks and dikes of fine-grained granites, aplites, and pegmatites. The granitoids themselves are universally altered by superimposed processes of sodium and potassium metasomatism.

The maximum depth of the foundation surface—650 m—is noted in the southern part of the field. There are swell-like protrusions and local paleovalleys with relative elevations of up to 250 m. The main morphostructural elements of the basement are oriented in a northwestern direction.

The upper part of the deposit section is represented by Upper Jurassic-Lower Cretaceous rocks of the Bayantumen formation, formed under continental conditions. It forms a monoclinal (6–10, rarely 25°) strata with a thickness of up to 800–1200 m, and is divided in section into lower, middle and upper subformations. Directly at the Nemer deposit, formations of the lower and middle subformations are represented.

A characteristic feature of the lower subformation is the repeated alternation of lavas of basic, less often acidic composition, alternating with horizons of terrigenous and tuffaceous-sedimentary rocks. At the base of the section of the lower subformation, a compositionally heterogeneous basal horizon of conglomerates (thickness up to 90 m) is developed, leveling the relief of the basement. A cover of basaltic andesites 100–250 m thick, overlying the basal horizon, is distributed over almost the entire area of the deposit and occupies about 30% of the volume of all volcanics. Higher in the section lie andesidacites, dacites, rhyodacites, their lava breccias and tuffs, combined into a single cover with a thickness of 60–160 m. Volcanic rocks of basic and intermediate composition are overlain by a horizon of sedimentary and tuffaceous-sedimentary formations, which belongs to one of the main lithological-structural levels of local uranium mineralization. The thickness of the horizon varies from 5 to 10 m above the swell-like projections of the underlying dacites and up to 60 m in paleoincisions. The complex lens-shaped-layered internal structure of the terrigenous-sedimentary strata with frequent interlayering, pinching out and sharp facies transitions of the lithological variations composing it led to the widespread development of intrastratal failures and gentle zones of fracturing. Sedimentary rocks throughout the mass are saturated with carbonized plant detritus. In the general stratigraphic section, this horizon is the boundary of the lower and middle subformations and corresponds to the lithological-structural level of ore bodies 2 and 3 of the Dornod uranium deposit. The total thickness of the lower subformation reaches 400–450 m.

The middle subformation within the deposit is represented by polyfacies felsic volcanics and their tuffs. The section begins with a cover of quartz-feldspathic porphyries and lava breccias with interlayers and lenses of tuffs with a consistent thickness of 40–60 m. In their appearance, they correspond to the facies varieties of oligophyric rhyolites. The overlying sequence of mixed-clastic, crystallithoclastic tuffs, tuff breccias and felsic ignimbrites is characterized by their frequent mutual transitions. The heterogeneous tuffaceous strata contains interlayers and lenses of tuffites and sandstones, which are ore-bearing on the northwestern flank of the deposit. The section of the middle subformation, and with it the entire upper structural floor at the Nemer deposit, ends with a cover of felsite and lava breccia with interlayers of mixed-clastic tuffs up to 200 m thick. The total thickness of the middle subformation reaches 150 m.

The stratified formations of the Bayantumen Formation are intruded by numerous subvolcanic intrusions of felsic porphyries.

The ore-bearing rocks of the Nemer deposit are terrigenous-sedimentary deposits (sandstones, silty sandstones, gravelites,

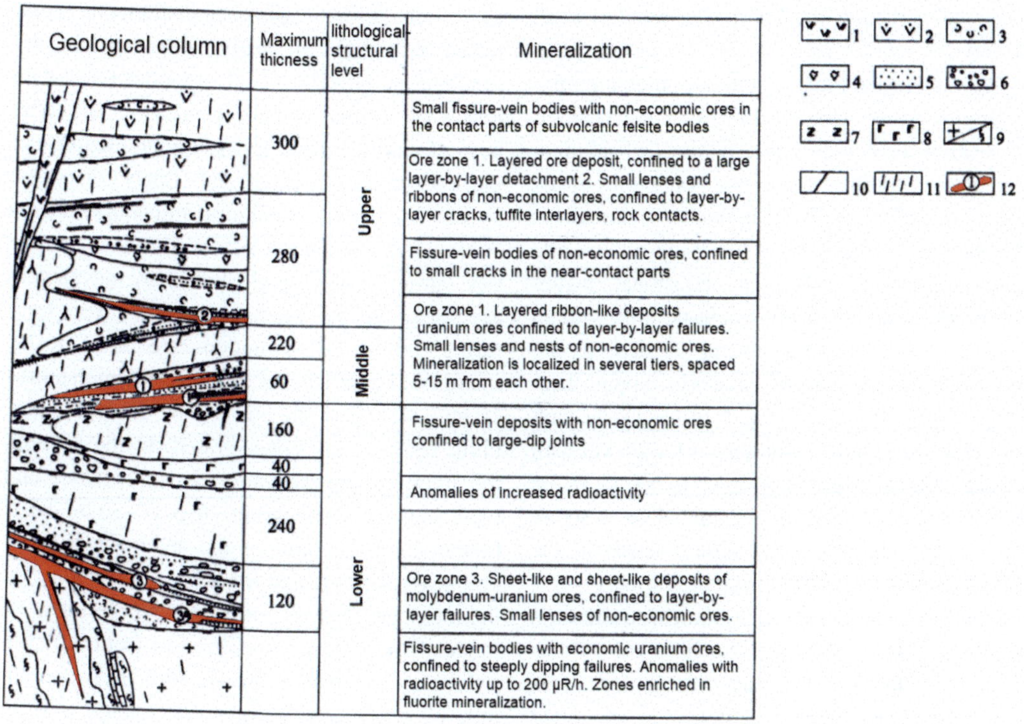

Fig. 4.27 Position of uranium mineralization of the Nemer deposit in the stratigraphic section (according to Mironov 2003): 1–8—late Mesozoic formations (J_3–K_1bt): 1—subvolcanic felsite; 2—felsic porphyry; 3—felsite tuffs; 4—quartz-feldspathic porphyries and their tuffs; 5—sandstones; 6—conglomerates; 7—dacites, rhyodacites; 8—basaltic andesites; 9—granites with xenoliths of metamorphites; 10—faults; 11—fracturing; 12—ore bodies

conglomerates), felsic tuffs and tuffites (Fig. 4.27).

Tectonics of the Deposit. The structural plan of the field is determined by systems of tectonic disturbances of northwestern, near-meridional and sublatitudinal strike. The largest is the Dagai fault zone, which crosses the Mardaingol tectonic block in a northwest (320–330°) direction and is clearly traced along a strike of 15 km. Steeply dipping (70–85°) large faults 8, 10, 10a and 11 of this zone, together with systems of feathering faults within the deposit, form a tectonically weakened strip up to 500 m wide. The main ore-controlling structural element of the zone is fault 10, which belongs to the category of large intra-block faults. In the southeastern part of the field, it consists of three branches—10, 10a, and 11, each of which is an independent tectonic suture with amplitudes of vertical displacements of up to 150 m. The total amplitude of movement of individual blocks along fault 10 varies from 160 to 300 m.

Along with widespread manifestation of steeply dipping tectonics, gently dipping faults are developed within the field along the contacts of rocks with different physical and mechanical properties, zones of layered fracturing near large steeply dipping faults, as well as interstratal and interfacies faults. Zones of gentle faults are the main ore-distributing and ore-bearing structures of the deposit and are accompanied by wide halos of hydrothermal alteration with the same type of mineral composition.

Conditions for Deposition of Uranium Mineralization and the Morphology of the Ore Bodies. Within the Nemer deposit, economic uranium mineralization is deposited in sedimentary-tuffaceous formations of the Bayantumen formation in the form of dough-like deposits, separated vertically by 110 m (Jamsrandorj et al. 2017a).

With the general manifestation of uranium mineralization in all rock complexes of the geological section, economic concentrations of uranium are confined to the stratigraphic boundaries of the basement and cover (basal horizon), sedimentary-volcanogenic formations of the lower and middle subformations, as well as to intrastratal disruption in the thickness of mixed-clastic tuffs with an acidic composition. Everywhere, ore deposits are controlled by a system of circularly dipping hectopic sutures 10, 10a, 11 and 8 of the Dagai fault zone and can be traced along them for 2.6 km with a width of the ore-bearing strip up to 600 m. In the vertical section, uranium mineralization is localized at three lithological-structural levels in the interval depths 130–320 m from the surface.

Within the Nemer deposit, industrial uranium mineralization is localized in sedimentary-tuffaceous formations of the Bayantumen formation in the form of dough-like deposits, separated vertically by 110 m.

A clearly expressed lithologic-structural control, due to the confinement of mineralization to porous (permeable) horizons of heteroclastic tuffs, tuffites, and terrigenous-sedimentary formations enriched in unified organic matter, predetermined the sheet-like morphology of the ore bodies.

Based on the nature of the geological and structural conditions of deposition, confinement to certain rock complexes and position in the vertical section of the deposit, ore-bearing zones I, II and III were identified, containing, respectively, 41, 37, and 22% of the total uranium reserves (Figs. 4.28 and 4.29).

Ore-Bearing Zone I unites all manifestations of uranium mineralization of the average lithological-structural level, confined to the horizon of terrigenous-sedimentary deposits at the border of the lower and middle subformations. Economic uranium mineralization is developed in the southeastern part of the deposit and is controlled by steeply dipping faults 10, 11 and 11a in a northwestern direction, which extends for 1100 m. The ore-bearing horizon is a terrigenous-sedimentary horizon, composed of often alternating sandstones, silty sandstones, gravelites and conglomerates with lenses of tuffites and felsic tuffs. In all varieties of sedimentary rocks, accumulations of unified organic matter and finely dispersed pyrite are noted. Uranium mineralization is confined to the boundaries of the facies varieties, along which gentle faults are developed, accompanied by intense layer-by-layer fracturing. Interfacial disruptions are most clearly manifested near steeply dipping tectonic sutures 10 and 11, where the bulk of the economic uranium reserves are concentrated. In the section, the uranium mineralization is multi-tiered and represents an ore-bearing strip 20–30 m thick, bounded by the base of the sedimentary horizon at the bottom and a gentle detachment 1 at the top. The thickness of individual ore intervals ranges from 0.3 to 3.4 m, grade—from 0.075 to 0.869%. Economic mineralization is accompanied by wide aureoles of non-economic ore. In ore-bearing zone I, two independent sheet-like ore bodies are present—Eastern and Western, having a subhorizontal occurrence. The eastern body extends along faults 10 and 11 for 600 m with a width of 140–200 m and has a trapezoidal shape in plan. The Western deposit is raised along fault 11 by 80 m relative to the Eastern body and is stretched in the form of a ribbon for 1000 m with a width of 100–200 m. The bulk of the uranium reserves are localized at a depth range of 120–230 m from the surface. Both ore bodies are characterized by an uneven distribution of uranium mineralization with a predominance of low-grade and average ores.

The average uranium grade in ore-bearing zone 1 is 0.219% with a thickness of 1.4 m. In addition to uranium, molybdenum is present with a grade that varies from 0.005 to 0.01%.

Ore-Bearing Zone II combines manifestations of uranium mineralization of the upper lithological-structural level, localized in the horizon of felsic tuffs of the middle subformation. Within the ore-bearing zone, two sheet-like deposits are present, controlled by steeply dipping tectonic sutures 10a and 8 with a northwest strike.

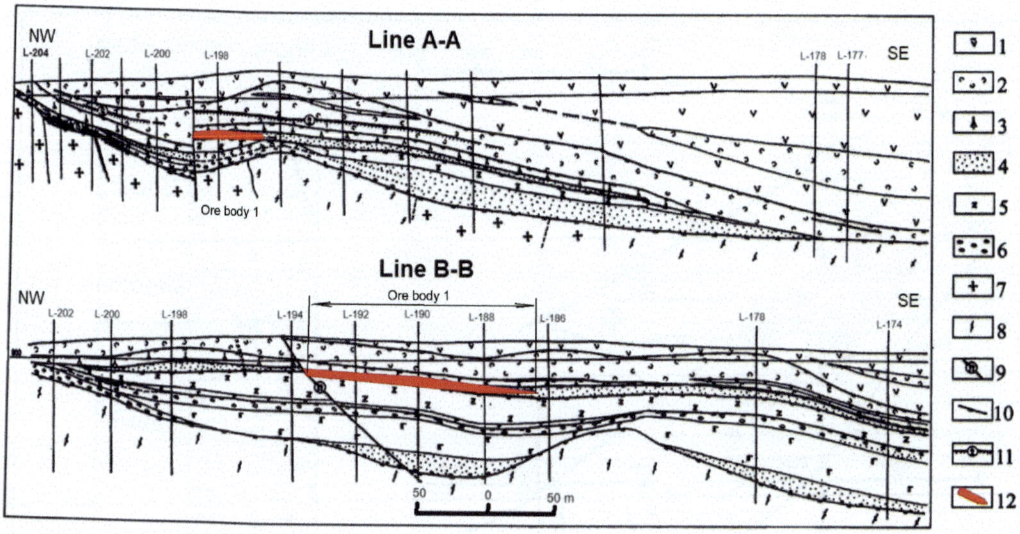

Fig. 4.28 Geological sections along lines A–A, B–B of the Nemer deposit (according to Mironov 2005): 1–6—sedimentary-volcanogenic formations of the Bayantumen formation: 1—felsite; 2—felsic tuff; 3—quartz porphyry; 4—tuffite, sandstone, conglomerate; 5—dacite; 6—conglomerate, conglobreccia; 7—undifferentiated granitoids of variable composition; 8—crystalline schists, marbles; 9—steeply dipping tectonic faults; 10—inter-formation disruptions; 11—interfacial ore-interfering faults; 12—ore bodies

The main ore-bearing structure is a gentle shear 2, which is accompanied by zones of intense fracturing.

The ore zone, located in the central part of the deposit, is traced in a northwestern direction along fault 10a for 400 m at a depth of development of economic ores of 130–23 m from the surface. The thickness of ore intersections ranges from 0.5 to 5.2 m, and the grade ranges from 0.124 to 0.377% with an average grade of 0.149%. Economic uranium mineralization is accompanied by an aureole of non-economic ore, more widely developed in the southeastern part of the deposit.

According to the local conditions of mineralization, the northwestern body is similar to the first ore body, but is smaller in size. It is also confined to the gentle fault 2 in the area where it adjoins the steeply dipping fault 8, which bounds the horst-shaped projection of the basement. Standard mineralization was studied by a network of 100 × 100 m wells. The thickness of ore intersections varies from 0.9 to 2.1 m, and the grade ranges from 0.062 to 0.163%. The average grade in the reserve block is 0.099%.

In general, ore-bearing zone II is characterized by a predominance of monoelement average and low-grade ores with a relatively consistent distribution both along strike and dip.

Ore-Bearing Zone III is located in the extreme northwestern part of the deposit and is controlled by tectonic sutures 10 and 8. Uranium mineralization is localized in basal conglomerates in the form of a sheet-like deposit measuring 100 × 400 m, which has a complex internal structure. The ore-bearing horizon, which belongs to the lower lithological-structural level of mineralization deposition, lies directly on the crystalline rocks of the basement and is represented (from bottom to top) by coarse clastic conglobreccias, boulder-block conglomerates, sandstones, tuff gravels and silty sandstones, filling a graben-like structure with a north-western orientation. Terrigenous sedimentary rocks are overlain by layers of acidic volcanics.

The main ore-bearing structural elements are gently dipping (10–15°) interfacial failures, zones of layered fracturing near tectonic suture 8, and areas of development of "dry" breccias

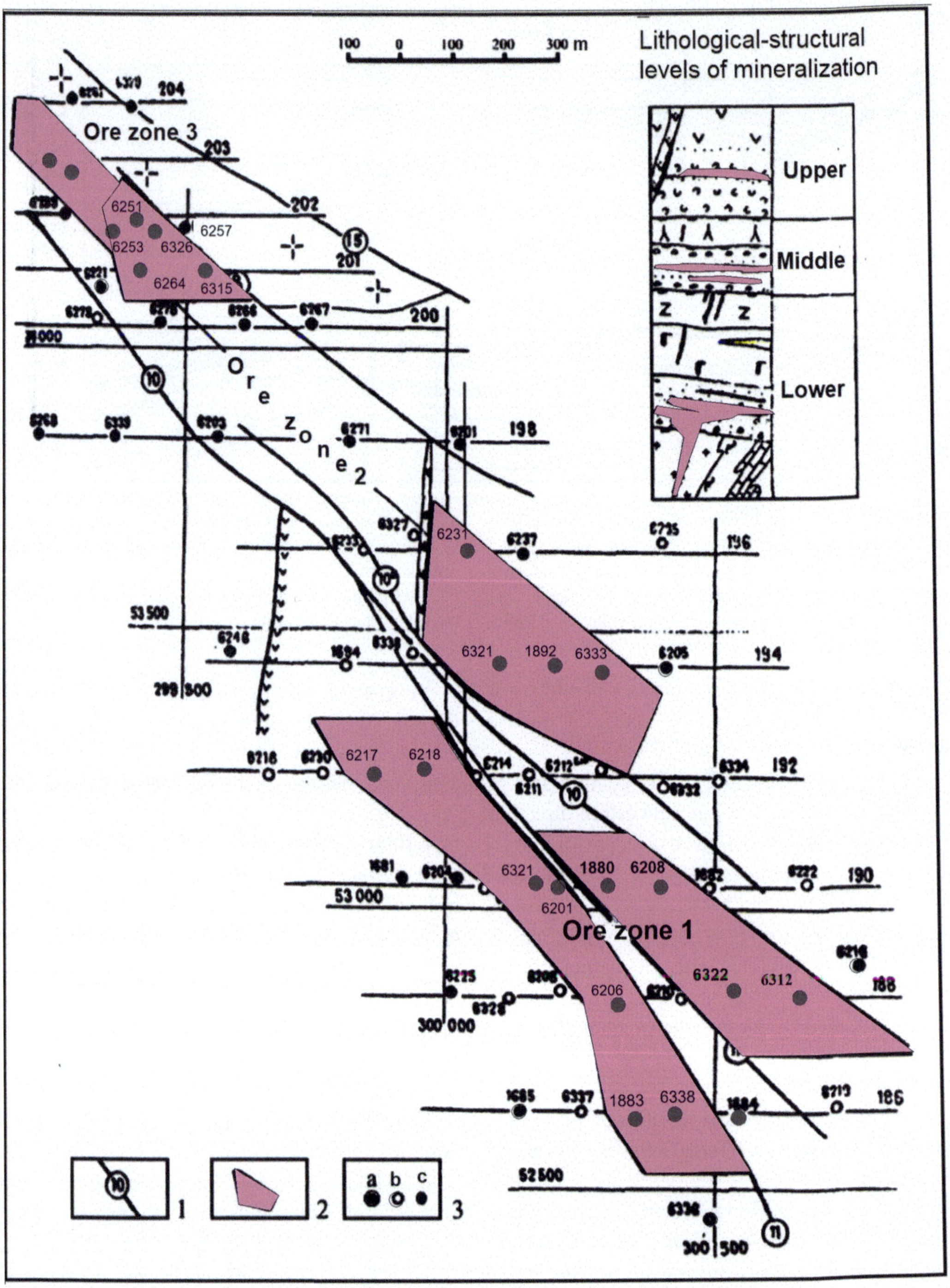

Fig. 4.29 Summary projection of ore deposits of the Nemer deposit onto a horizontal plane (According to Mironov et al. 2012): 1—steeply dipping tectonic faults; 2—contours of reporting ore-bearing blocks; 3—wells that discovered economic (a), non-economic (b) mineralization and barren (c)

above the swell-like protrusion of basement granitoids. Economic mineralization is developed in the depth range of 200–300 m from the surface and tends to gradually pinch out along the rise in a northwestern direction. In the southern part, the ore zone has the form of a flattened stockwork, characterized by a combination of small subhorizontal ore beds with steeply dipping vein-like bodies. To the northwest, the morphology of the zone simplifies to a distinctly sheet-like structure (Table 4.13). The thickness of economic ore intersections ranges from 0.3 to 9.6 m with a grade of 0.037–0.705%. The average uranium grade (taking into account the accepted cut off) is 0.153%.

The richest ores are localized in the southeastern part and are confined to zones of hydrothermal breccias.

Unlike other ore-bearing zones, ore-bearing zone III is represented by complex molybdenum-uranium ores. Molybdenum mineralization is closely related to uranium mineralization and in most cases does not extend beyond the uranium ore accumulations. No independent ore bodies with molybdenum ores have been identified within the deposit. Molybdenum is characterized by an extremely uneven distribution of concentration in different parts of the ore-bearing zone with a clear correlation with the uranium content. The highest molybdenum grades (up to 0.497%) are characteristic of rich uranium ores. Molybdenum grades in ore intervals range from 0.005 to 0.454% with an average of 0.185%.

Along with uranium mineralization localized in the rocks of the upper structural floor, small vein-like ore bodies were identified in the rocks of the crystalline basement at the Nemer deposit. So, at a depth of 280 m hole No. 6253 in exploration line 202, economic uranium mineralization with a thickness of 3.9 m and a grade of 0.083% in leucocratic granites was discovered. Economic mineralization is accompanied by a wide aureole of non-economic ore, localized in a steeply dipping zone of tectonic crushing and cataclasis with intense low-temperature hydrothermal-metasomatic changes. The ore was encountered 35 m below the roof of the foundation and cannot be traced to depth (Fig. 4.30). At deeper horizons, as well as in neighboring wells, manifestations of uranium mineralization are noted in the form of non-economic thin ore intervals and radioactivity anomalies present to a depth of 400 m (140 m from the roof of the foundation).

Mineral Composition of Ores and Alterations. According to the mineral composition and type of hydrothermal-metasomatic alterations in the host rocks, the ores of the Nemer deposit belong to the pitchblende-coffinitic mineral type in argillicites. There are several stages and stages of rock alteration. The first, pneumatolithic-hydrothermal stage is caused by the granitization of the pre-Late Mesozoic substrate. K-feldspathization, greisenization, and albitization are typical. Late Mesozoic pre-ore alteratios are associated with post-volcanic (autometamorphic and, possibly, fumoral-solfatar) processes. Two stages are clearly distinguished. The earliest high-medium-temperature hydrothermal-metasomatic is manifested by areal K-feldspathization, beresitization, propylitization, silicification, albitization with an increase in their intensity in subvolcanic varieties of acidic volcanics.

The second stage of this alteration is caused by the processes of low-temperature acid leaching, which manifested itself in the field area in widespread hydromicaization of sedimentary-volcanogenic rocks. A characteristic feature of hydromica formation is the predominant development of hydromicas and mixed-layer minerals with a subordinate role of kaolinite, montmorillonite, chlorite and quartz. The halos of these alterations usually extend beyond the boundaries of the deposit.

The ore stage appeared at the Nemer deposit in three successive stages: pre-ore, actual ore and post-ore. Low-temperature pre-ore alterations resulted in the formation of quartz, chlorite (chamosite), fluorite, carbonate, hydromica, montmorillonite, kaolinite and hematite. The noted mineral association is distributed locally and gravitates mainly to ore-hosting zones. The sizes of halos of pre-ore alterations are

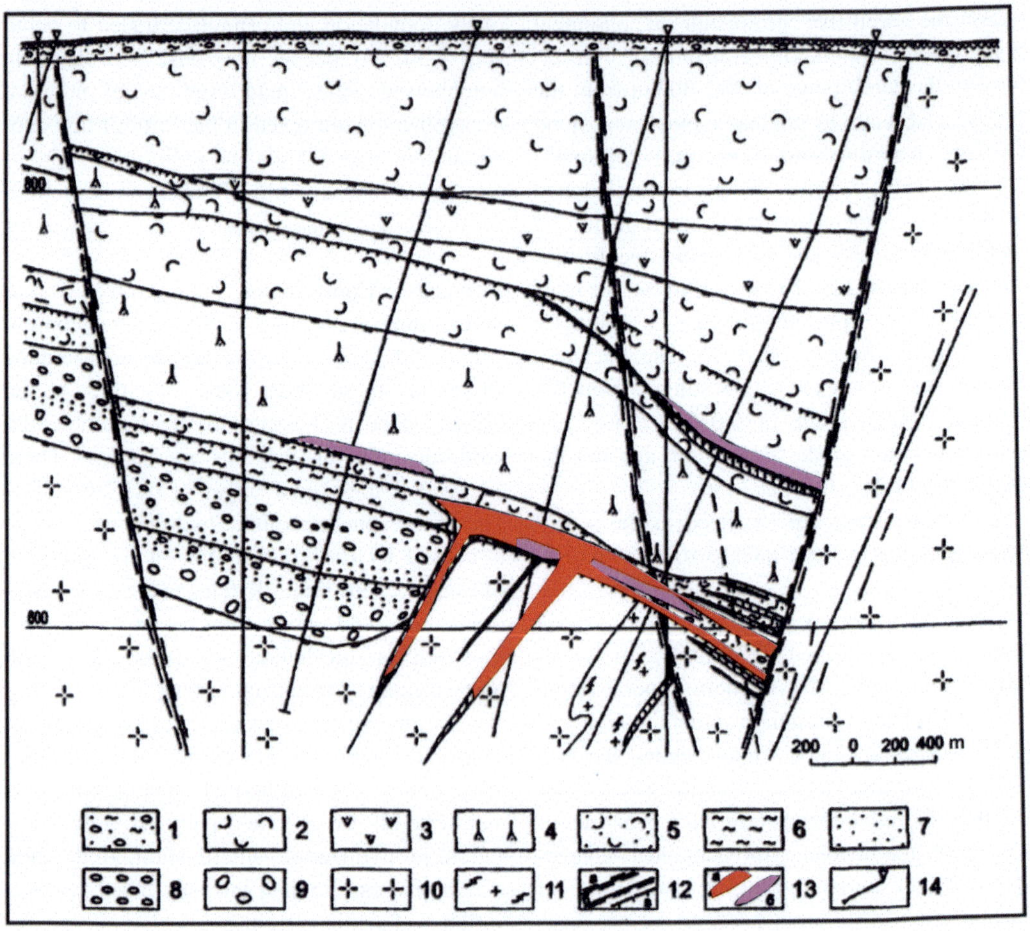

Fig. 4.30 Geological section along line 201 of the Nemer deposit (according to Mironov 2003): 1—modern deposits; 2—sanidine trachyrhyolites; 3—felsic porphyry; 4—dacites, andesidacites; 5—tuffs of felsites, rhyolites; 6—siltstones, siltstones-sandstones; 7—sandstones, gravelites; 8—small pebble conglomerates; 9—large-block conglomerates; 10—granites; 11—granite-gneisses, gneisses; 12—faults: steeply dipping zones of crushing and brecciation, main (a) and secondary (b), gentle faults (c); 13—ore bodies with economic uranium (a) and non-economic (b) mineralization; 14—boreholes

usually an order of magnitude larger than the sizes of ore bodies. In the section, these alterations do not extend beyond the ore bodies, with the exception of montmorillonite and kaolinite, which are mapped in the form of narrow linear halos along the ore-controlling steeply dipping tectonic sutures. The halos of near-ore hematitization are closest to the uranium mineralization, but do not extend beyond the contours of the ore bodies and give the ore zones a specific reddish-pink color.

The ore stage is recorded by the deposition of uranium minerals—coffinite and finely dispersed pitchblende in association with sulfides: pyrite, galena, molybdenite, marcasite, arsenopyrite, sphalerite, and occasionally chalcopyrite. Molybdenum mineralization with varying intensity is observed in almost all ore bodies of the Nemer deposit, including basement rocks. However, economic accumulations were established only in body 3.

The post-ore stage of mineral formation is characterized by the widespread development of thin veinlets of quartz, carbonate, fluorite, and kaolinite. In the oxidation zone, which occurs at the deposit along steeply dipping faults to a depth of 400 m, the ores contain uranophane, betauranotyl, and in isolated cases curite, which form thin veins, crusts, and small accumulations. Along with uranium minerals, brown iron hydroxides and sooty deposits of manganese are present.

The age of hydrothermal uranium mineralization, according to geological data, is Early Cretaceous. The absolute age of uranium ores, established in the IGEM laboratory using the lead isotope method, is 136–153 million years (J_3–K_1).

In the economic ores of the Nemer deposit, the main productive minerals are coffinitite, pitchblende, and molybdenite.

Pitchblende-coffinitic ores with sulfides form sheet ore bodies in tuffaceous-sedimentary and volcanic rocks of ore deposits in ore-bearing zones 1 and 2. Coffinitite here sharply predominates over pitchblende. The amount of sulfides in the ore is 1–2%. Nasturanium-coffinitite ores with molybdenite form sheet-like ores that transform into vein-like ore bodies in ore-bearing zone 3. Molybdenum mineralization is closely related to uranium and in most cases does not extend beyond the uranium ore intervals. In the overall balance of mineral reserves of the deposit, molybdenum-uranium ore aggregates account for 21%. Uranium ores are primarily composed of an aggregate of fine and fine-grained, and to a lesser extent, finely dispersed uranium minerals with sinuous and rounded boundaries. Unit sizes range from 0.001 to 2 mm. The textures of the ores are disseminated, dispersed-disseminated, spotted-disseminated, nested-disseminated, veined-disseminated, less commonly banded disseminated and earthy. Breccia and cement textures are developed in rich ores. The structure of the ores is colloform, corrosive, emulsion, granular, looped, reticulate, spherulitic, entangled fibrous and lamellar.

Chemical Composition of Ores. In terms of chemical composition, the uranium ores of the Nemer deposit are of the aluminosilicate type with a low content of acid components. Molybdenum, lead, zinc and silver present in ores in high concentrations are of no economic importance. In the ores of ore-bearing zone III, molybdenum is of economic importance (Table 4.14).

Geochemical Characteristics of Rocks and Ores. In the geochemical field, the deposit is manifested by the widespread development of halos of uranium and satellite elements (lead, arsenic and molybdenum) of uranium mineralization. Primary uranium halos are significantly larger in size than the ore deposits, and their configuration is determined by the morphology and geological and structural features of the localization of uranium ore deposits. The halos of all indicator elements, as a rule, have an elongated shape and are controlled by fault tectonics. On the surface, contrasting halos exceed the size of the ore deposits by 2–3 times. The most extensive and contrasting are the halos of uranium. Within the deposit along strike, they are present more than 500 m from the ore deposits. The halos of indicator elements individually are inferior in all respects to the halos of uranium. Multiplicative halos look more contrasting. Secondary halos are comparable in size to the primary ones, but are displaced relative to each other.

Uranium, lead, arsenic and molybdenum in the central part of the deposit have a high level of accumulation and are related by a positive correlation. The maximum intensity halos of indicator elements are mapped in areas of development of rich uranium ores.

Reserve Status. According to a preliminary assessment, the Nemer uranium deposit was determined to be a small deposit in terms of reserves with average ore quality. The overall estimate of the deposit's reserves, taking into

Table 4.14 Chemical composition of Mardaingol deposit ores, % (Mironov et al. 2012)

SiO_2	69.36	CaO	0.46	F	0.08
Al_2O_3	15.35	MgO	0.73	CO_2	0.74
Fe_2O_3	0.72	MnO	0.05	Stotal	0.186
FeO	1.30	K_2O	3.23	Loss on ignition	4.61
TiO	0.43	Na_2O	3.12		
P_2O_5	0.11	H_2O	2.15		

account the predicted resources of category P, is 4000 tons. The identification and assessment of the reserves of the Nemer field make it possible to increase the overall estimate of the Mardaingol block from 5500 to 6000 tons (by 8%).

The ores of the Nemer deposit are monometallic and complex. Complex molybdenum-uranium ores are established only at the lower lithological-structural level of mineralization and account for 21% of the total reserves of the deposit. Average grades in complex ores: uranium—0.153%, molybdenum—0.130%.

It is believed that the mineral resource base of the Dornod uranium ore cluster can be increased by 15–20 thousand tons with additional exploration of already explored deposits. The Dornod and Gurvanbulag fields are prepared for economic development. In addition to uranium mineralization, deposits of polymetals, gold and fluorite have been studied within the Dornod ore district.

4.6 Kharaat Deposit

The Kharaat uranium deposit is located in the southern part of the Choir depression. The deposit covers an area of about 30 km^2.

Administratively, the Kharaat deposit site is located in the Middle Gobi aimag, 320 km southwest of the city of Ulaanbaatar, 50–60 km west of the trans-Mongolian railway.

The Kharaat deposit was discovered by Russian geologists in 1988 when conducting a 1:25,000 scale aerial gamma spectrometric survey in the area. During the same year, the discovery was preliminary assessed by drilling and geotechnological properties of the ores samples were studied in laboratory conditions. However, further exploration of the deposit was stopped due to the cessation of funding for geological exploration work for uranium in Mongolia.

Exploration of the deposit commenced again in 1994, within the framework of the joint Mongolian-Russian-American enterprise project “Gurvansaikhan”. Exploration consisted of a wide range of geological studies, including delineation and hydrogeological drilling, and core sampling. Metallurgical and hydrogeological studies in wells and laboratory and field experiments on uranium leaching, and analytical studies.

The Choir depression is located in the northern part of the Gobi region. It is a linear depression structure of submeridional strike with a length of 150 km and a width of 10–25 km. Geomorphologically, it is a gently rolling plain with absolute elevations of 1100–1150 m, surrounded by mountainous elevations that exceed the depression by 100–200 m.

The depression is filled with weakly lithified and unlithified deposits of the Lower and Upper Cretaceous with a total thickness of about 1500 m. In the composition of the Lower Cretaceous formations belonging to the Zuunbayan formation, two facies successive complexes are distinguished—variegated and gray-colored. The first is represented by poorly sorted gravelites, small pebble conglomerates, sandstones and siltstones, developed along the edge of the depression and belonging to proluvial-deluvial deposition. The second, gray-colored complex corresponds to alluvial and lake-marsh formations, which include clays, clayey sandstones, sands with carbonaceous detritus, lenses and layers of brown coal with scattered iron sulfides. Gravelites and conglomerates are of subordinate importance here. Several lignite deposits are localized in the Lower Cretaceous deposits, the largest of which, the Shivee-Ovoo deposit, is currently being exploited.

Upper Cretaceous sediments, thickness which varies between 5 and 40 m, are deposited in small areas in the marginal parts of the depression. They are represented by variegated (often yellow-colored) coarse-grained sand and gravel formations, sometimes cemented at the base by limonite-goethite material.

Based on the results of geological exploration work, 11 ore deposits were identified at the Kharaat deposit, two of them (No. 1, 2), the most significant in terms of uranium resources, were explored in 1996. Exploration drilling of wells, in deposits No. 1 and 2, were on a grid of 100–200 × 50–100 m with in-fill drilling up to 50×25 m. The remaining deposits were assessed by drilling on a grid of 400–800×100–200 m. Drilling depths varied from 20 to 80 m and averaged 40 m. During the entire period of work in the field, about 40,000 linear meters of exploration wells, more than 1200 m hydrogeological wells and over 1500 m technological wells were drilled.

The Kharaat ore deposits are concentrated in a river paleovalley stretching along the Choir depression for more than 20 km (Fig. 4.31).

Mineralization is primarily located in the central part of the valley, occupying a segment of 5 km in length. Uranium mineralization is localized in sediments of channel and floodplain facies (Figs. 4.32, 4.33 and 4.34), which are represented by alternating inconsistent horizons of gravelites, fine to coarse-grained sands, silts, clays and their transitional varieties. As a rule, these are loose, weakly lithified rocks containing large amounts of carbonaceous fragments and dispersed organic detritus, as well as fine dissemination of pyrite and marcasite.

Clays and silts are of subordinate importance, their thickness varies from several tens of cm to 5 m. At the surface, all of the rocks are oxidized. The oxidation zone is characterized by a complex structure and irregular thickness, varying from 1–2 to 25–30 m. According to the nature of the changes, it is divided into two subzones: the upper with continuous limonitization of rocks (surface oxidation) and the lower with sporadic limonitization in permeable rocks (ground oxidation). Ores are concentrated mainly in areas of pinching out oxidation zones or within oxidation zones in the remains of weakly oxidized and non-oxidized gray sediments.

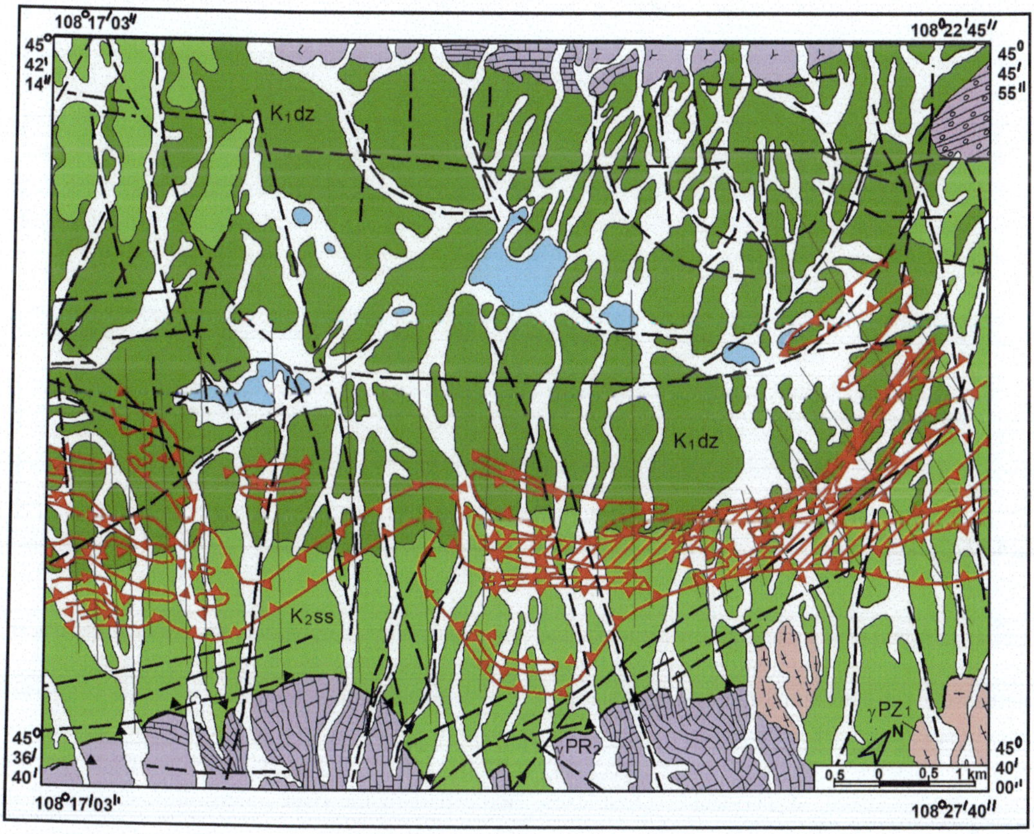

Fig. 4.31 Geological structure of the Kharaat deposit (Budunov and Jamsrandorj 2012)

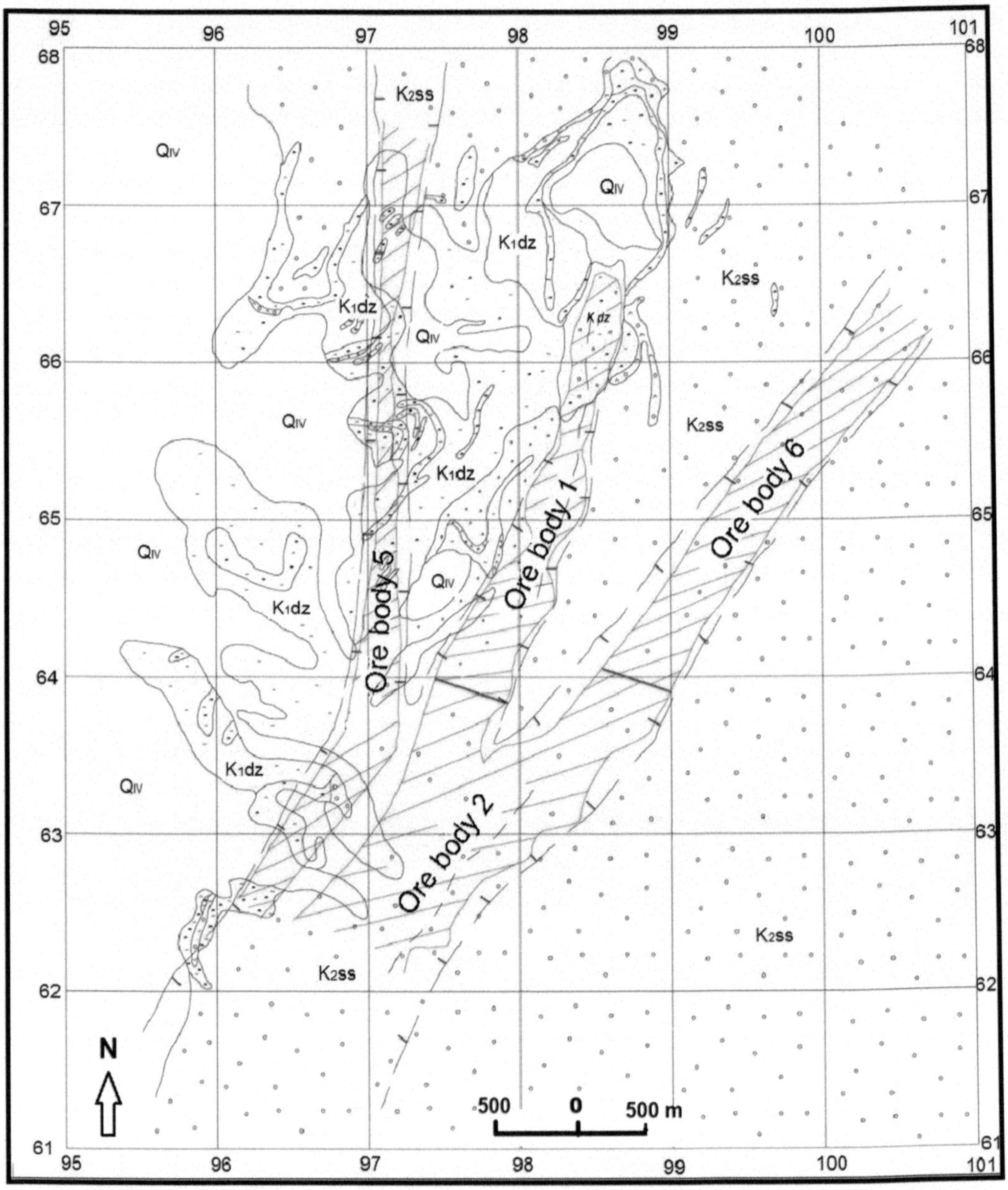

Fig. 4.32 Location of ore bodies at the Kharaat deposit (Budunov and Jamsrandorj 2012)

The ore deposits have a complex ribbon-lens-like and sheet-like morphology and are elongated in the submeridional direction. Economic mineralization is framed by a wide aureole of out of balance ores. Most of the ore accumulations are localized in permeable complexes (Figs. 4.32, 4.33 and 4.34).

The main parameters of mineralization of explored deposits are given in Table 4.15.

Uranium mineralization is formed at one absolute hypsometric level in the interval of 1060–1120 m. Spatially, it is confined to the axial part of the river paleovalley, expressed in

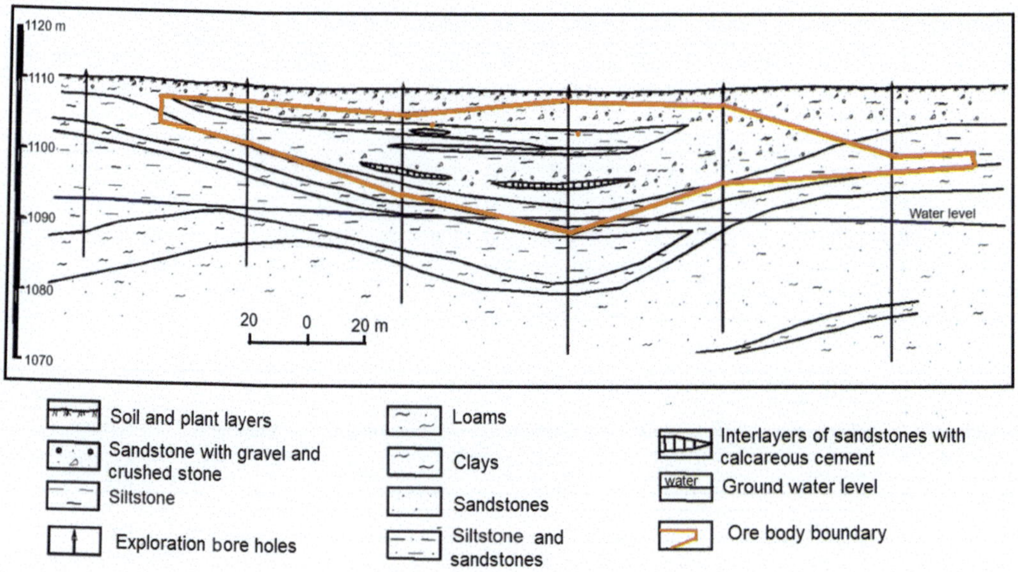

Fig. 4.33 Section through ore body 1. Line RL-58 (Budunov and Jamsrandorj 2012)

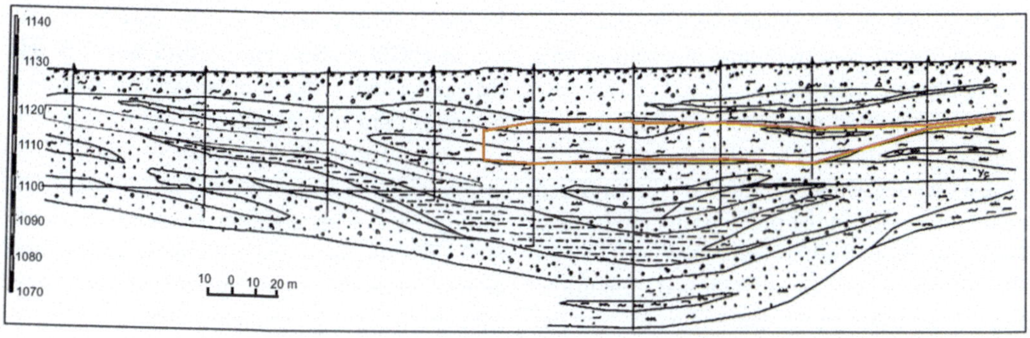

Fig. 4.34 Section through ore body 2. Line RL-24. See legend Fig. 4.34 (Budunov and Jamsrandorj 2012)

the form of an asymmetrical V-shaped subsidence of the paleorelief.

The ores of the Kharaat deposit are classified as aluminosilicate, non-carbonate. The organic carbon grade is 0.65–1.25%.

Two mineralogical and geochemical types of uranium mineralization are found at the deposit: "oxide", where uranium is contained in a tetravalent form, and "mica" with hexavalent uranium. The "oxide" type makes up the main part of the deposit. It is represented by finely dispersed uranium oxides (pitchblende) with subordinate amounts of coffinitite and ningioite. "Mica" type mineralization, developed mainly in the zone of surface and ground oxidation, was formed due to the transformation of primary "oxide" ores. Secondary mineralization is represented by phosphate uranium micas and uranium hydroxides.

Uranium mineralization at the deposit is accompanied by increased concentrations of cerium, lanthanum, scandium, yttrium, ytterbium, rhenium and selenium (Table 4.16). According to quantitative spectral analysis of core samples, the following contents of these elements in the ore (g/t) are noted.

Table 4.15 Dimensions of ore bodies at the Kharaat deposit (Budunov and Jamsrandorj 2012)

No. ore bodies	Depth (m) occurrence roofs from surfaces from–to average	Length on the strike (m)	Wide (m) from–to average	Thickness (m) from–to average	Uranium grade (%) from–to average
1	0.15–22.4 6.5	3000	100–600 300	0.6–42.5 10.2	0.01–0.247 0.036
2	4.9–36.6 14.5	2700	300–1300 800	0.7–25.5 6.0	0.01–0.147 0.030

Table 4.16 Rare earth grade (Budunov and Jamsrandorj 2012)

Scandium	Rhenium	Selenium	Yttrium	Lantanium	Ytterbium	Serium
6.5–33.2 11.0	0.3–4.7 0.6	1.0–359 22.0	7.9–158 31.0	17.0–195.5 53.0	0.9–105.6 2.9	85–850 105

Note In the numerator—fluctuations in the grades of elements from–to, and the denominator is the average

The contents of associated useful components generally slightly exceed the background ones. Their industrial concentrations in uranium ores have not been established.

During exploration work in the field, hydrogeological studies and full-scale metallurgical tests were conducted. Most of the explored uranium reserves in the ore deposits are located above the groundwater level and only about 20% of them are in a saturated condition. Significant lithological heterogeneity of the ore-bearing strata predetermined the complex hydrogeological conditions at the deposit. The water conductivity of rocks varies widely from 6–50 to 200–250 m^2/day. The filtration coefficient ranges from 1 to 10 m/day. Low permeable rocks and aquicludes are noted. Based on research results, the hydrogeological conditions in the field were determined to be favorable for its development using a in situ leaching method.

The chemical composition of the groundwater is sulfate with a mixed composition, where the magnesium cation predominates. The waters are salty with a concentration of 3–4.5 to 7 g/l and have high total hardness (40–70 mEq/l).

The study of geotechnological parameters of uranium leaching from ores was conducted both in laboratory and in field conditions. Laboratory studies were conducted in the following labs; Krasnokholmsky production-geological organization (PGO Uzbekistan), VIMS and VNIIHT (Moscow) and in the Priargunsky Gas Chemical Complex (Krasnokamensk).

Laboratory studies established a fairly high processability of ores using weak solutions of sulfuric acid with uranium recovery exceeding 80%. The uranium concentrations in solution, depending on the initial content in the ore, range from 40 to 100 mg/l or more.

Metallurgical research in field conditions was carried out in two stages within the first ore deposit.

At the first stage (1994), a two-well experiment was conducted for the purpose of preliminary assessment of geotechnological parameters in field conditions to decide on the feasibility of pilot tests.

As a result of the work carried out, the fundamental possibility of sulfuric acid in situ leaching of uranium in both saturated and anhydrous conditions was established. In the saturated part of the ore deposit, the maximum concentrations of uranium in productive solutions were 165 mg/l, with average values from the beginning of the experiment being 80 mg/l.

This was the first time that an experiment in underground uranium leaching was carried out in anhydrous rocks. Such experiments have not been carried out anywhere before. At the first stage of the experiment, artificial flooding of the ore zone was carried out by pumping water into wells and raising the level, and subsequently—leaching of uranium according to the usual scheme. As a result of the experiment, the possibility of implementing lateral filtration of the solution was decided and industrial concentrations of uranium in productive solutions were obtained, which reached 85 mg/l.

In 1996, geotechnological studies of the second stage were conducted at the Kharaat deposit in a semi-industrial complex. A special experimental technological complex was equipped which included two mining zones in the saturated and anhydrous parts of the ore deposit. The processing plant, provided technological processing of productive solutions, desorption of uranium using ammonium carbonate salts and obtaining finished product in the form of wet AUTC crystals (ammonium uranyl tricarbonate).

In 1997 and in 2008–2011 additional exploration work was conducted on the flanks of the Kharaat deposit and in adjacent areas. As a result new ore zones and uranium ore deposits were identified. The results of this exploration indicate real prospects for increasing the uranium reserves of the Kharaat deposit.

4.7 Dulaan Uul Deposit

The Dulaan Uul uranium deposit is located 60 km northwest of Ulaanbadrakh soum. The deposit belongs to the Zuunbayan ore district of the South Mongolian metallogenic province (Jamsrandorj et al. 2019).

Stratigraphically, the Dulaan Uul deposit is associated with the Sainshandin formation of the Upper Cretaceous in the Unegt depression. The Upper Cretaceous sediments have become widespread, which are underlain by Upper Jurassic sediments of the Sharil formation. Having studied the deposits of the Sainshandi formation, French geologists Fiette and Olivier Parise divided it into upper and lower strata, corresponding to the subformations of the stratigraphic classification of Mongolia.

The upper part of the Sainshand Formation (K_2ss^2) is the main uranium-bearing deposits, consisting of coarse- and fine-grained sandy rocks of the channel facies, having a thickness of 200–300 m. In the upper section, the thickness of clays and siltstones of yellow, red and gray colors increases and lignite interlayers with a thickness of up to one cm and contains iron nodules.

The conditions for the formation of rocks of the upper subformation of the Sainshandin formation began with the river alluvial system and gradually moved to oxbow lakes with an overdominance of deposition. Channel sand rocks are less developed.

During geological exploration at the Dulaan Uul deposits, the formations of the Sainshand formation were divided into 4 lithological variations, which the Mongolian Stratigraphic Commission recognized as the minimum unit of the stratigraphic unit of Mongolia—bed (member) (Fig. 4.35).

- Bed 1 or base layer (U1): interbedded mudstones and siltstones, coarse sands with an admixture of fine- and medium-grained sands and gravels. Average thickness 40 m.
- Bed 2 or lower layer (U2): coarse sands with thin interbeds of clayey siltstones. Thickness 60–80 m. Research in the paleospheric laboratory of Bernard University established that these deposits were formed in open lacustrine conditions with rich vegetation and were attributed to the Lower Cretaceous (Aptian-Albian).
- Bed 3 or middle layer (U3): sandy-clayey unstable heterogeneous sediments of channel and floodplain facies. The thickness of the clay layers increases upward. The total thickness of the unit ranges from 60 to 110 m.
- Bed 4 or upper unit (U4): consists primarily of fine-to-medium-grained sandy channel facies units. The thickness of the sandy components decreases upward and the thickness

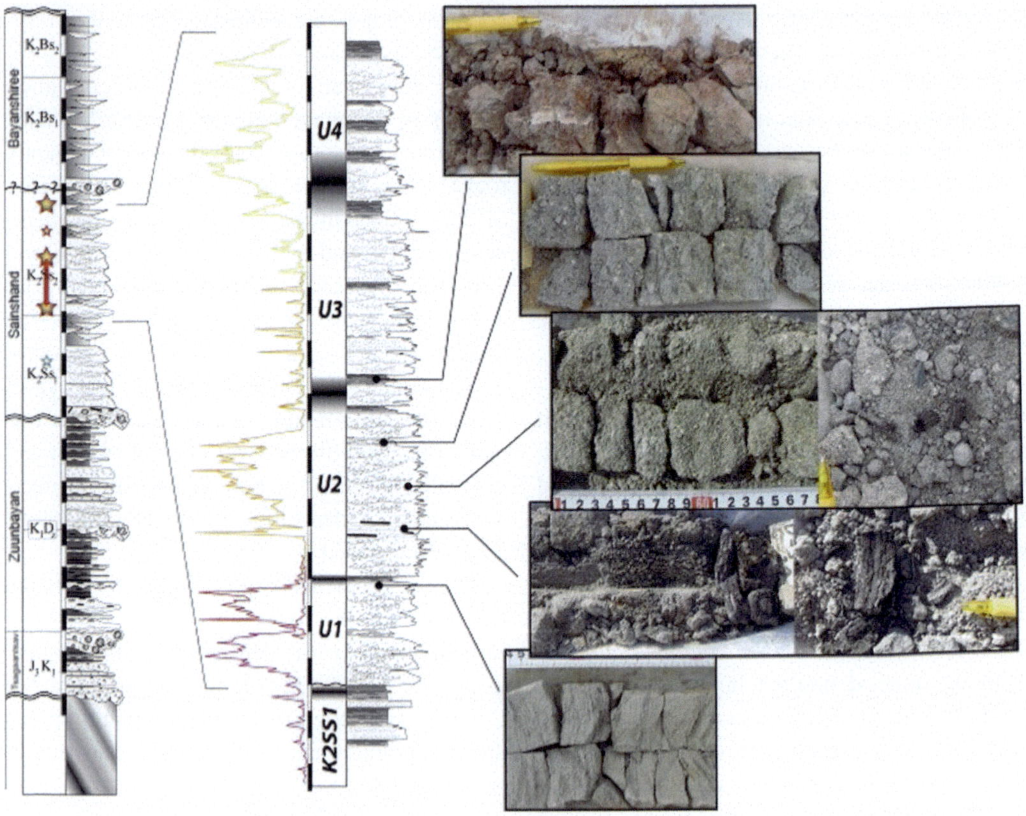

Fig. 4.35 Stratigraphic units of the upper member of the Sainshand formation (Enkh-Amgalan et al. 2011)

of the deposits of the nominal facies increases. The thickness of the member varies greatly and in the central part of the Unegt depression is 70 m, in its northern section it reaches 110 m, and within the Myagmar body it is about 40 m.

In terms of the structure of ore deposits, the quality of mineralization and the size of sheet-like and lens-shaped bodies, the Dulaan Uul deposit is a complex geological structure. The ore bodies at the deposit are: Myagmar, Australia, South-Western and South-Eastern (Fig. 4.36). The Myagmar ore body is located in unit-4 and the rest are located within bed-2.

Myagmar Ore Body: Located in unit 4, located closer to the surface. Up to 15 blocks with a thickness of 1.5–18.5 m have been defined. The shapes and sizes of the ore bodies, common at the contacts of the redox zones are shown in the example of the exploration line-73 section (Fig. 4.37).

The Australian Ore Body: Localized mainly in member-2, occupies an area of about 695.7 thousand m^2 and has a shape elongated in a latitudinal direction. The thickness of the eleven (11) ore blocks varies from 1.5 to 8.5 m. Several ore blocks were identified in bed-3.

South-Western Ore Body: Bed 2 is the ore-bearing unit. The total area is 1702 thousand m^2, stretching from north to south. The thickness of the oreblocks ranges from 1.6 to 7.5 m. Single ore blocks were identified in bed-3.

South-Eastern Ore Body: The main ore-bearing horizon is bed-2. The smallest ore blocks in the ore deposit occupy about 8 thousand m^2. The

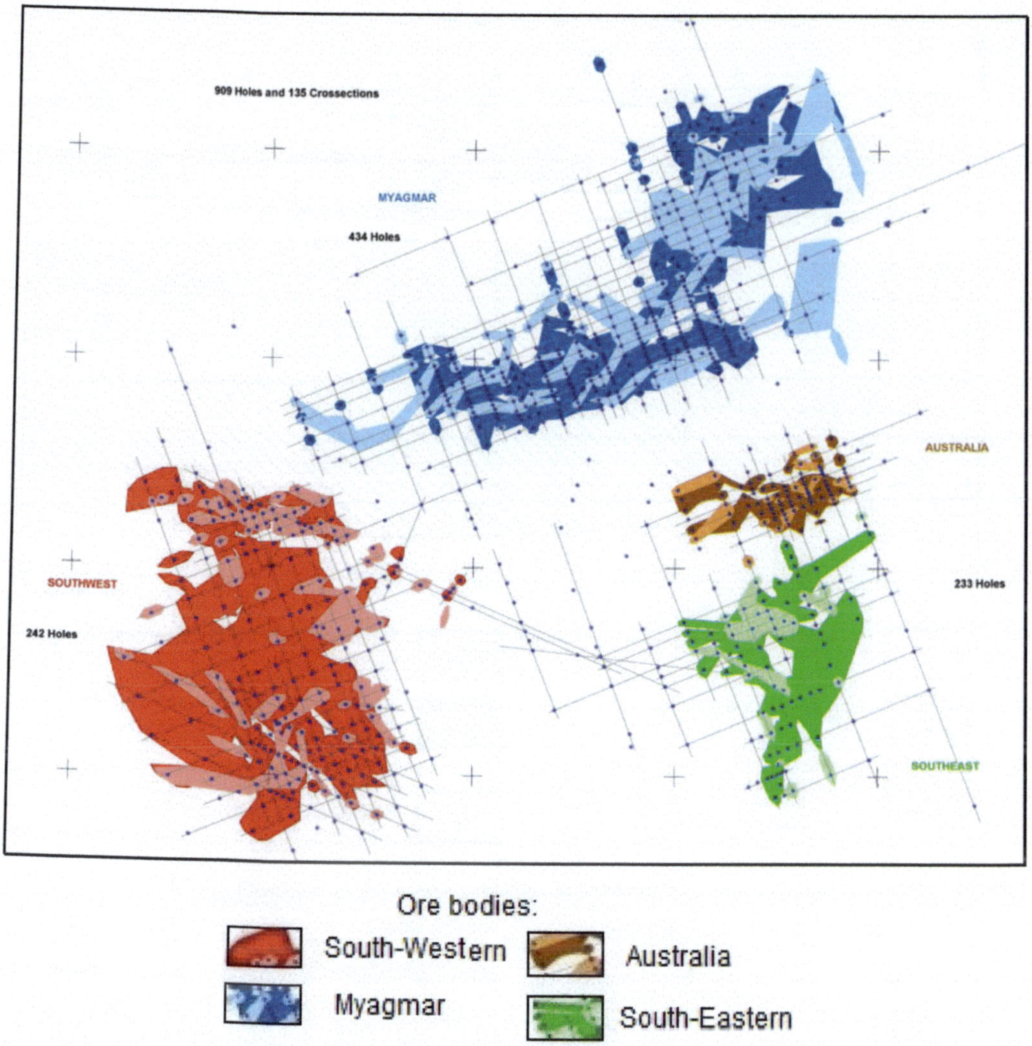

Fig. 4.36 Layout of ore bodies of the Dulaan Uul deposit (Jamsrandorj et al. 2013)

ore deposit has a thickness of 1.55–20.0 m and there are approximately 6 ore blocks. About 4 blocks are localized in bed-3.

The redox character of groundwater is not depend on rock facies and, for example, in the oxidizing environment of unit-2, both oxidizing and reducing waters are recorded, in unit-4, both oxidizing and reducing waters are also recorded within the reducing environment. Based on this data it can be concluded that the current state of hydrochemistry, and the placement and relationship of redox fronts indicates that there is no definite connection between them, therefore they have been preserved only as relics or "fossils". In this regard, the deposition of uranium occurred only due to redox processes.

Uranium mineralization is found in pyrites, organic material and silicates, confirming their epigenetic origin. Uranium is concentrated in fragments of epigenetically modified lignites, often pyritized, in titanium and iron oxides in a framework structure, rarely adsorbed in lignites, and in uranium sulfates (schroeckingerites), which emit ultraviolet rays during fluorescence.

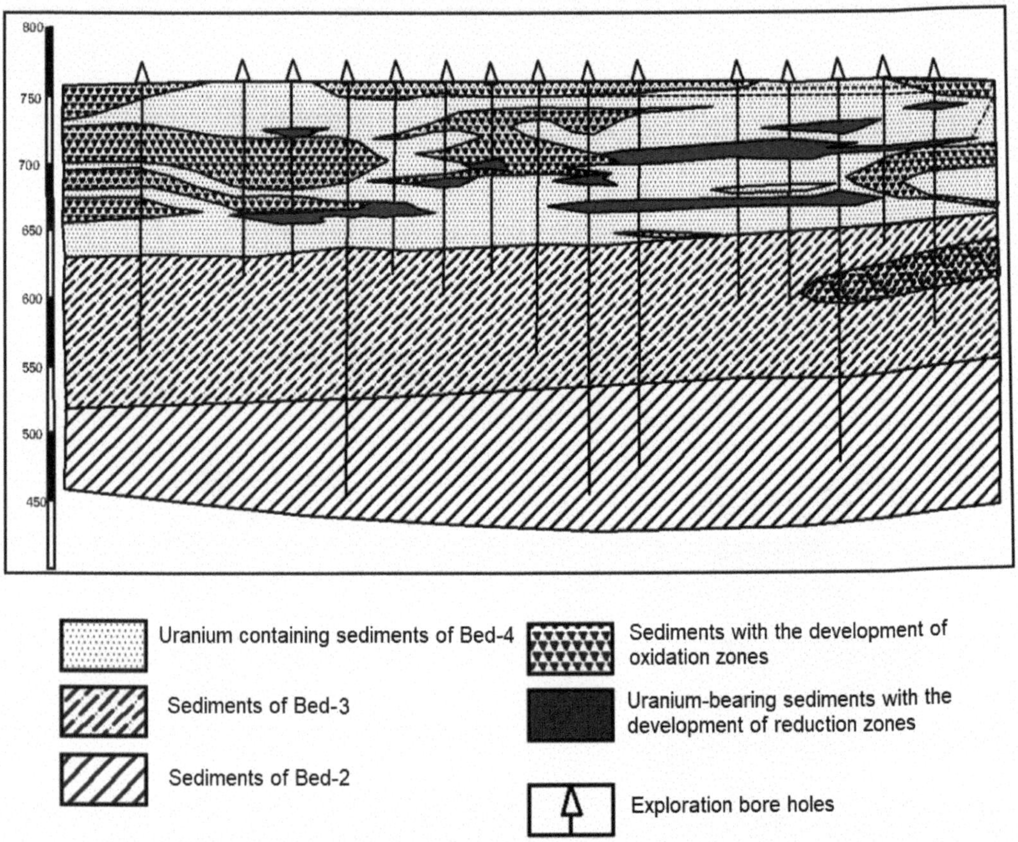

Fig. 4.37 Section of uranium mineralization of bed-4 in the Myagmar ore body (Jamsrandorj et al. 2013)

Uranium sulfate is formed in the presence of water, and this indicates that, in this case, the water contains free uranium.

The Dulaan Ul deposit is a representative of a hydrogenous uranium deposit of the "roll-front type", formed at the boundaries of oxidizing and reducing rocks. When searching for deposits of this type, hydrogeological, sedimentological, geochemical and structural studies play an important role.

Hydrogeological Conditions, Genesis of the Deposit

Based on the results of hydrogeological and hydrochemical studies, it was assumed that mineralization in the upper part of the Sainshandin formation and redox processes occurred in several stages:

Stage I Oxidized waters passed through the organic-rich sediments of the Sainshandin Formation with reducing potential from the west, southwest to the east, southeast, where redox fronts (at the end of the Cretaceous) were formed in beds 1, 2 and 4.

Stage II Due to the uplift (Neogene) of the southeastern margin of the Zunbayan depression, a regime of oxidized flows was established in a north-northwest direction along the Zuunbayan fault, separating the Unegt from the Zuunbayan depressions. This regime was the source of the formation of a secondary front directed from the southwest to the northeast, accompanied by the appearance of low-grade mineralization and the redistribution of previous uranium mineralization.

Stage III The current phase of groundwater movement, in eastern part of deposit in gray-green color, opens the hydrogeological rarefaction funnel and sets the current flow temperature.

According to the study by Benedicto, phase A of groundwater movement is probably associated with a tectonic (Quaternary?) structure directed at an angle of 135° to the southeast. It is assumed that such structures regulate modern landforms. The movement of groundwater has a partial influence on the mineralization of sediment outcrops in unit 4 and its partial migration.

This preliminary conclusion requires clarification in future. The waters in the upper part of the Sainshangdin Formation have a sodium chloride composition and contain sulfate compounds. The waters are highly mineralized (solid grade 1–1.5 g/l), slightly alkaline (pH 7.5–8.5) and generally have a reducing character. A relatively oxidizing environment is observed in surface waters and some areas of bed-4.

Within the field, there are well-cemented layers less than 1 m thick and layers with limited material, which are associated with radiometric anomalies (> 3800 pulses/sec). This may be due to the fact that this organic or carbonaceous material, acting as a reducing agent, came from oil fields. When measuring cores from wells in the Tsagan Els oil field, no increased radioactivity was detected. But uranium-bearing deposits lie above oil deposits and it can be assumed that reducing carbonaceous substances could come from underlying oil-bearing horizons in the form of gases.

In the Unegta depression, dispersed uranium mineralization is intermittently concentrated in unit 4, which contains abundant organic material. In the underlying layers, redox fronts with uranium are observed in sands without organic materials. The sizes and shapes of these fronts follow modern landforms, and their morphology is partly subject to the direction of large faults of northern and southern orientation. It can be assumed that some semi-permeable faults significantly influence the formation of individual local ore bodies and deposits.

Reducing factors migrating from oil deposits in the Zuunbayan depression improved the conditions for uranium accumulation, and therefore the uranium content in ores increased.

These redox fronts appeared in Tertiary times, when crustal compression and relief formation occurred. The tectonic inversion of the Zuunbayan fault and the formation of the Zuunbayan anticline divided the Zuunbayan depression into several blocks and led to the formation of oil only in the Zuunbayan part of the depression. In this regard, in the Unegt and Zuunbayan depressions, different conditions for uranium concentration according to redox conditions appeared.

Mineral Composition. In highly mineralized zones, uranium occurs relatively "freely" than in low-mineralized areas, where uranium is contained in silicates and other minerals, which led to the following conclusion:

- In cracks and intergranular spaces, uranium mineralization is intense (UO_2>80%), and in some heterogeneous phases it contains elements such as Si, Ca, Ti, Fe, Al, Na. This mineralization is associated, with pyrite, and organic material and leads to epigenetic changes (Fig. 4.38).
- Lignite fragments pyritized and epigenetically replaced by uranium (Fig. 4.39a).
- Titanium and iron oxides of the frame structure are partially replaced by uranium (Fig. 4.39b).
- Uranium is less commonly adsorbed on lignites.
- Uranium sulfate schreckingerite, which reflects light under ultraviolet fluorescence rays.

Uranium sulfate forms in a short time in the presence of water, indicating the presence of free uranium (Fig. 4.39).

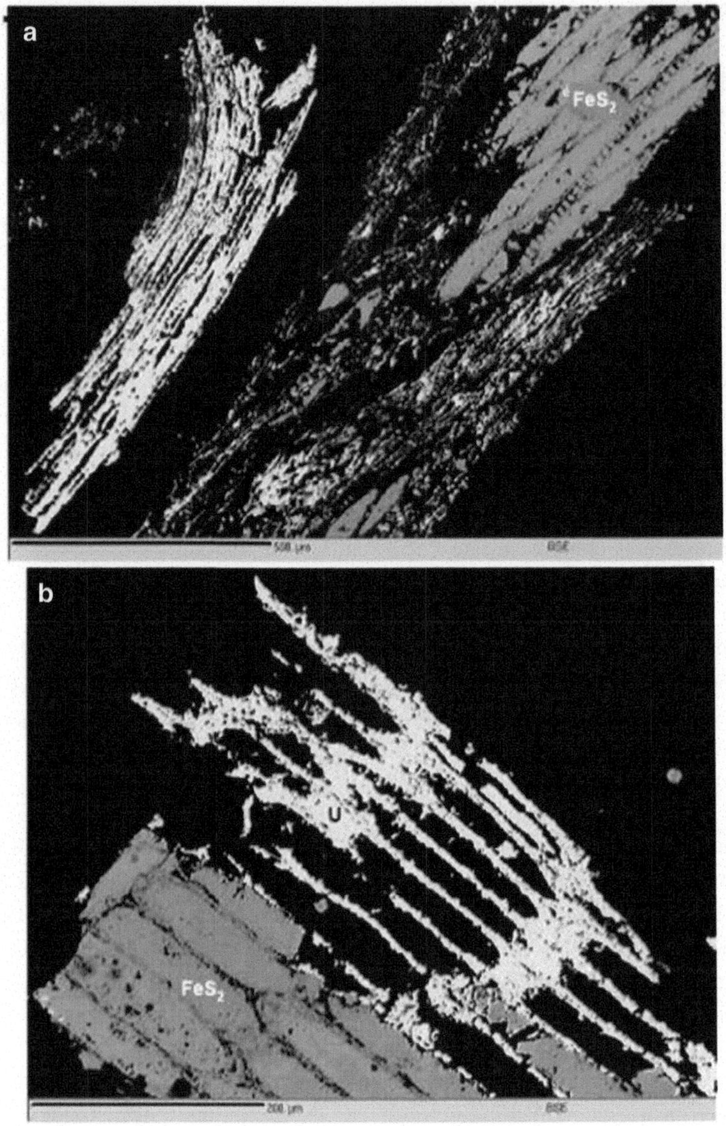

Fig. 4.38 Uranium mineralization in pyrite (light-uranium)

4.8 Zoovch Ovoo Deposit

The Zoovch Ovoo deposit is located in the Zuunbayan uranium district in the Zuunbayan depression, where deposits of the Upper Cretaceous Bayanshiree formation are widespread (Fig. 4.40).

The thickness of the sand layer within the main uranium deposit ranges from 35 to 40 m, and its composition is homogeneous and has medium-sized sandy grains and small organic remains. Redox boundaries in the stratification of the sedimentary layer are characterized by gray-yellow sand for the oxidation zone, grains of gray feldspar and coal fragments in the reduced zone. Pyrite and kaolinite in the oxidizing layer are present in association with rare hematite grains.

Reduced sandstone has higher radioactivity. Geological and hydrogeological conditions of

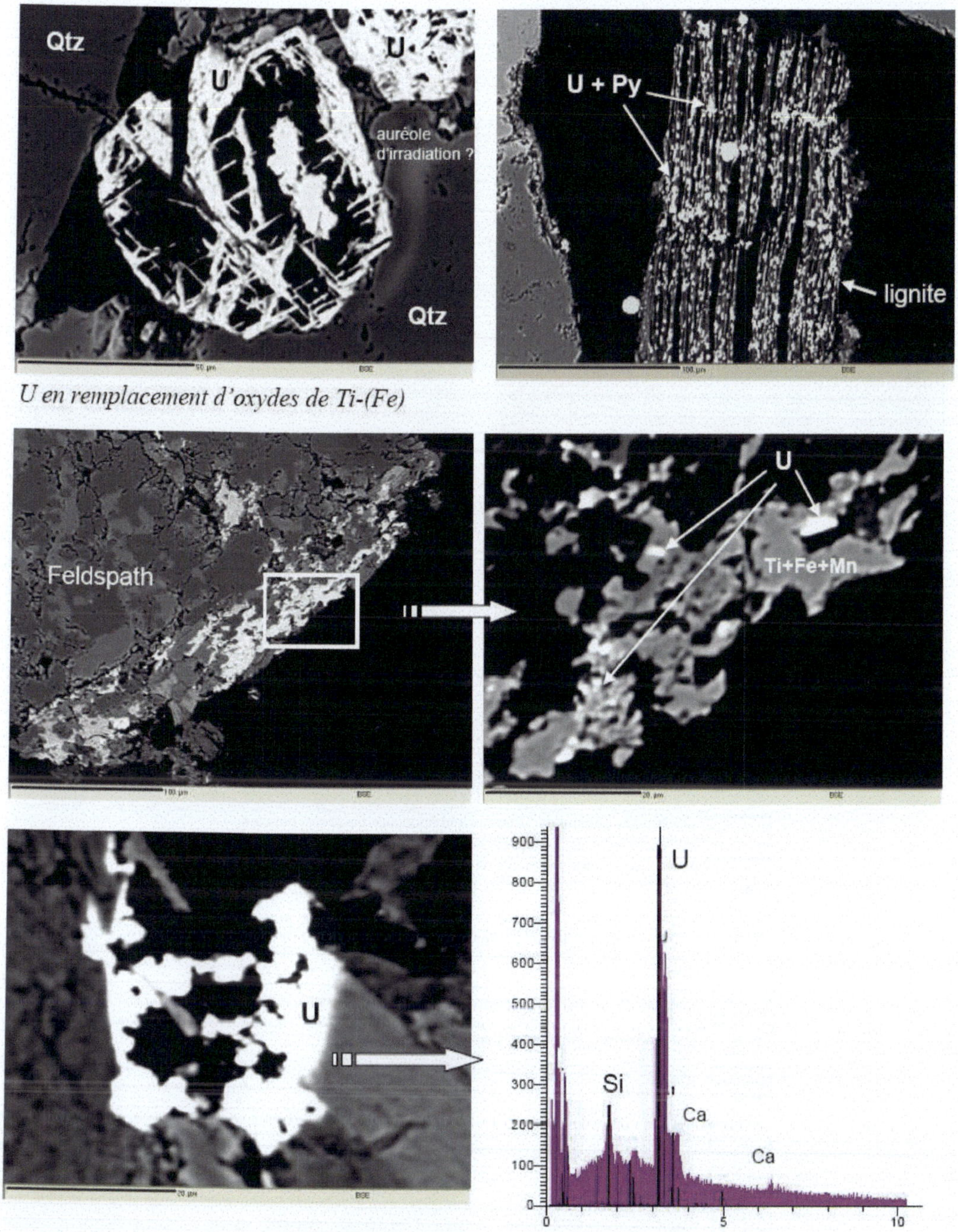

Fig. 4.39 Epigenetic uranium-substituted pyrite, iron–titanium oxides

the area played an important role in the accumulation of uranium.

Electron microscopic studies of ore samples showed that uranium occurs in lignite in the form of micrograins together with micrograins of iron and quartz.

As can be seen from the above analysis, uranium is relatively mobile and amenable to acid

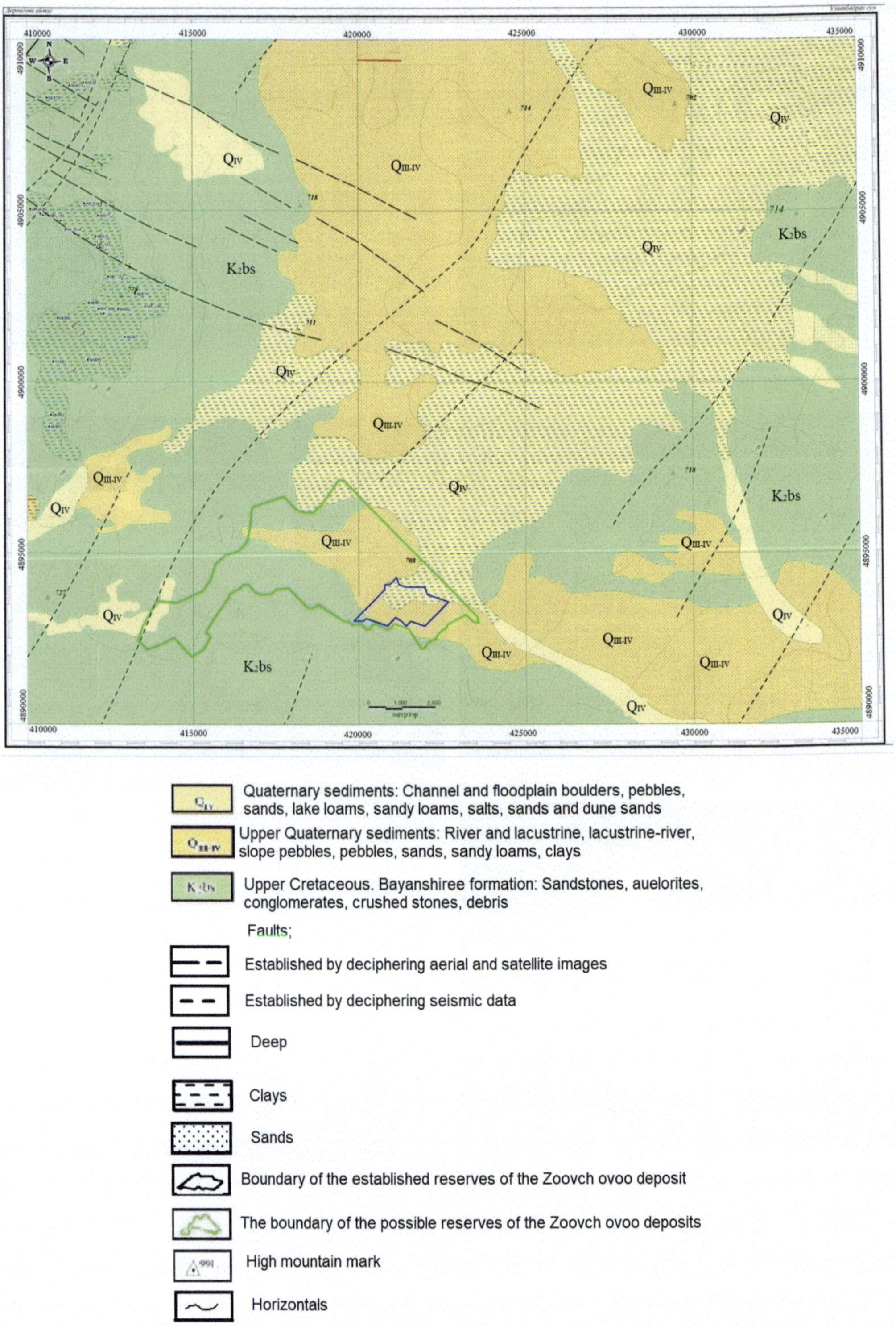

Fig. 4.40 Geological map of the area around the Zoovch Ovoo deposit (Enkh-Amgalan et al. 2012)

leaching. Kozhegobi held the majority of uranium prospecting and exploration licenses in the Zuunbayan depression and played a major role in the discovery of the Zoovch Ovoo (67,706.77 t U), Dulaan Uul (7054.71 t U) and Umnut-east-south continaution of Dulaan Uul (4746.14 t U) uranium deposits in the East Gobi province.

These deposits were discovered as part of the Sainshand project, implemented by the French Orano group as a result of joint work by French and Mongolian geologists in 2002–2018.

Geological Structure of the Deposit. At the Zoovch Ovoo deposit, uranium accumulates in gray sandy sediments dispersed and unevenly distributed in organic material of the Upper Cretaceous Sainshand formation. In this section of sedimentary rocks, interlayers and lenses of aleuropelites of gray, greenish-gray and red colors are noted (Fig. 4.41).

In the Zuunbayan depression, where the Zoovch Ovoo deposit is located, the Kozhegobi Company studied the full composition of the sediments of the Sainshand formation and assumed the possibility of dividing it into upper and lower horizons, but according to results of recent work on sedimentology, this was not justified. According to the classification of Cretaceous deposits of the Gobi province, the Lower Cretaceous section ends with gray-colored deposits of the Khukhteeg formation. Higher up in the section, sediments of the Sainshand formation, occurring with a stratigraphic unconformity, became widespread (Fig. 4.42).

Below the stratigraphic unconformity, red-colored weathering surfaces ranging in thickness from several m to tens of meters are often observed in the gray clayey deposits. In the Zuunbayan depression, a sequence of red-colored clayey deposits, unknown in other stratigraphic units, was intersected by many wells and is located between gray clays and sands. In the Zuunbayan depression the Upper Cretaceous deposits begin with a thick layer of clayey sediments, then, in well ZOOV_0024, gray clayey deposits of the Khukhteeg formation were found at an interval of 162–227 m. Higher up the section in an interval of 67–162 m, clayey sediments with interlayers and lenses of sands with a thickness of a few cm to one meter were seen. The color of the deposit is red, brownish-red and greenish-gray. The same section was encountered in well ZOOV_0023, drilled in the central part of the depression. The rocks of the Khukhteeg formation were exposed at a depth of 246.3 m and overlain by red-colored sediments with a thickness of 124 m.

Most of the Zuunbayan depression is composed of red-colored sediments of the Bayanshiree formation with a sandy-siltstone-pelite composition. The thickness of the red

Fig. 4.41 Sediment outcrops of the Sainshand formation (Enkh-Amgalan et al. 2012)

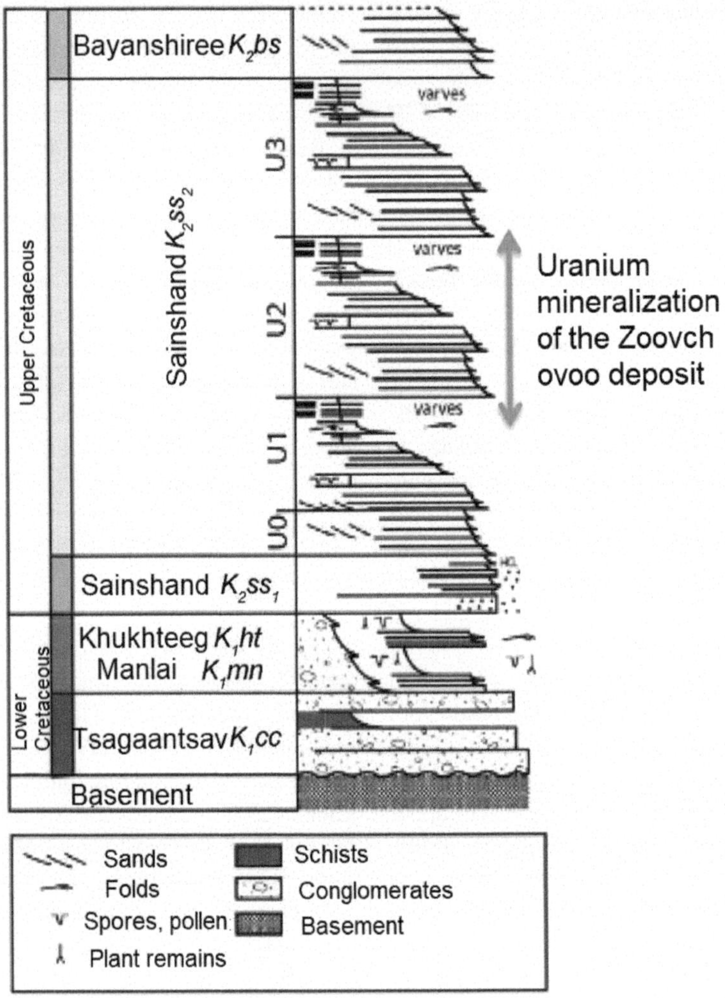

Fig. 4.42 Schematic stratigraphic column of the Zuunbayan basin (Enlh Amgalan et al. 2012)

deposits ranges from tens of meters in the eastern and southeastern parts of the depression to 400 m (well ZOOV_0001) in the central part to the east of Bayanshiree-Uul. There is a clear disconformity between the deposits of the Bayanshiree and Sainshand formations. The change in the boundaries of these deposits is identified by the appearance of epigenetic zones of limonite and mixed red- and yellow-colored deposits. The deposits of the Bayanshiree formation complicate the search due to the fact that they almost completely cover the ore-bearing deposits, overlying and screening them. When there is a thin cover of exposed rocks of the Sainshand formation, radioactive anomalies are noted.

In the Zuunbayan depression, outcrops of sandstones of the Sainshandin formation, weathered and covered with black iron-manganese ocher, are rarely observed, but outcrops of rocks of the Bayanshiree formation are often exposed—light-colored sandstones with spots of red clay and carbonate conglomerates, forming locally high pointsin the area.

Sedimentological research in 2015–2016 in the area of the Zoovch Ovoo deposit by Lia RADOUANT and Roger MATTHIEU identified

4 different stratigraphic horizons among the uranium-bearing deposits of the Sainshand formation according to lithological and physical characteristics. These horizons were compared with standards identified by the Mongolian Stratigraphic Commission and were named beds U-0, U-1, U-2, U-3, U-4.

If in the neighboring Unegt depression uranium mineralization is confined mainly to members U-1, 2, 3, and at the Zoovch Ovoo deposit uranium is concentrated mainly in member U-2 (Fig. 4.43). This is one of the important criteria for distinguishing uranium deposits.

According to drill hole data, the main ore bodies are localized in bed-2, and to a lesser extent—in bed U-1.

Following are features of the rocks of the U-2 bed:

1. The roof of U-2 is covered with the bottom of the U-3 bed or the U-4+3 bed.
2. Interlayers of gray or variegated siltstones appear in the section.
3. With the appearance of gray or variegated facies, radioactivity increases, reaching an anomalous level.
4. Due to changes in the intensity and nature of oxidation, limonitization in bed U-2 turned out to be a constant phenomenon. In these deposits, oxidation occurred throughout, and unoxidized gray-colored deposits are rare.

In the U-2 bed, sandy material predominates and there are interlayers of siltstone up to a few meters thick. The thickness of sedimentary rocks along the periphery of the depression is 35–40 m, and in the central part they reach 80–90 m. The thickness of uranium-bearing sandy rocks is 35–40 m, lithologically their composition is constant, coarse sand outside the oxidized areas contain scattered organic residue.

Weakly carbonated rocks are rare; the thickness of these layers is up to 2 m. Carbonates were formed along with the oxidation of rocks. For example, in well ZOOV_0041 in the upper section of bed-2 below the oxidation zone in

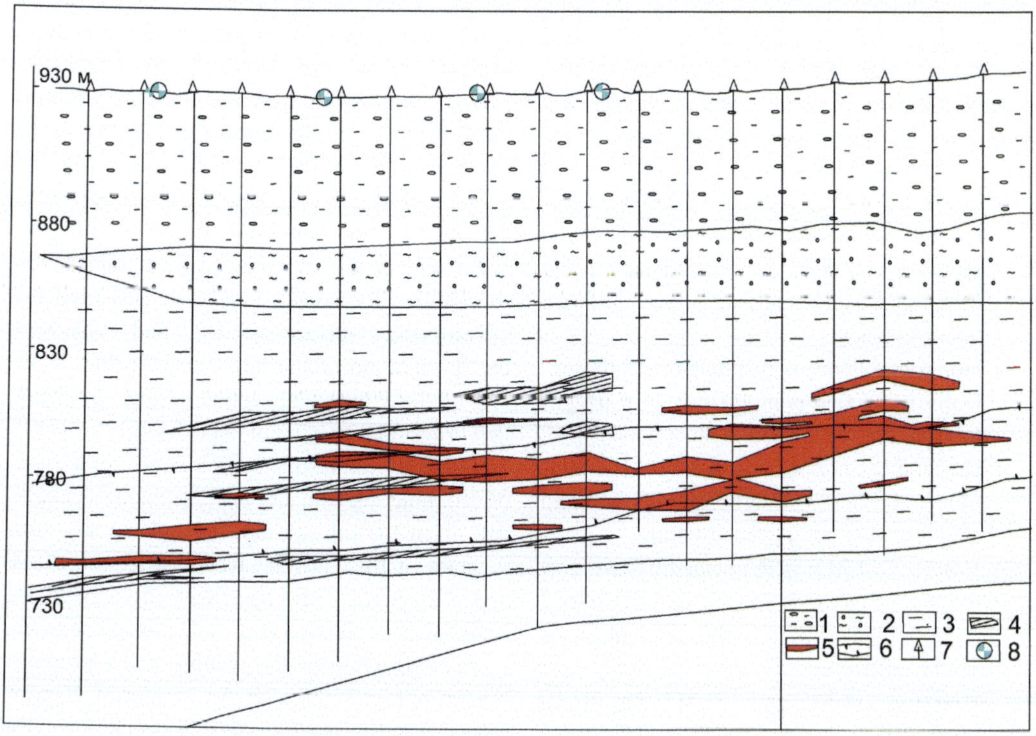

Fig. 4.43 Section along line 6400 of the Zoovch Ovoo deposit (Enkh-Amgalan et al. 2012): 1—deposits of the Bayanshire formation; 2—deposits of member-1; 3—deposits of member-2; 4—clay layers; 5—uranium mineralization; 6—oxidation zone; 7—wells; 8—faults observed on the surface

the interval of 144–146 m, sandstone layers with carbonate cement and radioactivity were encountered.

Bed U-2, compared to bed U-1, has an increased thickness of siltstone (up to 5 m) and a total thickness of 7–10 m. Thin layers of sand are noted in bed U-1. Epigenetic oxidation in beds U-1 and U-2 differ from each other due to facial features. In this system, the role of the upper waterproof horizon is played by red-colored clayey deposits of the Sainshandin formation, and the lower aquiclude is the red-colored sediments of the tops of the U-2 bed with a thickness of 6–16 m. The upper aquitard screens the uranium layer from the general oxidation zone.

- The direction of development of the oxidation zone in the U-2 bed is from the south and southeast to the northwest and north. The shape of redox fronts has a bizarre morphology and depends on the internal structure of the host rocks.
- For the Zoovch Ovoo deposit, bed U-2 is the main uranium-bearing stratigraphic unit and is distinguished by a difference in litholic composition and grain size. According to the decipherability of geophysical characteristics, it is divided into 5 horizons: from bottom to top M-1, M-2, M-3, M-4 and M-5.
  - Horizon U2-M5: is delineated by two correlated surfaces within the deposit;
  - Horizon U2-M4: is distinguished by lithology, characterized by relative unconsistency;
  - Horizon U2-M3: is distinguished lithologically because it contains layers of irregular granular clayey sediments (lacustrine);
  - Horizon U2-M2: also is distinguished lithologically, being underlain by U2-M3 and characterized by uneven structure;
  - Horizon U2-M1: is delineated by two correlated surfaces, it can be divided into 2 parts, separated by discontinuous intervals of clay-siltstone;
  - Horizons U2-M5 and U2-M1: are stratigraphically enclosed and provide marker horizons. Horizons U2-M2; U2-M3 and U2-M4, are identified by lithological characteristics.

Based on stratigraphic, sedimentological studies and the study of tectonic reconstruction, it was established that these deposits are lacustrine sediments of a clayey-siltstone facies.

Structural Conditions of the Deposit. The Unegt and Zuunbayan depressions filled with continental sediments extending from the northeast to the southwest were formed as a result of the expansion of the earth's crust in the Upper Jurassic-Lower Cretaceous. The development of rifting reached a peakin Berriasian-Hauterivian and Albian, where sediments of maximum thickness were deposited in the Lower Cretaceous, which were also subjected to intense tectonic activity. It was believed that due to rifting, the northeastern fault No. 60, separating the Unegt and Zuunbayan depressions, was activated, and Upper Jurassic–Lower Cretaceous sediments formed on the basement with an angular unconformity. The surface of this unconformity was noted by Benedicto and others as the "D1" deformation. After the formation of these deposits, at the end of the Lower Cretaceous period, compression from the northeast to southwest, formed anticlinal structures in the Lower Cretaceous deposits. This area was in an erosional regime until the Upper Cretaceous, when the Zuunbayan depression began to fill with sediments of the Zuunbayan and Sainshandin formations. Benedicto and Prost classified this unconformity surface as "D2" and considered the reactivation of the left-slip fault #60, at the end of the Upper Cretaceous rifting, an important criterion for uranium deposition.

Based on the results of field work, processing of drill hole log data and interpretation of seismic data from the oil industry, an updated structural diagram of the Zoovch Ovoo district (Fig. 4.44) and a diagram of the basement structure of the Zuunbayan depression (Fig. 4.45) were compiled.

A comparison of the results of the propagation of redox fronts with the structural diagram shows that the northwestern boundaries of the redox fronts and the ore-controlling faults "F" have a spatial connection. These faults played a semi-shielding role and influenced the direction of groundwater flow to the southeast.

In Beds U-1 and U-2, facies movement in a northeastern direction has been established. It manifests itself in the fact that in the northeast direction there are thicker layers of sand with a small number of beds of interbedded sand and clay. Also in this direction the capacity of the above packs are increasing. These mostly coincide with existing faults. Therefore, sedimentation and faults formed simultaneously, suggesting that facies movement of rocks is controlled by tectonic factors (Fig. 4.45).

The Tsaagan Els oil field is located in the northeastern part of the main fault and from here, theoretically introduced reducing fluids and gases, which rose or moved along the fault in a southerly direction, creating a reducing environment.

Uranium Mineralization. The Zoovch Ovoo deposit is a large uranium deposit in Central Asia in terms of reserves. Ore bodies and ore quality are controlled by redox fronts developed in the ancient valley. The ore bodies have lens and sheet-like shapes. Uranium mineralization is distributed in a northeastern direction for more than 20 km with a width of 5 km. Numerous ore layers are confined to the bottom of the section and are separated from each other by weakly mineralized layers (Fig. 4.42). The reserves have been calculated for underground leaching and the uranium grade of the reserve blocks is 0.0087–0.045%, with an average of 0.0226%, which is comparable to other deposits in Mongolia.

The thickness of uranium-bearing layers ranges from 1.0 to 29.7 m, and the thickness of the largest layer of the 1st block averages 6.61 m.

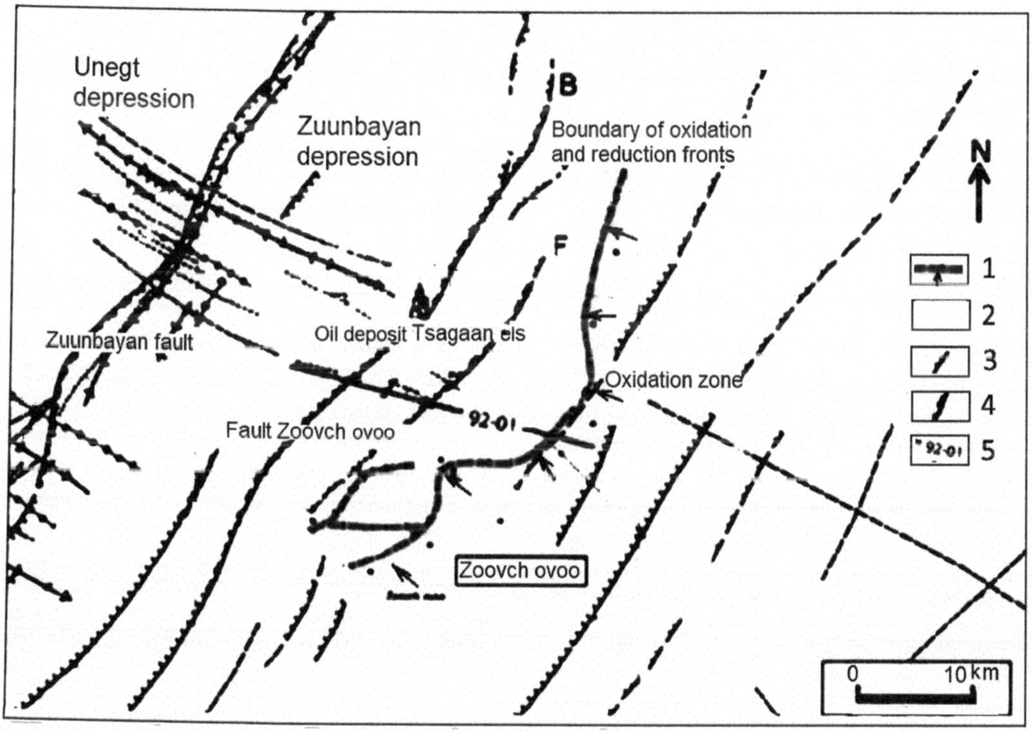

Fig. 4.44 Updated structural diagram of the Zoovch Ovoo district (Enkh-Amgalan et al. 2012): 1—boundary of redox zones; 2—oxidation zones; 3—simple faults; 4—faults with displacement; 5—seismic line 92_01

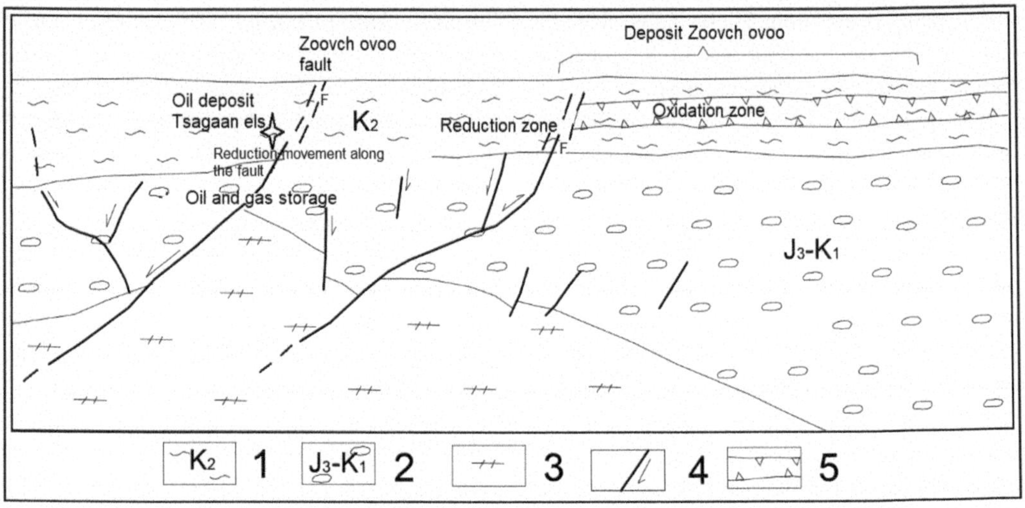

Fig. 4.45 Section through deciphered seismic line 92_01 (Enkh-Amgalan et al. 2012): 1—Upper Cretaceous deposits; 2—Jurassic-Cretaceous deposits; 3—Proterozoic basement; 4—faults; 5—oxidation zones

Features of Uranium Mineralization. Research has established that uranium minerals are more common in fine-grained facies than in coarse-grained facies. The primary ore mineral is coffinitite and is slightly enriched in phosphorus and calcium. Coffinite is fine, found together with pyrite and oxides of iron and titanium.

Main Forms of Uranium Mineralization

- under an electron microscope, the sizes of grains and minerals are very small, but sufficient for the formation of uranium oxides, pyrite micrograins and layered lignite fragments;
- cracks in clayey facies filled with late "pure" oxides (> 80%) of uranium along with quartz and pyrite;
- coffinate within the skeletal structure of ilmenite or along the edges of titanomagnetite crystals contains calcium and phosphorus;
- coffinite and phosphocofinite are observed on mica planes and less frequently in clays;
- cofinite with calcium and phosphorus are established in pyrite microcracks localized in clays in the form of clausthalite microinclusions;
- uranium-containing minerals—uranothorite, monazite, zircon, xenotime, etc. They are observed in the form of accessory inclusions.

Uranium is contained in micropyrites in the form of woody fragments. These fragments under the electron microscope (EDS) contain uranium, iron, and in small quantities silicon, aluminum, magnesium and sodium. Elements such as molybdenum and selenium have no economic significance, but are important as geochemical properties. For example, in a reducing environment, uranium and molybdenum are deposited together, but in an oxidizing environment they are observed sporadically. This pattern needs further studied.

Sources of Uranium. An important pattern has been established, namely that the economicaccumulation of uranium is recorded in the hypsometrically lowest part of the depression, where groundwater transportation is located. Uranium is mobile in surface conditions and can also be transported by surface rainwater and concentrated at the redox barrier. For the Zoovch Ovoo deposit, the main source of uranium may be the granitiode massifs surrounding the depression from the southeast and south. Granitoid magmatism occurs in several stages. The largest of the stages is the Middle-Late Paleozoic magmatism. Also noted here, along with granitoids of the Devonian, mid-late Carboniferous and Permian, are the Mesozoic small massifs of granitoids. Based on the results of a gamma spectrometric survey on a scale of 1:200,000 of Late Paleozoic granitoids, it was established that they contain numerous anomalies of uranium with a grade of more than 4 g/t. The northeastern sides of the Unegt and Zuunbayan depressions have greater prospects for uranium content compared to the northern and western sections.

Genesis of the Deposit. French geologists (Kozhegov) believe that the Zoovch Ovoo deposit is representative of the well-known (IAEA) "roll-front type" deposit, and that this is one of the names for sandstone type uranium deposits in sedimentary strata.

4.9 Khairkhan Uranium Deposit

The Khairkhan deposit is located in the northern part of the depression. The deposit is located 100 km south of the aimak center of Mandal-Gobi and 150 km west of the Kharaat deposit. The terrain is flat, semi-desert, accessible to vehicle traffic at any time of the year and in any direction. The nearest railway station, Choir, on the Ulaanbaatar-Beijing highway is 250 km to the north. In general, the natural conditions of the field site are favorable for geological exploration and production work.

The Khairkhan deposit was discovered by geologists of the Gurvansaikhan joint venture in 1996 when conducting autogamma spectrometric surveys on a scale of 1:10,000 and exploratory drilling in the depression. In 1997–98 prospecting, appraisal and exploration work was carried out at the field site using drilling, geophysical logging, hydrogeological and technological studies, as well as various types of sampling and laboratory analyses. Drill hole exploration was laid out on a grid of 100–200 × 50–100 m. The depth of the wells was 50–200 m and a total of 65,000 linear meters were drilled, which included about 1700 linear meters of hydrogeological wells and 600 linear meters of technological wells.

The district is located in the Khairkhan depression, which is located in the south of the Middle Gobi aimag of Mongolia.

The Khairkhan depression is a basin with a flat topography, cut in the northern edge by numerous gullies and ravines.

The first specialized studies for uranium in the depression were carried out in 1973 by Russian geologists. During this period, an airborne gamma spectrometric survey on a scale of 1:25,000 was carried out here. As a result of this work, 3 uranium anomalies and a large number of halos of elevated uranium and thorium occurrences were identified. However, these occurrences had not received a reliable ground-based assessment. Since 1996, geological exploration work for uranium by the Gurvansaikhan joint venture has been carried out in the Khairkhan depression.

The geological structure of the Khairkhan depression (Figs. 4.46 and 4.47) is similar to the structure of the Choir depression. It is composed primarily of carbon-bearing gray deposits of the Zuunbayan formation (K_1). In the south of the depression, these deposits are overlain by variegated sediments of the Sainshand Formation of the Upper Cretaceous.

The Khairkhan deposit is controlled by a latitudinal river paleovalley with a length of about 25 km (Fig. 4.47). Economic mineralization is established in the lower part of the valley over a length of 12 km; the highest-grade block (Central) is located in the mouth of the paleovalley and has a dimension of 2.0 × 1.5 km. Ore deposits are concentrated in the channel, floodplain, and coastal-lacustrine facies, represented by heterogeneous sands, gravelites, conglomerates, siltstones, and clays (Fig. 4.48).

The sediments are gray-colored and contain a significant amount of carbonized organic matter, interlayers of brown coal, and fine dissemination of sulfides. The thickness of river sediments varies from 30–70 to 120–150 m in the mouth area. Surface and ground oxidation of sediments is prolificto a depth of 50–80 m;

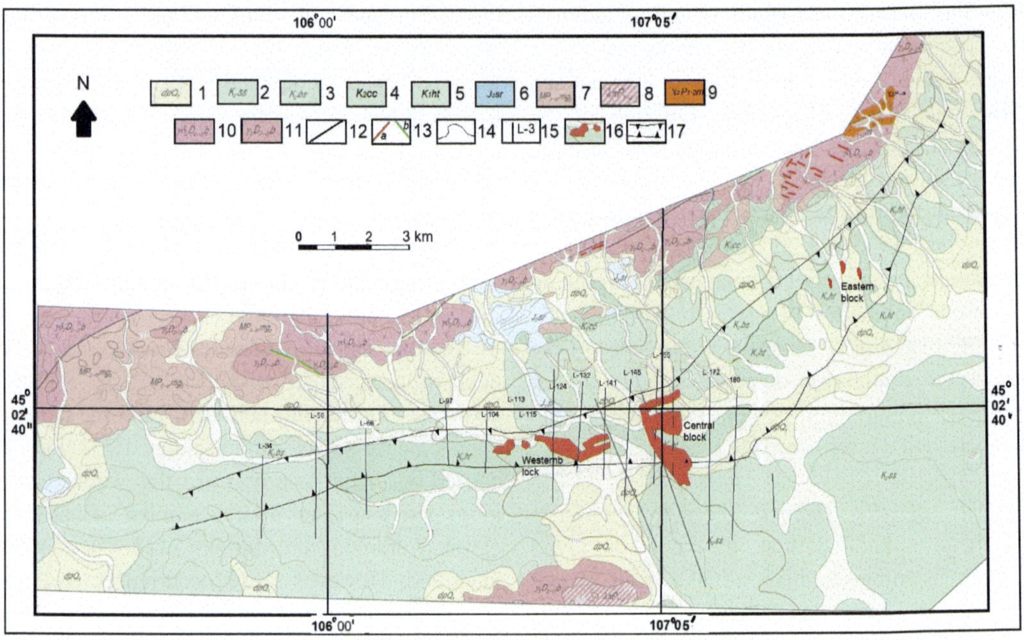

Fig. 4.46 Geological map of the Khairkhan deposit area (Budunov and Jamsrandorj 2013): 1—modern deposits. Sands, sandy loams, clays, loams, pebbles; 2—Upper Cretaceous. Sainshind formation. Variegated sands, gruss, pebbles, with layers of clays and siltstones; 3—Upper Cretaceous. Bayanshire formation. Clay, sandstone, aulerolite, conglomerate; 4—Upper Cretaceous—Tsagaantsav formation. Vulcanogenic member. Basalt, andesite basalt; 5—Lower Cretaceous. Khukhteeg formation. Sands, sandy loams with layers of red clays, gray alluvial pebbles, sands, clays, alluvium of lake shores, lacustrine, alluvial-lacustrine gray clays, siltstones; 6—Upper—Middle Jurassic. Sharil formation. Andesite porphyries, trachydacites, dacites, basaltic andesites and rhyolites; 7—Mesoproterozoic. Estasi-steny. Upper member. Limestones, crystallized marbles, quartzite layers, metamorphic schists; 8—Permian. Gvadelup. Subvulcan complex. Syenite porphyry, trakhyandesite; 9—Permian. Mandalgovi complex. Biotite-hornblende granites and alkaly granites; 10—Late Devonian. Buyant complex. Tird fase. Leucogranites and biotite granites; 11—Middle Devonian. Buyant complex. Second fase. Granodiorites, plagiogranites, biotite-hornblende granites, rarely diorites; 12—faults; 13—dykes: a—granites; b—dyorites; 14—geological boundary; 15—exploration lines and their numbers; 16—uranium ore blocks and their name; 17—buried channel

individual tongues of limonitization along permeable horizons penetrate to a depth of 170 m, and the oxidation processes acquire a substratal character. The primary mineralization is located in areas where oxidation zones pinch out and transition to red-colored oxidized formations and gray-colored non-oxidized sediments. These areas have a high reduction potential due to significant concentrations of organic residues and diagenetic sulfides.

The central ore block, where the main reserves of the deposit are concentrated, is limited from the west by the Carbonate fault, and from the east by the Meridional fault. The Carbonate fault has a NNW strike. It is characterized by the carbonatization of sediments, which gives the appearance of strong lithified rocks. They form pillars, lenses, and nests ranging in size from a few meters to a few tens of meters. No ore concentrations have been found in them.

The Meridional fault is distinguished by a number of indirect signs: first of all, due to a sharp drop in the level of ground water, there is a change in the facies composition of the sediments from channels and floodplains from the west to essentially lakes from the east. The role of faults in uranium deposition, have not been studied, but it has been established that the

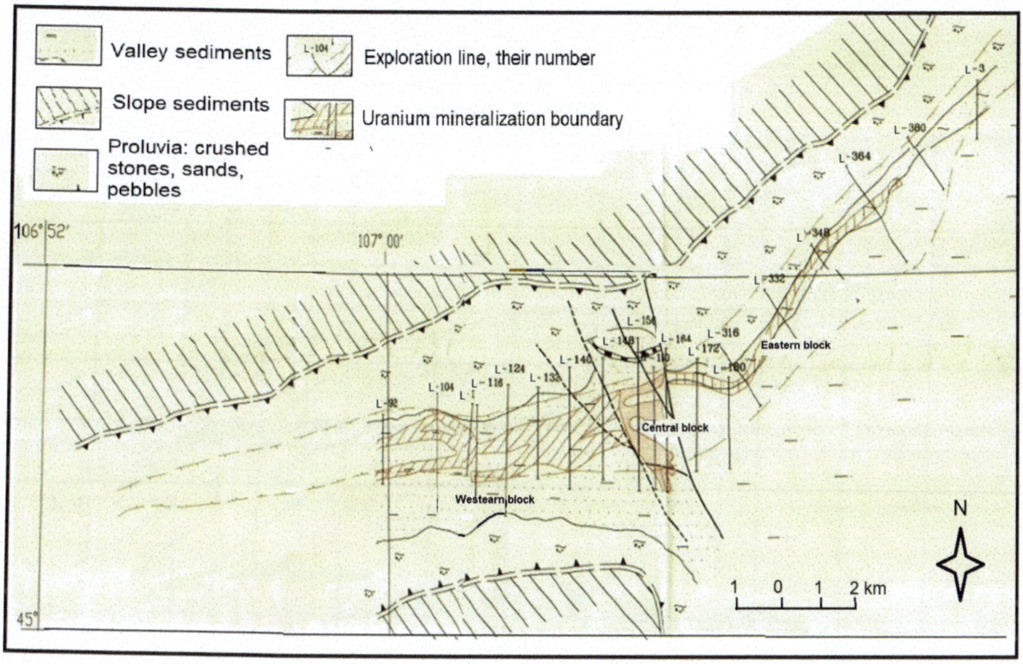

Fig. 4.47 Geology sketch of the Khairkhan deposit (Budunov and Jamsrandorj 2013)

highest grade mineralization, with productivity of up to 70 kg/m^2, is concentrated along the fault. This may be due to the role of fractures as hydrodynamic, lithological, and geochemical barriers to the movement of oxygen and uranium-enriched groundwater. It is not to exclude the concept that the fault could serve as a conduit through which additional gaseous reducing agents (hydrogen sulfide, hydrocarbons, etc.) came from the deeper horizons to the upper layers of the sediments (zone of surface and soil oxidation).

Geological exploration work at the Khairkhan deposit was primarily concentrated in the Central ore block. Here, according to exploration results, 3 ore deposits were selected: Northern, central and southern, in each of which there are 3–5 levels of mineralization, which extend from the surface to a depth of 200 m. On every level, there are 1–5 ore bodies localized. All of them have a complex tape-, lens-, sheet-like-, nest-shaped morphology, and column-shaped uranium roll deposition is noted. The size of the ore bodies vary; from the first hundred to 1200 m, the width is from 50 to 600 m, and the thickness varies between 0.5 and 22.0 m. Uranium grade is extremely inconsistent, generally, it ranges from 0.010 to 0.100% however, there are high grade zones with uranium concentrations of 0.305–0.619% noted in some of the drill holes with varying thicknesses.

Ore bodies are located in sub-horizontal deposits with a slight inclination to the east. The deposits are composed of aluminosilicate and carbonate-free ores. Uranium mineralization is represented mainly by pitchblende, with a minor amount of coffinite. Secondary uranium mineralization of otenite and uranophane is noted in oxidized rocks. Primary associated mineral with uranium is molybdenum, at a grade of 0.05%. Economic concentrations of Mo, in uranium mineralization, have not been established.

According to these studies, the hydrogeological conditions of the reservoir are determined to be complex, but generally are favorable for ISR extraction. Mineralization is located below the level of groundwater, mainly in permeable rock. The conductivity of ore-bearing deposits varies

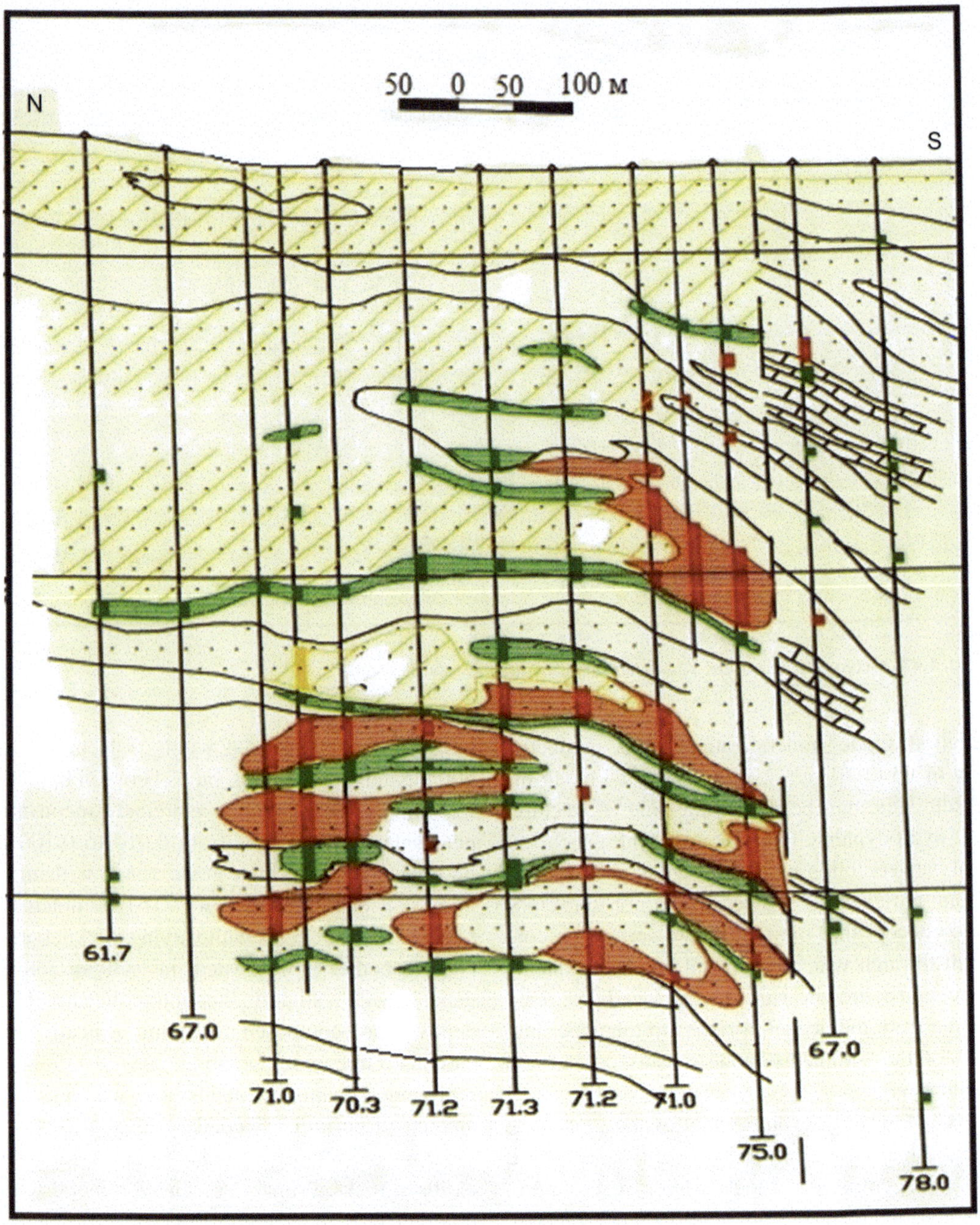

Fig. 4.48 Section along line 158 of the Khairkhan uranium deposit (Budunov and Jamsrandorj 2013)

widely from 10 to 370 m^2/day. The filtration coefficient is 1–10 m/sut.

In 1998, a leaching test was carried out at the Khairkhan site. 5 bore wells were constructed for the test; 1 injection well and 4 extraction wells. 4 holes were equipped for monitoring. The test was carried out in a closed loop operation for sorption and desorption of uranium. The duration of the test was 3.8 months.

The results of the in situ leach test were positive, indicating uranium from the Khairkhan deposit can, technologically, be extracted using

Table 4.17 Comparison of the parameters of deposits Kharaat and Khairkhan (Budunov and Jamsrandorj 2013)

Compatible parameters	Kharaat	Khairkhan
Thickness of floodplain and channel gray-colored uranium bearing sediments (m)	50–100	100–250
Depth of oxidation zones (m)	5–40	50–200
Number of layers or ore-bearing levels and depth of occurrence (m)	1–2 5–40	3–5 20–220
Uranium average grade (%)	0.036	0.066
Shape (form) of ore layers	Tape-, sheet-like	Tape, lens, sockets, roll and pillar
Proportion of reserves located in permeable sediments below the groundwater level (%)	20	78

this technique. The average concentration of uranium obtained during the test was 74 g/l. Based on the results of exploration, the reserves, suitable for leaching, in category B+C in the Central ore block were calculated. New ore deposits were discovered by exploration drilling on the eastern and southern flanks of the deposit, and the projected resource in the P_1 category is estimated at 10,000 tons, which is a significant increase of reserves.

Thus, the Khairkhan deposit has relatively positive test results and better uranium recoveries than compared to the Kharaat deposit, which is the object of priority development (Table 4.17).

4.10 Gurvansaikhan Deposit

This deposit was discovered by geologists of the Gurvansaikhan joint venture in 1996 during autogamma spectrometric surveys on a scale of 1:10,000 and prospecting and reconnaissance drilling in the area. In 2005, prospecting and exploration was conducted in in the field, and continued through 2008–2009. As a result of this work, two mineral occurrences were identified at the deposit, each of which contained 2 ore deposits 10 km apart. In 2010–2011 exploration was conducted by drilling on three ore bodies, Nos. 1, 2 and 4. Research was also carried out here to study the hydrogeological conditions of the ore; the permeability of ore-bearing sediments and other parameters necessary to determine the suitability for in situ leaching. In 2011, laboratory leaching tests of the ore was conducted.

The Gurvansaikhan deposit is located within the Late Mesozoic depression of the same name, composed of terrigenous-sedimentary coal-bearing deposits of the Lower Cretaceous. A significant part of the area is overlain by younger Upper Cretaceous, Neogene-Quaternary and modern sediments (Fig. 4.49).

Both ore zones of the deposit are located near the southeastern edge of the depression, composed primarily of Early Paleozoic granitoids containing xenoliths and roof pendant of metamorphic formations of Early and Late Proterozoic age. Rocks of the crystalline structure are broken by numerous tectonic disturbances, characterized by crushing, brecciation, and increased fracturing, which contributed to the significant destruction of crystalline for mations and the leaching of uranium and other components from them.

The ores are composed of proluvial-deluvial sediments of alluvial fans from the foothills, alluvial deposits of the channel and floodplain facies and lacustrine and lacustrine-marsh sedimentary deposits.

Sediments of the proluvial-deluvial piedmont-fan train are deposited in the southern part of the depression. This strip of sediments has a width from a few hundred meters up to 1.5–2 km. These deposits are represented by red-colored, primary oxidized, unsorted clayey-sandy-crushed stone sediments with inclusions

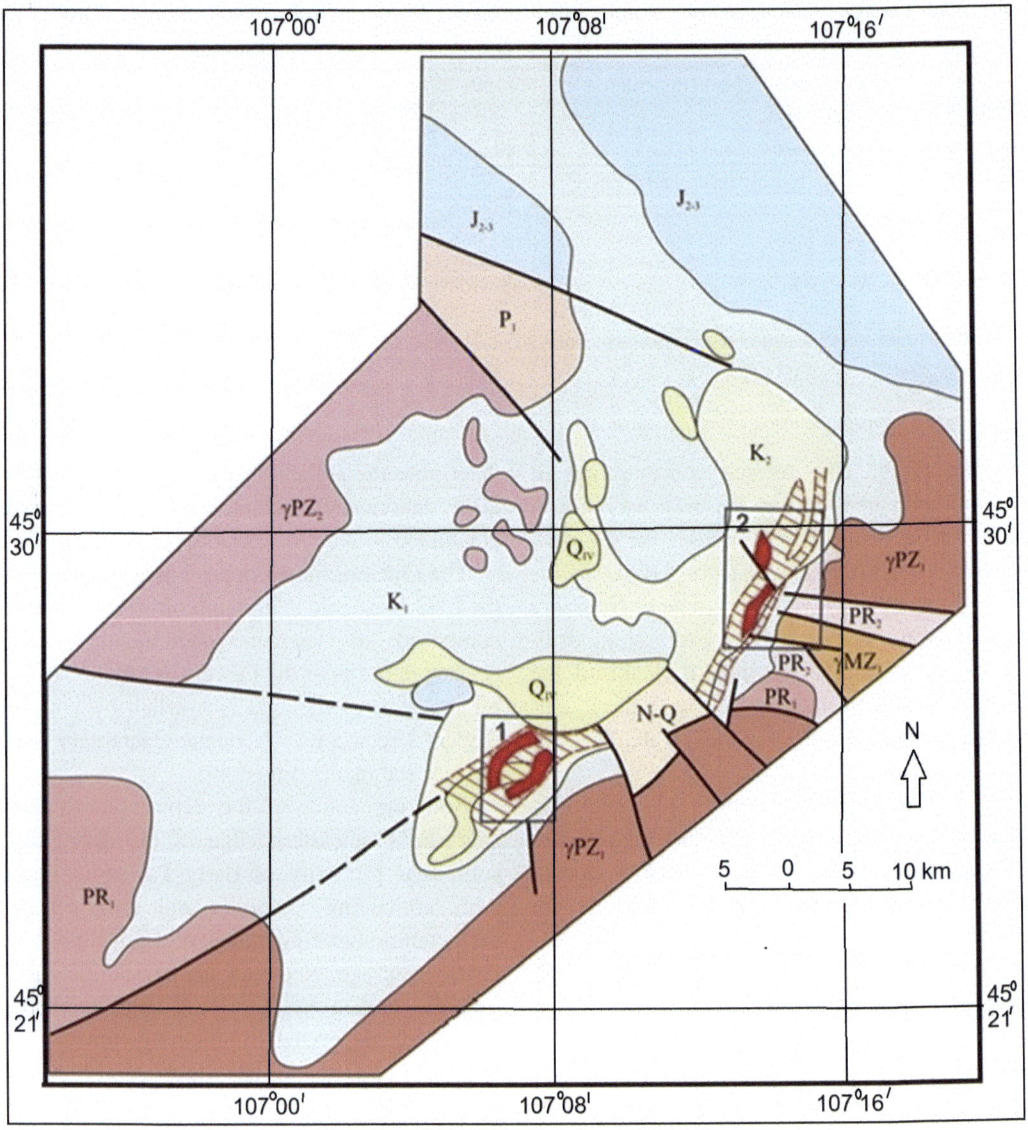

Fig. 4.49 Geological scheme of the Gurvansaikhan depression (Budunov and Jamsrandorj 2016): 1—modern lake-marsh, alluvial deposits (sands, silts, clays); 2—Neogene-Quaternary deposits (unsorted pebbles, sands and clays with crushed stones and boulders); 3—Upper Cretaceous gray deposits (variegated pebbles, sands and clays); 4—Lower Cretaceous gray deposits (sands, silts, clays, clayey sands, coal layers); 5—Middle, Upper Jurassic (andesites, dacites, andesite—basalts, rhyolites, tuffs); 6—Lower Permian (rhyolite-porphyry, andesite-porphyry, tuff); 7—Upper Proterozoic (shales, limestones, quartzites); 8—Lower Proterozoic (crystalline schists, gneisses, amphibolites). Intrusive rocks: 9—Mesozoic (leucocratic granites, granite porphyries, granisyenites, syenites); 10—Middle Paleozoic (biotite, biotite-amphibolite granites); 11—Early Paleozoic (granodiorites, plagiogranites, granite gneisses); 12—faults: a—established; b—possible; 13—radiometric anomalous zones established by airborne gamma spectrometry; 14—uranium ore bodies; 15—ore areas: 1—southern; 2—eastern

of large fragments and blocks of crystalline framing rocks. Ore concentrations in them have not been established. The predominant deposition in the ore zones are alluvial deposits of channel and floodplain facies. In the northern area they spread from the side of the depression to its center. In the southern area they form a large river paleovalley, which stretches along the depression from southwest to northeast. The channel facies of these deposits consists primarily of mixed-grained sands with pebbles and gravels. They contain layers and lenses of silts and clays. Floodplain sediments are more complex consisting of frequent interlayered horizons, of varying thickness and extent, consisting of sand, silt and clay with transitional differences. These deposits are formed in river, lake, lake-marsh, and oxbow conditions. The rocks contain carbonized organic remains and interlayers of brown coal. In these deposits, as a rule, a contrasting redox environment is created, which creates favorable conditions for the formation of uranium ores. In the southern area, these deposits are mainly ore-bearing (Figs. 4.50 and 4.51).

Lacustrine sediments are distributed primarily in the central part of the depression. They are represented by dark gray and black clays and lignites containing small layers and lenses of silts and sands. These sediments are established in the western part of the eastern ore zone, where they create a contrasting lithological, geochemical and hydrodynamic barrier to the movement of oxygen and uranium-enriched groundwater, thereby creating favorable conditions for uranium deposition.

In the area of the deposit, Lower Cretaceous sedimentary rocks are, in places, overlain by Upper Quaternary and modern sediments of lacustrine and lacustrine-marsh types. Most often they are represented by variegated and red clays, sometimes highly saline and transformed into impassable salt marshes.

Faults known in the ore zones are of a hidden nature, they are established only by a sharp change in the lithological composition of sedimentary deposits. Their individual lineaments are identified when interpreting space-aerial photographs. The faults have different directions and are mainly of a fault-slip nature.

In the central part of the depression, between the Southern and Eastern ore zones, a large graben-like structure of Neogene-Quaternary age has been mapped. It is filled with large blocks of mixed boulder-sand-clay deposits, formed by powerful mudflows or, possibly, glaciation. The structure has a northwest strike and its presence is confirmed by drill hole data from 2005 to 2008. The width of the graben is 4–6 km. Perhaps, during the Neogene-Quaternary period, this structure "broke" the once single ore-bearing zone extending along the southeastern side of the depression. The negative impact of transverse young fault structures is that they have probably led to the fact that to date only small isolated ore deposits have been preserved in the Gurvansaikhan depression.

Almost throughout the Lower Cretaceous sediments epigenetic processes of surface and ground oxidation are manifested. According to drill hole data surface and ground oxidation are evident from the surface to a depth of 50–60 m and in some cases up to 100 m. Oxidation is most intensively manifested in permeable deposits of channel facies, developed less intensely in mixed-layer floodplain sediments and very weakly in lacustrine clays, where only surface oxidation is observed to a depth of 1–3 m. In sedimentary deposits, enriched in carbonized organic matter and in coal layers, sulfidization is noted. It is primarily encapsulated in small-lens.

In general, the geological structure of the Gurvansaikhan deposit is similar to the structures of the well-known economic uranium deposits Kharaat and Khairkhan. The similarity is noted in the nature and composition of the sedimentary section and the manifestation of epigenetic processes. The only difference between them is the structure of the ore zones. At the Gurvansaikhan deposit, tectonic processes are more intense, which significantly complicated hydrodynamics and limited favorable conditions for ore deposition, which led to the formation of small and unproductive ore deposits.

The Gurvansaikhan deposit is a sandstone type uranium deposit of the "paleovalley" type,

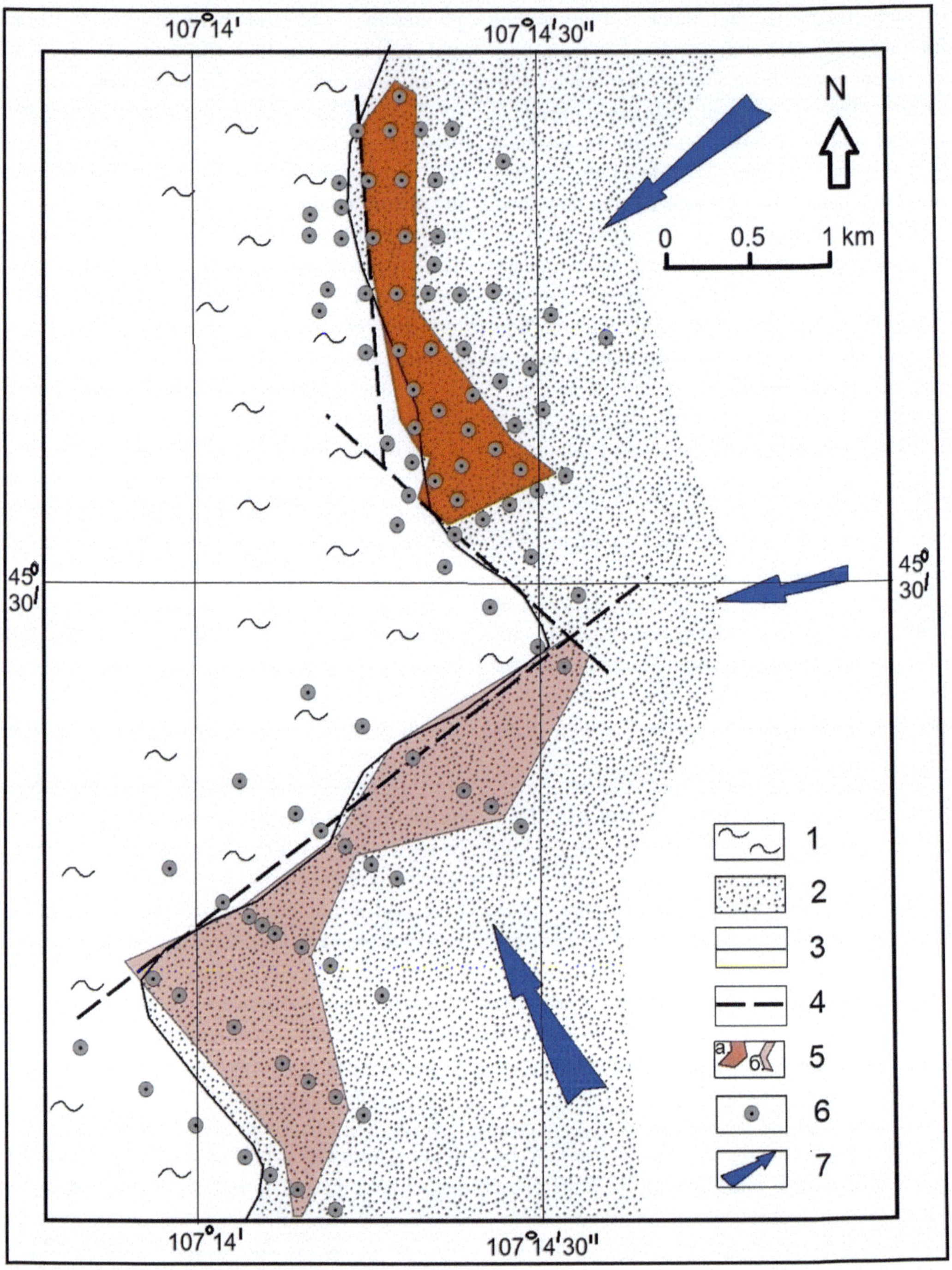

Fig. 4.50 Scheme of distribution of lithological facies of the eastern area (Budunov and Jamsrandorj 2016): 1—clayey facies enriched in coalified material; 2—sandy facies; 3—lithologic-facies boundaries; 4—expected faults; 5—ore bodies: a—with established reserves; b—estimated resources; 6—wells and their numbers; 7—direction of groundwater flow

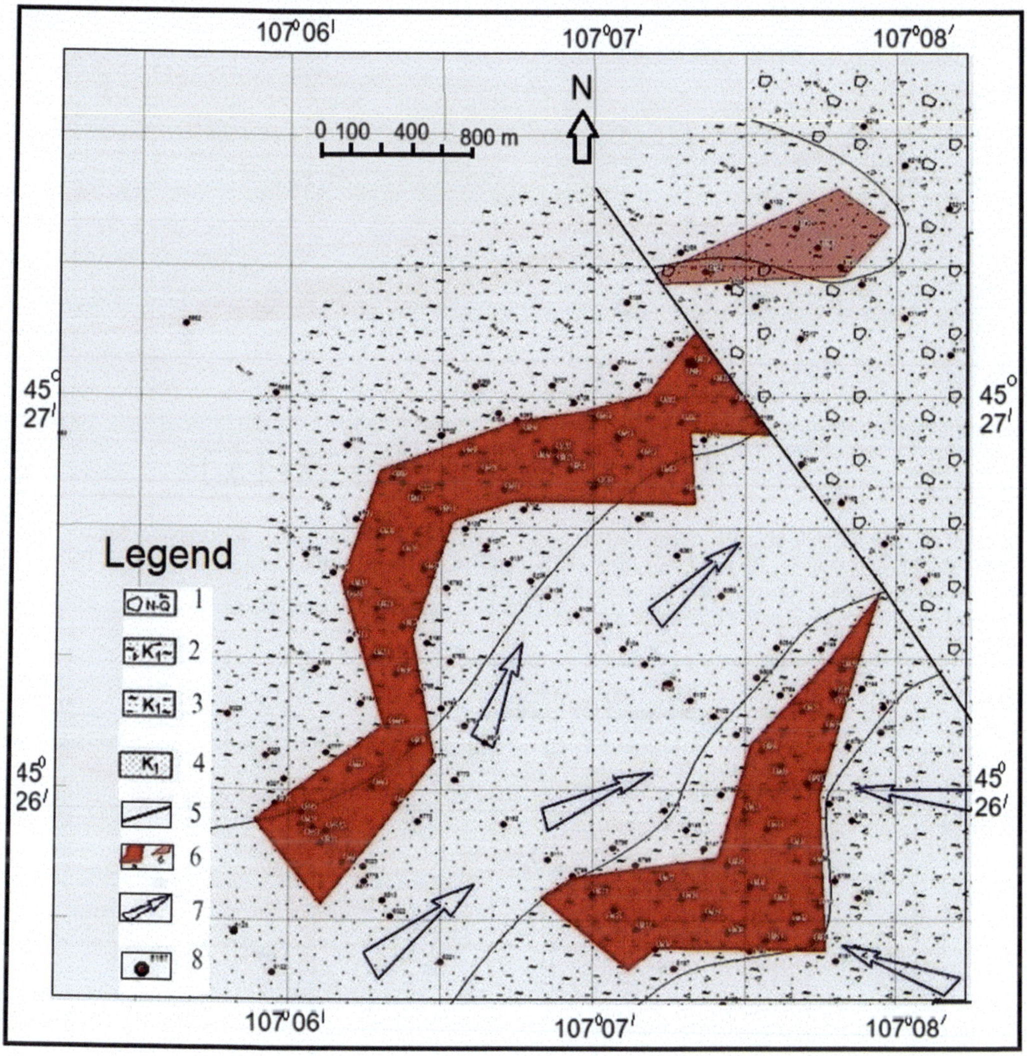

Fig. 4.51 Scheme of distribution of lithological facies of the southern area (Budunov and Jamsrandorj 2016): 1—proluvial facies: variegated debris, sands, clays; 2—proluvial facies: colorful, poorly sorted, primary oxidized gruss, sands, clays of mountain slopes; 3—alluvial-lacustrine and paleovalley facies: gray clays, silts and sand layers with thin layers of coal and organic remains; 4—alluvial-channel facies: gray sands with thin layers of silts and clay lenses; 5—faults; 6—ore bodies: a—with established reserves; b—estimated resources; 7—direction of groundwater flow; 8—wells and their numbers

associated with zones of surface, ground and substratal oxidation of the sediments. Ore deposits are located in gray-colored coal-bearing Lower Cretaceous deposits and are controlled by river paleovalleys. They are localized in loose terrigenous sedimentary deposits of channel and floodplain facies. Mineralization is associated with zones of surface and ground oxidation of sediments and is located both within these zones and in areas where they pinch out. The highest concentrations of uranium are observed near faults, which serve as barriers to the filtration of oxygenated, uranium-enriched groundwater. Mineralization is found in mixed-layers of sandy-silty-clayey deposits, enriched in carbonized organic matter and containing layers of coal.

Ore bodies have a lens-like, sheet-like morphology (Figs. 4.52 and 4.53).

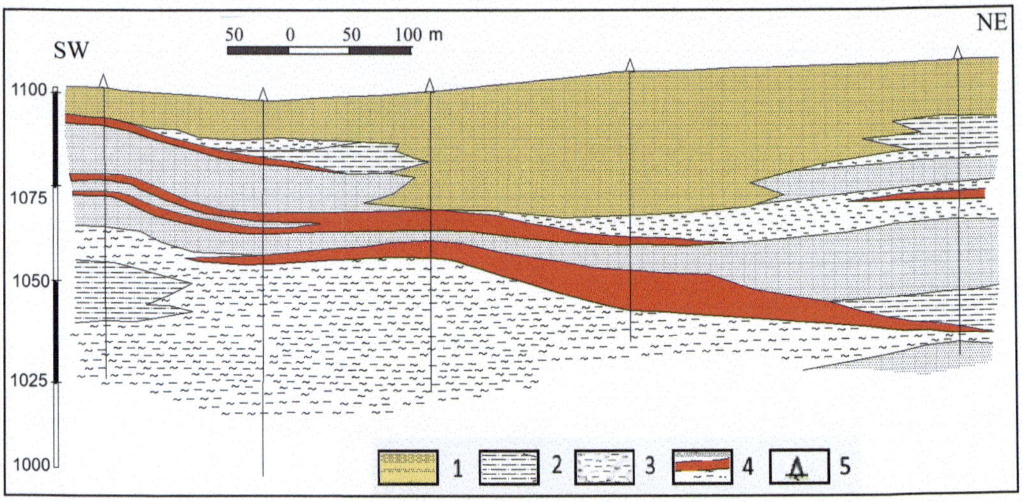

Fig. 4.52 Section through line 210 of the eastern area (Budunov and Jamsrandorj 2016): 1—oxidized sediments: sandy loam and loam with pebbles and crushed stones, redox sediments; 2—pebbles, sands, sandy loams; 3—sandy loam, loam; 4—uranium ore body; 5—drill holes

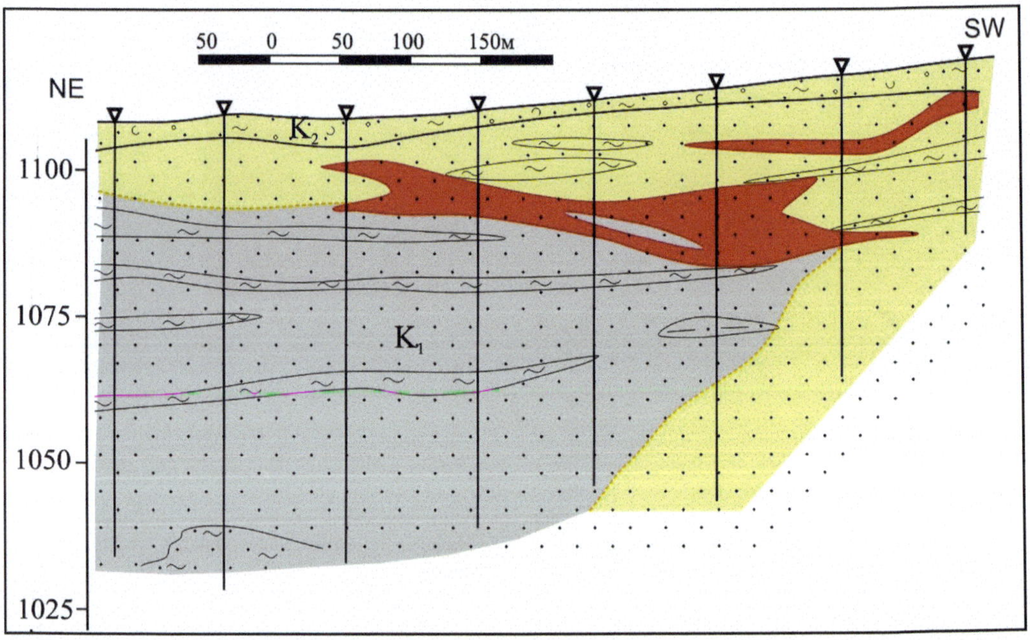

Fig. 4.53 Cut through line 13 of the southern area (Budunov and Jamsrandorj 2016). Ore body II. For the legend see Fig. 4.53

They lie subhorizontally at a depth of 3–70 m. The lengths of the deposits vary from 1500 to 3200 m, the widths from 100 to 900 m and their thickness varies from 0.6 to 12.9 m. The ore grades are predominantly poor. The uranium grades range from 0.01 to 0.099%; in some sections, richer concentrations of up to 0.244% are noted. Within the deposits, 1–3 levels of mineralization are observed.

The ores of the Gurvansaikhan deposit are aluminosilicate and non-carbonate. Uranium mineralization is composed primarily of pitchblende, with coffinite present in minor amounts. In oxidized rocks, secondary uranium mineralization is observed in the form of otenite and uranophane. The ores are predominantly uranium; in some cases, they contain increased concentrations of Mo, Pb, Zn, Cu, as well as rare earth elements. However, due to low concentrations they have no economic importance.

Hydrogeological conditions at the Gurvansaikhan deposit are favorable for in situ leaching of uranium. The ore deposits are located primarily below the groundwater level. According to the results of hydrogeological studies, carried out within the explored ore deposits, their filtration properties are acceptable for in situ leaching. To study the metallurgical properties of the ore, bench laboratory tests were carried out on the leaching of uranium with sulfuric acid solutions. The results proved the high processability of ores for leaching. The recovery was over 97% when using working solutions with an acid concentration of 20 g/l. The results obtained from the leach tests coincided with the results from similar studies conducted at the Kharaat and Khairkhan deposits.

4.11 Ulziit Deposit

The Ulziit deposit is located in the East Gobi region of Mongolia in the territory of the Sukhbatar aimag. It was discovered in 2011 by geologists of the joint Mongolian-Russian-American joint venture Gurvansaikhan.

The Ulziit deposit is located in the Ulziit depression, one of the largest Late Mesozoic depressions in the Mongolian Gobi. This depression is an erosional-tectonic graben-like structure with a northeastern strike, controlled by the primary Mongolian lineament orientation. The length of the depression is 200 km and its width varies from 10 to 25 km.

The uranium mineralization is a sedimentary deposit of the Ulziit depression and has been known since the late 1950s. Ground-based prospecting and exploration has been conducted since its discovery. In some areas, ore content was assessed by drilling to depths of 100 m. As a result of this work, several uranium ore occurrences and radioactive anomalies were identified. The ore zones were given a negative assessment due to the small scale of the ore bodies and low uranium grades.

In 1997, the Gurvansaikhan joint venture resumed exploration on the Ulziit deposit. An autogamma spectrometric survey, of the most promising areas, was conducted, as well as a small amount of shallow verification drilling (50–80 m depth) and geologic mapping. As a result, the widespread contamination of sedimentary deposits on the surface with uranium was confirmed, and according to drill hole data, the intense development of oxidation at depth was established, and based on this data single ore intercepts met the standard parameters for in situ leaching. In 1998, due to a sharp drop in uranium prices, the work of the Gurvansaikhan joint venture was stopped and didn't resume until 2005. Exploration in the Ulziit depression resumed in 2008 and resulted in the discovery of the Ulziit deposit.

The formation of the Ulziit depression occurred at the final stage of the Late Mesozoic orogene as a result of vertical block movements of the earth's crust. The regional tectonic structures that determine the position of the depression are the faults included in the system of the Main Mongolian Lineament, which determined its general morphology and northeastern orientation. Younger faults with a transverse NW strike divided the Ulziit depression into a number of blocks of different sizes, characterized by varying amplitudes of vertical movement and, accordingly, different thicknesses and compositions of the terrigenous-sedimentary deposits.

The Ulziit depression, like other well-known uranium-bearing Late Mesozoic depressions in the Mongolian Gobi, is composed of Lower Cretaceous deposits (Zuunbayan Formation) and Upper Cretaceous—Paleogene sediments. Quaternary formations are widespread and are represented by eluvial sediments a few meters thick, and in the valleys of dry riverbeds proluvium up to 20 m thick is developed. The deposits of the Zuunbayan Formation are ore-bearing. Their composition, according to lithological and facial characteristics, is comprised of the accumulation of sedimentary rocks of the foothill-fan complex and sediments of the alluvial-lacustrine valley. In the modern section, these accumulations lie in the northwestern side and are represented by a complex alternation of rocks of clay-sand-gravel-pebble-boulder composition. The section is characterized by the presence of large blocky fragments up to 2–3 m in diameter. The thickness of the deposits exceeds 200 m. The alluvial complex is represented by two facies zones: channel and floodplain-marsh. The section of the channel zone is dominated by sands of different grain sizes; gravelites, and conglomerates are less common. The thickness of sediments in this zone, in the area studied, is about 200 m, and the sandy facies comprises more than half. Sediments of the floodplain-marsh zone are represented primarily by siltstones and, to a lesser extent, clays. Sandy facies are developed locally and, according to observations in the coal quarries in the region, lie in incisions of narrow lens-shaped bodies with a thickness of a few meters to 10–15 m. Layers and interlayers of brown coal are characteristic in the section. Lacustrine accumulations are characterized by a homogeneous structure and are represented predominantly by finely dispersed clays, with occasional carbonaceous clays.

Oxidation in the rocks of the sedimentary cover, which are associated with hydrogenous accumulations of uranium, is intense. According to the research results, surface oxidation and zones of formational limonitization are identified in the Ulziit depression. Oxidation is distributed over a large area, beginning in the upper part of the section and, depending on the terrain, has a thickness from a few meters up to 30–40 m. In the area of groundwater discharge in the southeastern segment of the depression, oxidation has not been established. In most cases, there are no elevated concentrations of uranium in the surface oxidation zone, and only occasional anomalous increases and some thin ore intersections with low grades. It should be noted, that all known ore occurrences, ore concentrations reaching uranium content of up to 0.1%, occur spatially among rocks oxidized from the surface.

Zones of sheet limonitization appear only in the northwestern and near-side sections of the basin. These zones develop from the crystalline frame deep into the basin at a distance of 4–10 km. Near the basement outcrops, the section of the sedimentary cover is often completely oxidized from the surface to a depth of about 200 m, representing a single zone of limonitized rocks. In the pinch-out areas, there is a division of a single zone into tongues, which, developed along the most permeable rocks. This resulted in a clear layered morphology with the thickness of individual oxidation zones from 10–20 up to 40–50 m. The ore deposits of the Ulziit deposit are associated with the pinch-out of oxidation zones, as well as many known radioactive anomalies and ore intersections located at depths from 50–70 up to 150–250 m. This includes deposits that meet the economic requirements for in situ leaching.

The Ulziit deposit is located in the central part of the Ulziit depression, near its northwestern side. The crystalline frame of the depression is composed primarily of granitoids of different ages, as well as Jurassic and Lower Cretaceous volcanic rocks. The crystalline frame closest to the deposit is composed of Middle-Upper Jurassic granites and granosyenites, characterized by a high radiogeochemical background and elevated uranium grades. This factor is very important for sandstone type of uranium mineralization. Granitic rocks containing a significant proportion of free uranium are the most favorable source for its entry into a sedimentary basin.

The sedimentary cover at the deposit site is composed of Cretaceous terrigenous deposits, the thickness of which is a few hundred meters. There are two primary lithologic differences in the structure of the sedimentary strata. The first includes variegated alluvial-proluvial and colluvial-deluvial formations. The second is a gray-colored coal-bearing lake-marsh formation (Fig. 4.54).

Deposits of the first association are represented by coarse, poorly sorted, weakly rounded boulder-block deposits, conglomerates and conglobreccias of granitoids, as well as arkosic gravelites and sandstones. They are distributed along the edge of the depression, and also form a significant alluvial cone, up to 5 km long.

Lacustrine-marsh, and to a lesser extent alluvial-lacustrine deposits of the second association are represented by gray-colored conglomerates, conglobreccias, sandstones, siltstones, enriched in carbonized organic remains, carbonaceous clays and thin layers and horizons of brown coals. The deposits of this lithological association have a clear tendency to increase toward the center of the depression and facies replace the variegated deposits along the edge. The section is characterized by frequent rhythmic interlayering of different lithological and genetic types of sediments, which also indicates their facies transitions.

Uranium mineralization is localized in permeable and weakly permeable terrigenous-sedimentary Cretaceous deposits. In general, mineralization is controlled by the boundary of facies replacement of the first lithological coarse clastic association with the second fine- and fine-clastic one and the replacement of secondary oxidized rocks with gray-colored non-oxidized rocks (Fig. 4.55).

The oxidized front develops along the most permeable sediments of the coarse variegated formation from the side of the depression, along the sediments of the alluvial fan, as well as along fracture zones near the faults.

Uranium mineralization is spatially confined to the contact of gray coal-bearing rocks with reducing potential with oxidized sediments. Multi-level ore bodies of complex lens-, ribbon-, and roll-shaped forms are distributed at depths from 10 to 180 m. The most significant mineralization is confined to the top of the gray-colored strata and to the western border of facies replacement of rocks. In the section, ore bodies occur in accordance with the boundary of the facies transition from variegated to gray-colored deposits, enriched in carbonized organic remains, often repeating the complex bay- and stream-like forms of facies transitions.

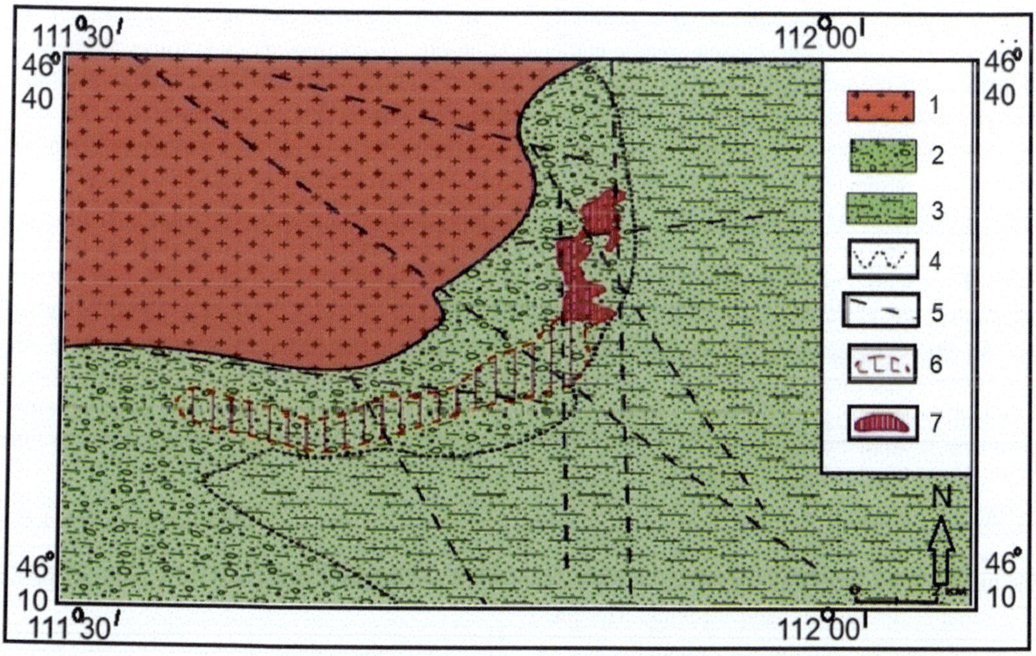

Fig. 4.54 Geological position of the Ulziit deposit (Grechukhin 2018): 1—Jurassic granites framing the depression. Cretaceous continental deposits: 2—large fan (conglomerates, conglobreccias, arkosic sandstones); 3—lacustrine-marsh coal-bearing deposits (gravelites, sandstones, siltstones, clays, brown coals); 4—boundary of facies replacement of fan deposits; 5—faults; 6—anomalous radiogeochemical zone in coal-bearing deposits; 7—uranium ore bodies of the Ulziit deposit

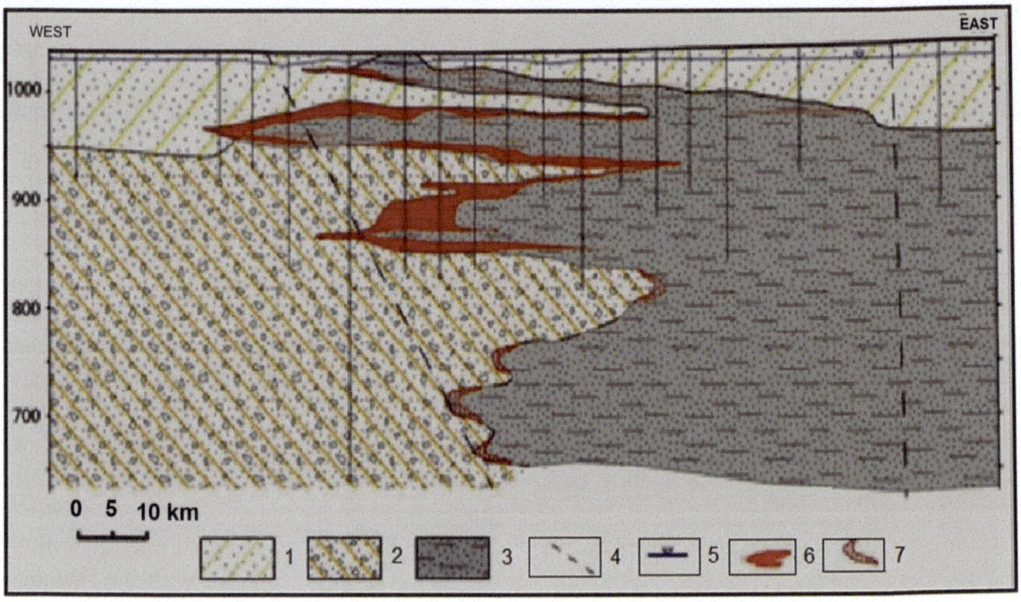

Fig. 4.55 Schematic geological section of the Ulziit deposit (Grechukhin 2018): 1—surface and ground secondary oxidation developing along permeable sandy-conglomerate chalk deposits; 2—bed-pore secondary oxidation developing along the Cretaceous coarse sediments of the fan; 3—gray carbon-bearing lake-marsh chalk deposits with a reducing geochemical environment; 4—suspected faults; 5—static groundwater level; uranium ore bodies: 6—explored; 7—prornozic

The contrasting reduction barrier is primarily due to the presence of carbonized organic matter in lake-marsh sediments and is partly associated with pyritization zones. The most contrasting uranium mineralization gravitates to linear zones along faults, where local manifestations of pyritization are also noted.

In plan, the ore bodies have a linear elongated shape with a northeastern strike. The most productive ore body lies at a depth of 25–40 and 50–90 m. In general, uranium mineralization covers the entire sedimentary section in the near-wall part of the depression and is developed to a depth of 180 m.

The ores of the Ulziit deposit are low grade, the uranium grade varies from 0.01 to 0.05% and averages 0.035%. The ores are aluminosilicate in chemical composition, carbonate-free and generally suitable for sulfuric acid leaching. The predominant part of uranium ore mineralization is represented by tetravalent uranium phosphosilicate (phosphorus-containing coffinitite), which is typical for similar sandstone type deposits in Mongolia and Transbaikalia in Russia.

In the Central Geological Laboratory of Mongolia (Ulaanbaatar), a sulfuric acid leach test was conducted and high recovery rates (86–97%) were obtained. This proves the fundamental possibility of extracting uranium from the Ulziit deposit using borehole in situ leaching.

The deposit's calculated reserves to date are estimated at a few thousand tons, which classifies it as small. However, there are good prospects that could significantly increase the resources through exploration along the flanks of the deposit and at depth, as well as in other promising areas in the Ulziit depression adjacent to the known deposits.

4.12 Dald Deposit

The Dald deposit is located 168 km northwest of the city of Tsetserleg in the Arkhangai aimag. The deposit was discovered by Soviet geologists in 1982 during reconnaissance work. Metallogenically, the deposit is located in the

Chuluut uranium ore district of the Butel-Khangai zone (Jamsrandorj 2019).

In 1982, the 19th team of the Mongolian Geological Survey Expedition (MGSE) identified anomalies 016 and 017 when detailing gamma spectrometric data. In 1984, during airborne geophysical work at a scale of 1:200,000 in northwestern Mongolia, the first MGSE team identified 7 radiometric anomalies. Exploration conducted at individual sites included 500 linear meters of geological investigation, 1490 linear kilometers of autogamma spectrometric surveys, and collection of 1460 lithogeochemical samples. In addition, electrical geophysical exploration was conducted on 4.6 km^2 and 22.1 linear meters of hand dug holes and 415 linear meters of drilling were completed (Russian URB-2A). As a result the Bayar, Sul, Daagiin Gol ore occurrences and a number of mineralized areas were identified.

As of 2009 the Mongolian company, Nuclear Energy Mongolia, received a license to explore the Bayar and Sul ore occurrences and conducted geological exploration for uranium. Nuclear Energy Mongolia conducted geological exploration from 2009 to 2012 on a scale of 1:50,000 on an area of 2700 ha that included a radiometric survey on an area of 2.36 km^2. They also diamond drilled 16,319.12 linear meters for exploration and assessment of ore reserves. As a result, these ore occurrences turned out to be a single deposit consisting of ore bodies Dalt-1, Dalt-2, Dalt-3 and Dalt-4 which were combined into one deposit, called the Dald deposit. In the area of the deposit, the Egindaban granitoid intrusive complex of Late Triassic-Early Jurassic rocks, with high levels of radioactivity, became widespread (Samovich et al. 1982).

The geological structure of the Chuluut uranium ore district includes Paleoproterozoic undifferentiated metamorphic rocks, Lower-Middle Devonian middle and acidic volcanics of the Nariin formation, Pleistocene loose, Middle and Late Pleistocene basalts and Holocene loose sediments (Fig. 4.56). Paleoproterozoic undifferentiated metamorphic rocks are represented by biotite, biotite-hornblende, amphibolite, quartz-plagioclase, biotite-amphibole gneisses, granite gneisses, migmatites, plagiogneisses, crystalline schists, quartzites and lenses and interlayers of limestones. In addition to metamorphic rocks, dikes of diorites, plagiogranites, and pegmatites of quartz-feldspathic composition are often found.

Rocks of the Nariin formation are distributed in the central part of the district in the interfluve of Shargachingol and Daagiin Gol and are represented by lavas, tuff lavas, rhyolite-porphyry lava breccias and tuff siltstones. The total thickness is about 1400 m.

Quaternary deposits are widespread within the region. They are important for uranium concentration. The valleys of large river systems and their watercourses and intermountain depressions are filled with loose sediments. According to their age range, they are divided into lower-middle Pleistocene, middle-upper Pleistocene and Holocene, and according to their genesis they can be divided into alluvial, alluvial-proluvial and proluvial-deluvial (Fig. 4.57).

Lower and Middle Pleistocene loose sediments have limited distribution and are recorded in the lower half of the river valley and on Shargachingol, on its southern residual terraces, up to 60 m above the river water level. Genetically they are proluvial and mixed with alluvial. The sediments consist of gravel, gravel, sand, sandy loam and loam. The thickness of these deposits increases from top to bottom along the river, from the sides to the floodplain, from the narrowed part to the expansion of the valley. Lithologically, the sediment composition is dominated by pebbles and pebbles with sandy filler; boulders make up 10–20%. Their thickness ranges from 5 to 20–30 m. Sediments of this age are the main uranium-bearing deposits of the Dald deposit.

Uranium-bearing deposits are overlain by intermediate-level Quaternary basalts (Devyatkin 2004; Byamba 2012). There are no lower-level Neogene basalts within the deposit. Samples from mid-level basalts on the river terrace of the Chuluut river are dated to be 200 + 20

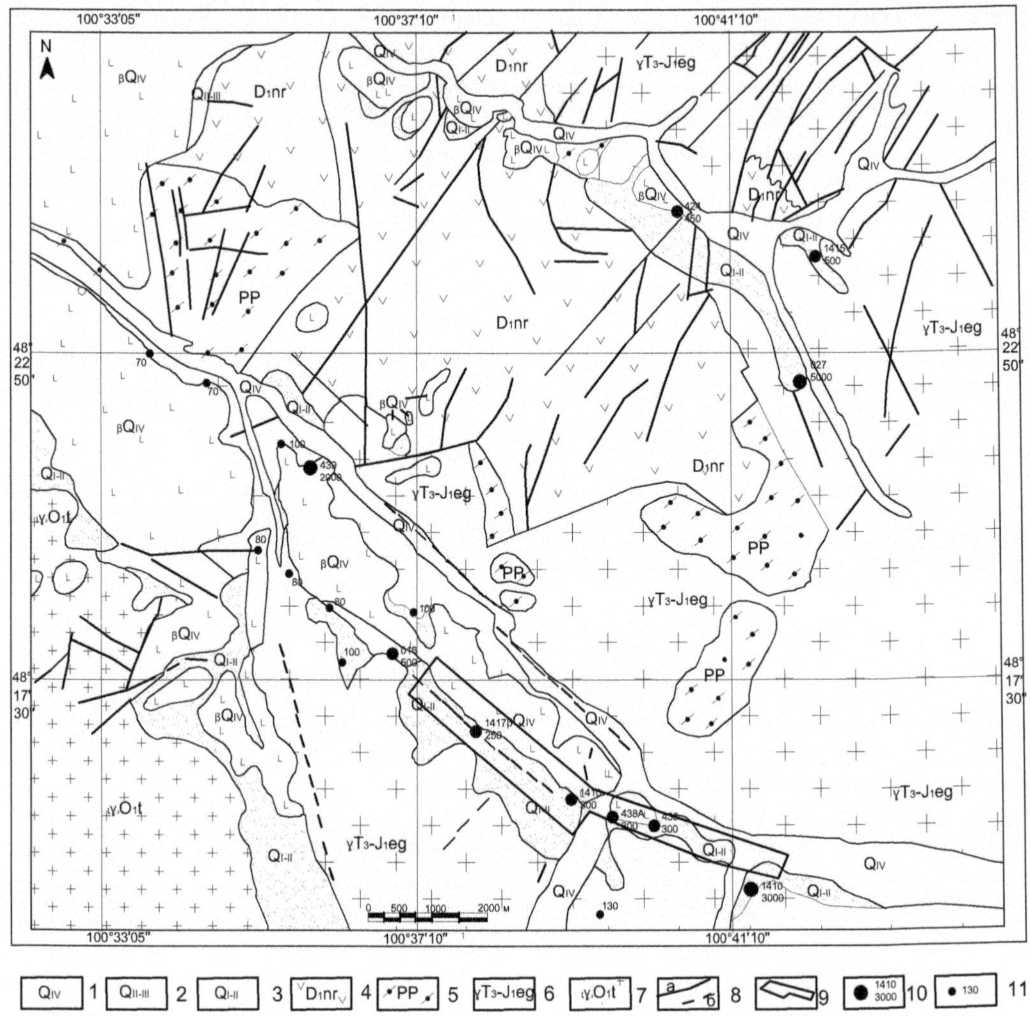

Fig. 4.56 Geological map and map of mineral resources of the Chuluut district (Jamsrandorj et al. 2019). 1—Holocene: alluvial clays, loams, sandy loams, pebbles, pebbles, sands, spongy olivine basalts; 2—middle, upper Pleistocene: alluvial-proluvial sands, sandy loams, clays, massive and porous basalts; 3—lower, middle Pleistocene: colluvial, colluvial-proluvial, alluvial-proluvial debris, sandy loam, loam; 4—Lower Devonian; Nariin formation: dacites, andesidacite porphyries, rhyolite porphyries and their tuffs, thin layers of siltstones; 5—Paleoproterozoic: gneiss-schist strata; biotite, biotite-hornblende, epidote-hornblende gneisses, crystalline schists with thin layers of marble; 6—Late Triassic—Early Jurassic; Egindavaa intrusive complex: biotite, biotite-amphibole granites, granosyenites, granodiorites; 7—Early Ordovician; Telmen intrusive complex: fine- and medium-grained leucogranites; 8—faults: a—established; b—expected; 9—Dald deposit; 10—radiometric anomalies: anomaly numbers—in the numerator, microR\h—in the denominator; 11—points with increased radioactivity, microR\h

and 280+30 thousand years old (Logachev et al. 1982), which corresponds to the Middle-Upper Pleistocene. Along the section, the Middle and Upper Pleistocene basalts have a thickness of 54–62 m. In the lower part of the section, they have a dense texture, and in the upper part they have a spongy texture. At the junction of the rivers Shargachingol and Daagiingol basalts form a high terrace-like plateau and even exhibit volcanic cauldrons.

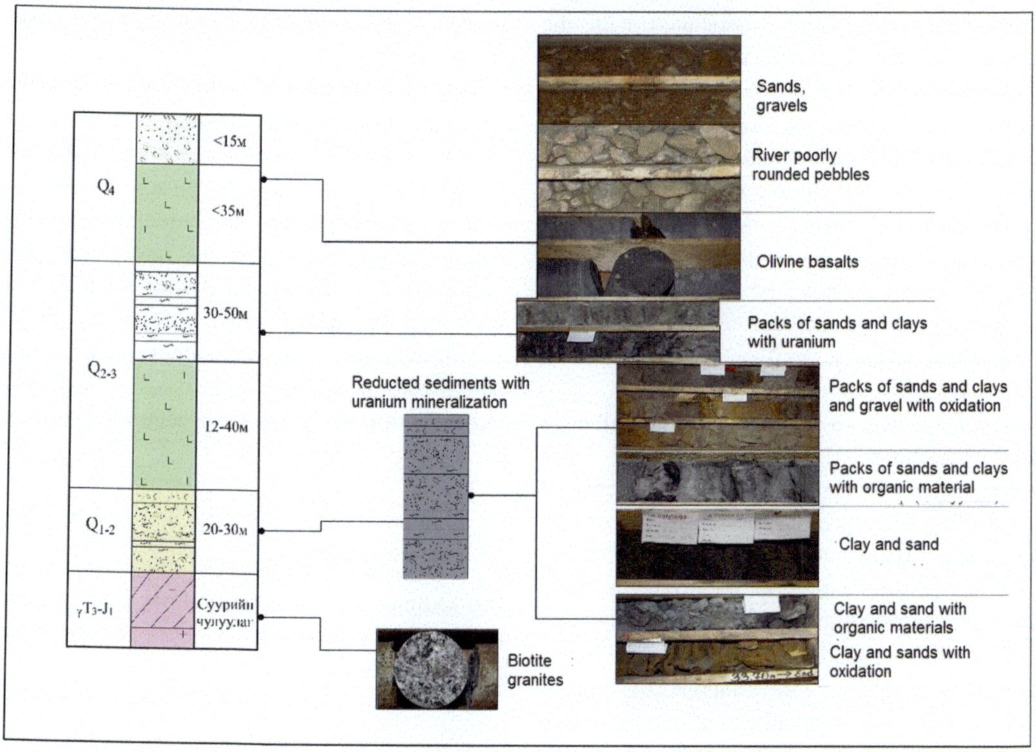

Fig. 4.57 Section of Quaternary sediments of the Dald deposit (Jamsrandorj et al. 2019)

Middle and Upper Pleistocene loose sediments 70–100 m thick consist of loams, clays and sands with inclusions of interlayers of sand, gravel and pebbles. These deposits are underlain by upper level basalts.

Within the deposit, Holocene sediments include upper-level basalts and unconsolidated sediments overlying them. The upper-level olivine basalts are gray and black in color, have a dense texture with plate-like separation, and are about 12–40 m thick.

Holocene loose sediments—alluvial, alluvial-proluvial formations. The alluvial variety is common in the valleys of the river. Chuluut, Azarga, Nomgon, Shargachingol, Daagiingol, and their tributaries. These are well-rounded boulders, pebbles, pebbles with cement consisting of sand, sandy loam, and loam. The raft of these valleys is subject to strong tectonic influence and has a block structure, due to which the thickness of loose sediments ranges from 15 to 30 m.

Alluvial-proluvial deposits are distributed in the sides and terraces of the river. Chuluut, Azarga, Nomgon, Shargachingol, and Daagiingol and consist of granite debris, semi-rounded pebbles, gravel, sandy loam, and loam.

Among the intrusive rocks, granitoids of the Telmen complex of the Early Ordovician and the Egindabaa complex of the Late Triassic-Early Jurassic became widespread. Granitoids of the Telmen complex form small massifs of mainly granite and leucogranite composition in the southern part of the district. The intrusive outcrops of the Egindabaa complex almost completely frame and underlie the Dald deposit. The composition of the intrusions is dominated by leucogranites; their age was dated by P. V. Koval K–Ar, Rb–Sr method (166–146–127 Ma). Uranium-bearing loose sediments lie directly on the intrusive massifs of the Egindabaa complex. The increased uranium grade of the Egindabaa intrusive complex was established in 1984 during a gamma spectrometric survey. Radioactive anomalies were spatially associated with the massifs of the Egindabaa complex, which was the basis for the assumption (Markov and Samovich 1984) that the uranium grade of the area is most likely associated with the granitoids of the Egindabaa granitoid complex.

The area has a widely developed fault system, and it played a role in the formation of the river valleys. Thus, the Shargachingol and Daagiingol valleys are formed along subparallel faults with a northwestern orientation (Fig. 4.58).

The Dald deposit is located in the valley of the Shargachingol river and consists of 4 ore bodies (Figs. 4.58 and 4.59). The Shargachingol river valley was traced to the northwestern section and is 10 km long with a width of 1.5 km. Uranium mineralization is localized in loose sediments consisting of interlayered fine- and medium-grained sands, crushed stone and loams with organic residue. Ore minerals are otenite, metaotenite, and to a small extent other uranium minerals, as well as uranium-containing rutile and pyrite.

Ore beds are located in lower and middle Pleistocene loose sediments. Holocene sediments overlie them conformably. The boundaries of the ore layers are separated by a pebble-sand horizon with boulders, which is clearly visible on the Dald-I ore body. In the area of the Dald-IV ore body, the wells show the presence of a basalt horizon on two levels (Pleistocene and Holocene).

Pleistocene sediments can be divided into 3 types: sands and sandy loams, clays and loams and pebbles and pebbles. Sands and sandy loams are heterogeneous, good differentiation is observed in fine-grained varieties, organic materials and sulfides are present, favorable for the deposition of uranium. Clays and loams are often black and gray, very intensely carbonized, and are often interbedded with layers of sand and sandy loam. Layers of clay, sand, pebbles and remnants of vegetation were noted. The clayey material probably played the role of a waterproof screen. In the loose sediments of large valleys, organic remains are noted, which created a reducing environment for the deposition of uranium.

The total thickness of the loose sediments filling the valley is about 30–40 m.

The oxidation and reduction process is widely observed in loose sediments. Oxidation is expressed by limonitization and hematitization. Reduction is represented by sulfidization, kaolinization, chloritization, montmorlinitization with organic material. Uranium bearing sediments are sometimes lithified and exhibit carbonatization. Three types of oxidation and reduction processes have been identified at the deposit:

a. Oxidation zone. Hematite and limonite are formed abundantly.

The sediments are colored yellow, light yellow and red-brown. This zone is developed above or below the reduction zone.

b. Oxidation and reduction zone. Interlayering of oxidized and reduced rocks. Hematite and limonite color the rocks in a red-auto-yellow, light yellow color; among them, restored rock packs are recorded. Uranium grades in this zone are relatively high. The zone extends above the eluvium or in clay-pebble deposits.
c. Recuction zone. It is painted in darker colors and contains a lot of carbonized material.

The uranium grade is directly dependent on the grade of carbonized materials. Pyrite crystals are often observed in the reduction zone

Dalt-I Ore Body. Extends 700–4000 m with a width of 200–500 m and has been explored with 157 drill holes. The average thickness of the uranium layer is 4.8 m and the maximum thickness reaches 15.9 m. The average uranium grade is 0.026%. The ore body lies under a cover of thin rocks.

Dalt-II Ore Body. Was drilled on a grid of 50×50 m with 36 holes. The length of the ore body is 450 m, the width on average is 200 m.

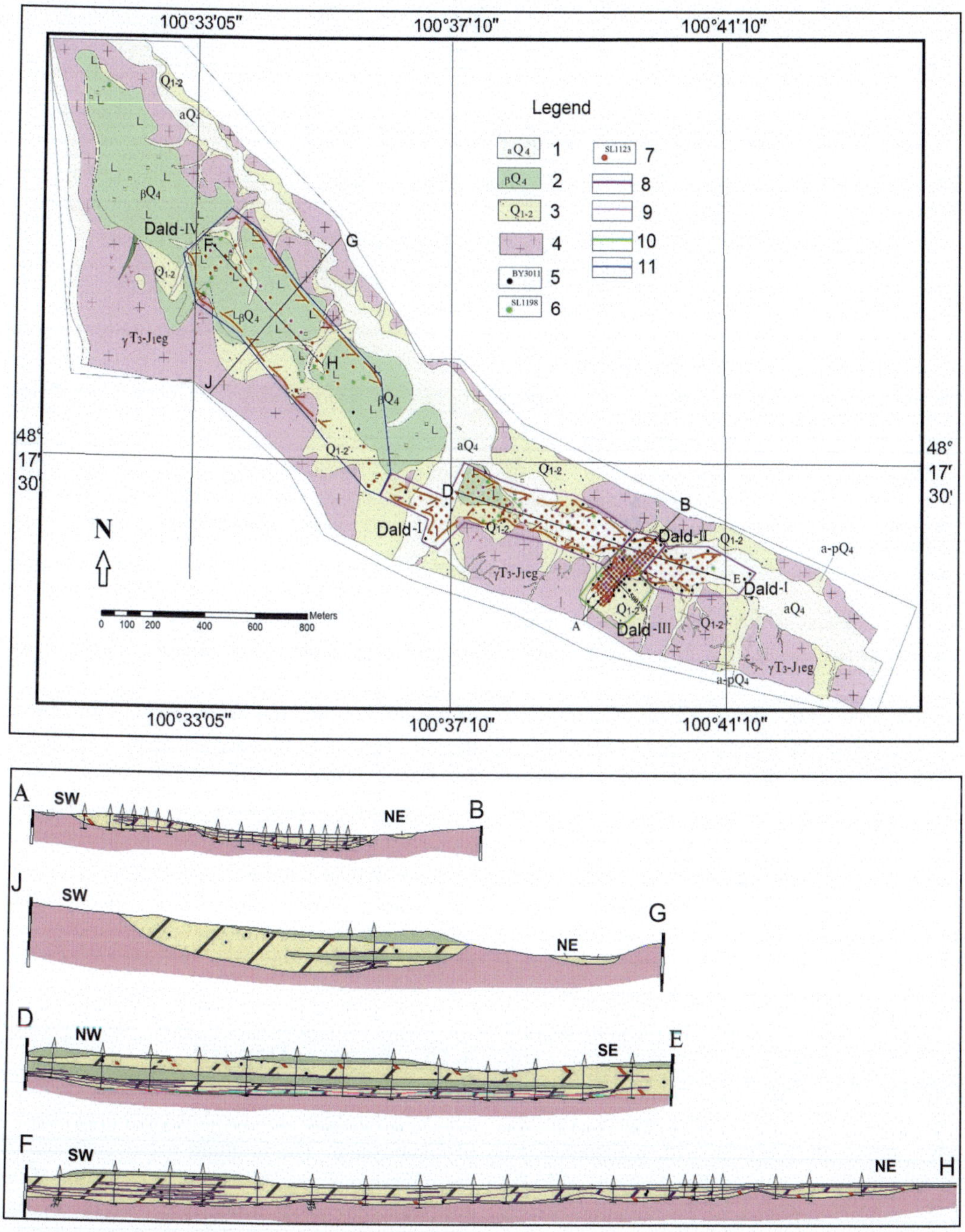

Fig. 4.58 Geological sketch of the Dald deposit (Jamsrandorj et al. 2019): 1—Holocene: alluvial boulders, pebbles, sands, loams, pebbles; 2—Holocene: spongy olivine basalts; 3—lower-middle Pleistocene: river and lake loams with sands and sandy loams; 4—Late Triassic-Early Jurassic: Egindabaa intrusive complex; biotite granites, subalkaline granites; holes with uranium grade: 5—less than 0.01%; 6—0.1–0.02%; 7—0.02–0.05%: ore bodies of the Dald deposit: 8—Dald-I; 9—Dald-II, 10—Dald-III, 11—Dald-IV. **a** Geological sections through the Dald deposit

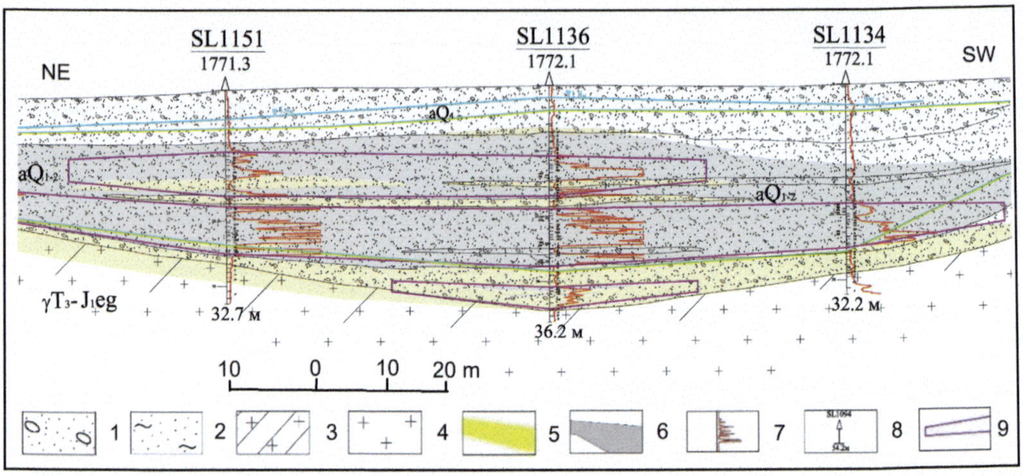

Fig. 4.59 Section through exploration line 500150 of the Dald deposit: 1—alluvial and lacustrine sands, sandy loams, and crushed stones; 2—loams containing sand and sandy loam; 3—weathered granites; 4—biotite granites; 5—oxidation zone; 6—recovery zone; 7—gamma-ray logging curves of bore holes; 8—borehole; 9—uranium ore bodies (Jamsrandorj et al. 2019)

And the thickness is 5.6 m. The uranium grade varies in the range of 0.01–1.25%.

Dalt-III Ore Body. Was drilled on a grid of 50×50 m with 90 holes over a distance of 400 m. The width of the ore-bearing layer is about 200 m. The uranium horizon consists of 2–3 layers, the thickness is 1.0–13.10 m, and averages 5.6 m. The uranium grade ranges from 0.01 to 0.5%, the highest uranium grade observed is 1.31% with a thickness of 0.3 m.

Dalt-IV Ore Body. It is a natural continuation of the Dalt-I ore body to the southeast and stretches for more than 3800 m with a width of 600 m. 51 holes have been drill with a more open grid than the other bodies. The ore body consists of 3 blocks with a thickness of 0.5–14.3 m. The depth of the ore body is 45.0–112.0 m from the surface. The uranium grade varies from 0.01 to 1.33%.

Ore Minerals: uranium-pyrite, uranium-rutile, uraninite, meta-otenite, non-ore minerals—calcite, gypsum, quartz, albite, illite, and dickite (Fig. 4.60).

Thus, the Dald deposit was formed under the influence of oxidation and the reduction processes occurring in the Pleistocene loose sediments in near-surface conditions, as indicated by the presence of minerals such as gypsum and calcite in the ore. Based on the data obtained, there is reason to classify the deposit as near-surface according to the IAEA classification. Outcrops of granitoids of the Egindaba complex with high uranium content are considered a source of uranium in loose sediments.

The Dald deposit is undoubtedly an economically viable deposit when developed and mined using in situ leaching.

4.13 Engerar Deposit

The Engerar deposit is located 160 km southeast of Ulaanbaatar. The uranium ore occurrence of the same name was discovered as a result of an airborne gamma spectrometry survey on a scale of 1:25,000, D. A. Samovich in 1984 gave the occurrence a negative assessment.

In 1985, the MGSE airborne geophysical team identified 17 radioactive anomalies, field checked them and came to the conclusion that only the Naidal ore occurrence had promise. The team made a recommendation to conduct exploration for uranium within the Cretaceous

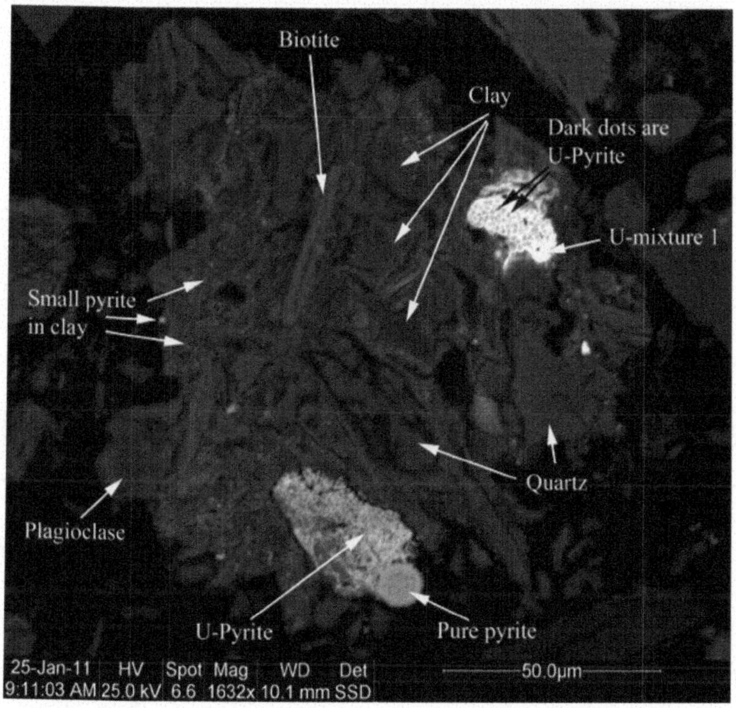

Fig. 4.60 Clay containing uranium minerals from the Dald deposit (Jamsrandorj et al. 2019)

depressions. Within the Shivee vulcano-tectonic structure, during 1985–1989, a Mongolian geological expedition conducted a 1:50,000 scale survey to assess the prospects for identifying ore occurrences of uranium, tungsten, fluorite, graphite and coal. Geologic field work was carried out at 13 sites; 6 of them are located in the Shivee VTS and surrounding area, 5 are in the endo- and exo contacts of the Bayanovoo granite massif, 1 is in the Tabznsubeet depression and 1 is in the Kerulen river basin.

The Mongolian company Lucria received a license for the area where the Engerar deposit is located in 2009. From 2009 to 2015 they conducted exploration to assess the uranium occurrence.

The Engerar deposit is located in the Shivee uranium ore district at the intersection of the Severogobi and Kharayrag-Choyr deep faults. E. G. Shishkin and L. P. Gavrilov identified the Niyalga and Maant mega blocks based on the results of geological, structural and geophysical surveys (Fig. 4.61).

The geological structure of the area is similar to the geology of the Dornod VTS. The basement is composed of Proterozoic metamorphic rocks, Permian volcanics and Lower Paleozoic and Mesozoic intrusions, which are intruded by Early Triassic-Late Cretaceous granites that are divided into two mega blocks (Fig. 4.63). The Niyalga megablock is bounded in the north by the Tsagaan Jargalan fault, and in the east by the North Gobi fault. These disturbances created the conditions for the formation of the Niyalga depression. Several small basement projections were noted in the central and southern parts of the area. In other locations in the depression, the foundation is lowered, the roof is located at a depth of 100–400 m. The foundation is covered by volcanic-sedimentary depositsof the Tsagaantsav and Bayanshire formations of the Upper Cretaceous, which correspond in age to the uranium-bearing Bayantumen formation in the Dornod uranium district. Deposits of the Bayanshire formation are common in the north and rhyolites and their tuffs of the Tsagantsav

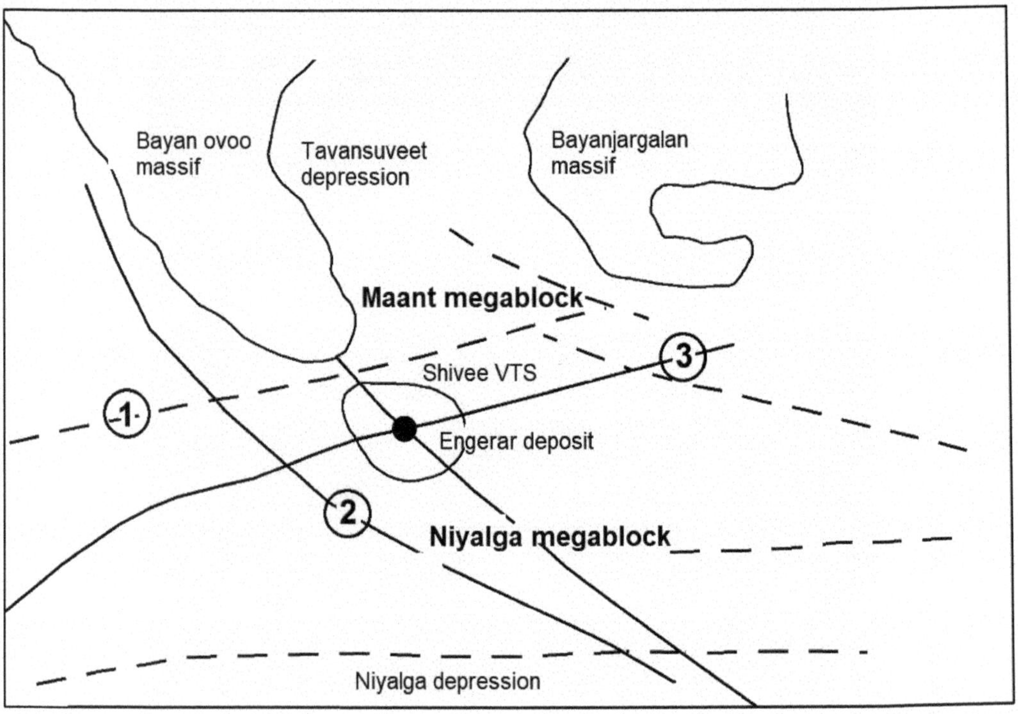

Fig. 4.61 Tectono-structural sketch of the area. Fault: 1—Tsagaan Jargalan; 2—Kharayrag-Choyr; 3—North Gobi

formation are widespread in the southern part of the Shivee VTS. Rhyolites and dacites cover an explosive facies consisting of pyroclastic tuffs and tuffaceous-sedimentary deposits at the top of the Tsagaantsav formation. They have a significant role in the deposition of uranium mineralization at the Engerar deposit. Rhyolites are fine- and medium-grained and massive with a porphyritic texture. Volcanic rocks have been subjected to low-temperature hydrothermal alterations: silicification, hydromicaization, montmorilinitization, etc. Mineralization (pyrite, arsenopyrite, fluorite, ilmenite) is dispersed along cracks in connection with hydrothermal processes. Intense radiometric anomalies associated with the dispersive distribution of uranophane in oxidized rhyolites was noted on the day surface in the trenches. Drill hole data notes that rhyolites of the Tsagaantsav formation are overlain by gray-colored conglomerates, sandstones, clays and shales with interlayers of lignites and brown coals of the Bayanshire formation with a thickness of about 90 m. This area of the deposit is divided by subparallel faults in a north-eastern direction (340°), and these faults are interpreted on cosmic and gamma spectrometric images (Fig. 4.62).

The deposits of the Bayanshire formation consist of interbedded gray siltstones, sandstones and conglomerates. Genetically they belong to lake formations. The sequence usually begins with basal conglomerates and includes shales with lignites and siltstones. At the top of the section, river sands and coarse-grained sediments predominate.

Fault tectonics are widely developed in the area, forming a zone of a junction of northwestern and northeastern regional faults. The faults are represented by zones of crushing and brecciation with quartz cement. Rhyolite and granite porphyry dikes are also found. The stratigraphic features of the Tsagantsav formation deposits at the Engerar deposit according to drill hole data are shown in Table 4.18.

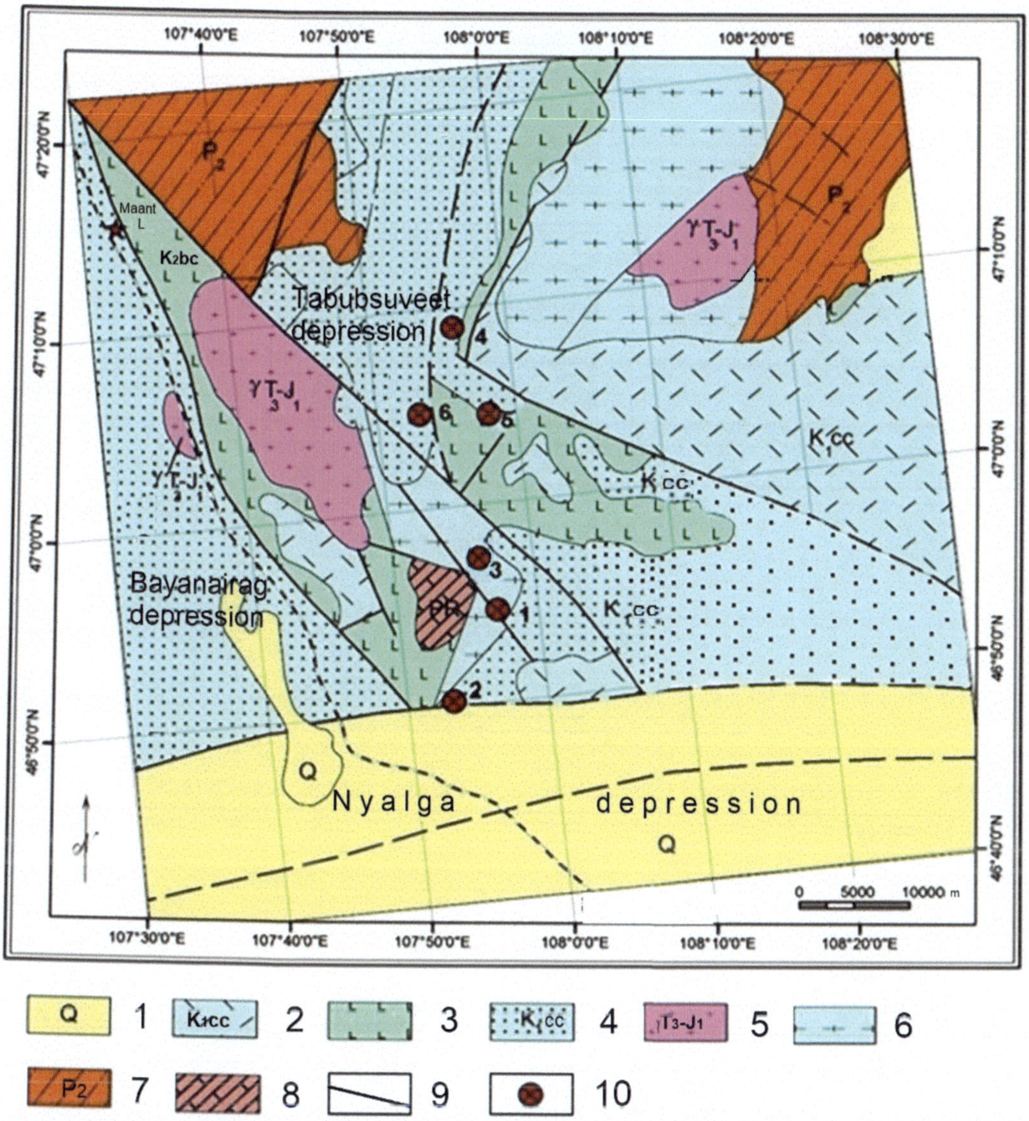

Fig. 4.62 Geological sketch of the Engerar deposit area (Samovich et al. 1982): 1—Quaternary deposits, rocks of the Tsagaantsav formation; 2—rhyolites, 3—basalts, 4—sandstones, shales, brown coals; 5—Late Mesozoic granitoids; 6—Early Paleozoic intrusions; 7—Permian deposits; 8—Proterozoic: limestones, shales; 9—faults; 10—uranium mineralization: 1—Naranbulag; 2—Elgen; 3—Engerar; 4—Naidal; 5—Durtai; 6—Tsaidamnuur

The deposits of the Bayanshire formation are divided into 6 stratigraphic horizons according to the drill hole data from 6 holes (Fig. 4.63).

The lower horizons of the Bayanshire formation include sediments in a reducing environment. These deposits may be of interest for future exploration of uranium mineralization, in depressions in the East Gobi region. There are plans to conduct exploration for uranium in the Tsaidam and Tabannsubeet depressions.

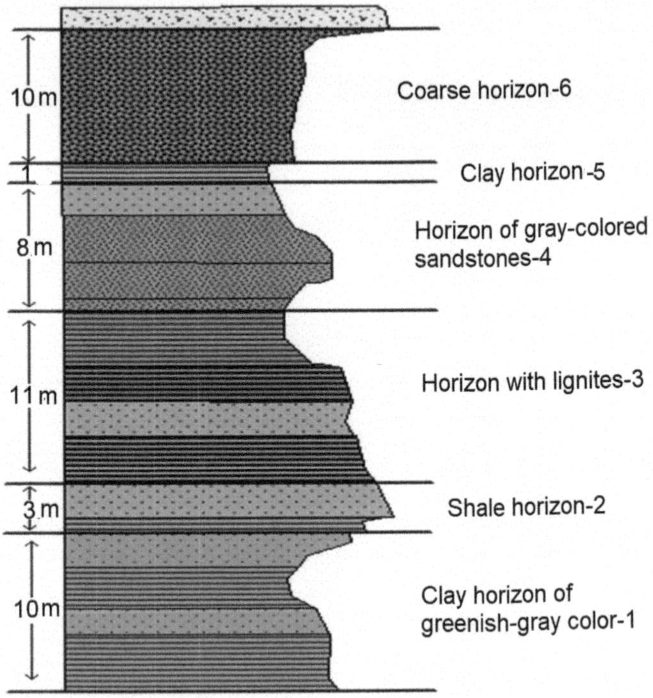

Fig. 4.63 Lithological column of the Bayanshire formation (Champan and Jamsrandorj 2018)

Table 4.18 Stratigraphic column of the Tsagaantsav formation (Champan and Jamsrandorj 2018)

Beds	Lithology	Thickness (m)
1	Volcanogenic-sedimentary rocks	10–35
2	Lithovitroclastic tuffs	25–60
3	Rhyolite lava with volcanic glass	10–30
4	Thin-bedded ash tuffs and rhyolites with flow texture	20
5	Massive and fluidal spheroidal rhyolites massive and with flow texture	300–350
6	Interbedded clastic lava and lava breccia with flow texture and spheroidal rhyolite	50

The Engerar uranium deposit is located in the southeastern part of the Shivee VTS in the fault zone with a northwestern direction. Mineralization of the deposit is localized in the upper part of the rhyolite section of the Tsagantsav formation, close to the lower contacts of sedimentary deposits of the Bayanshire formation (Figs. 4.64 and 4.65).

The Engerar deposit contains several industrial and non-industrial occurrences associated with rhyolites (Fig. 4.69). Uranium mineralization creates fracture zones in rhyolites and forms inclusions and scattered dissemination.

The near-ore alteration is represented by low-temperature silicification, argilitization, hematitization, and hydromica.

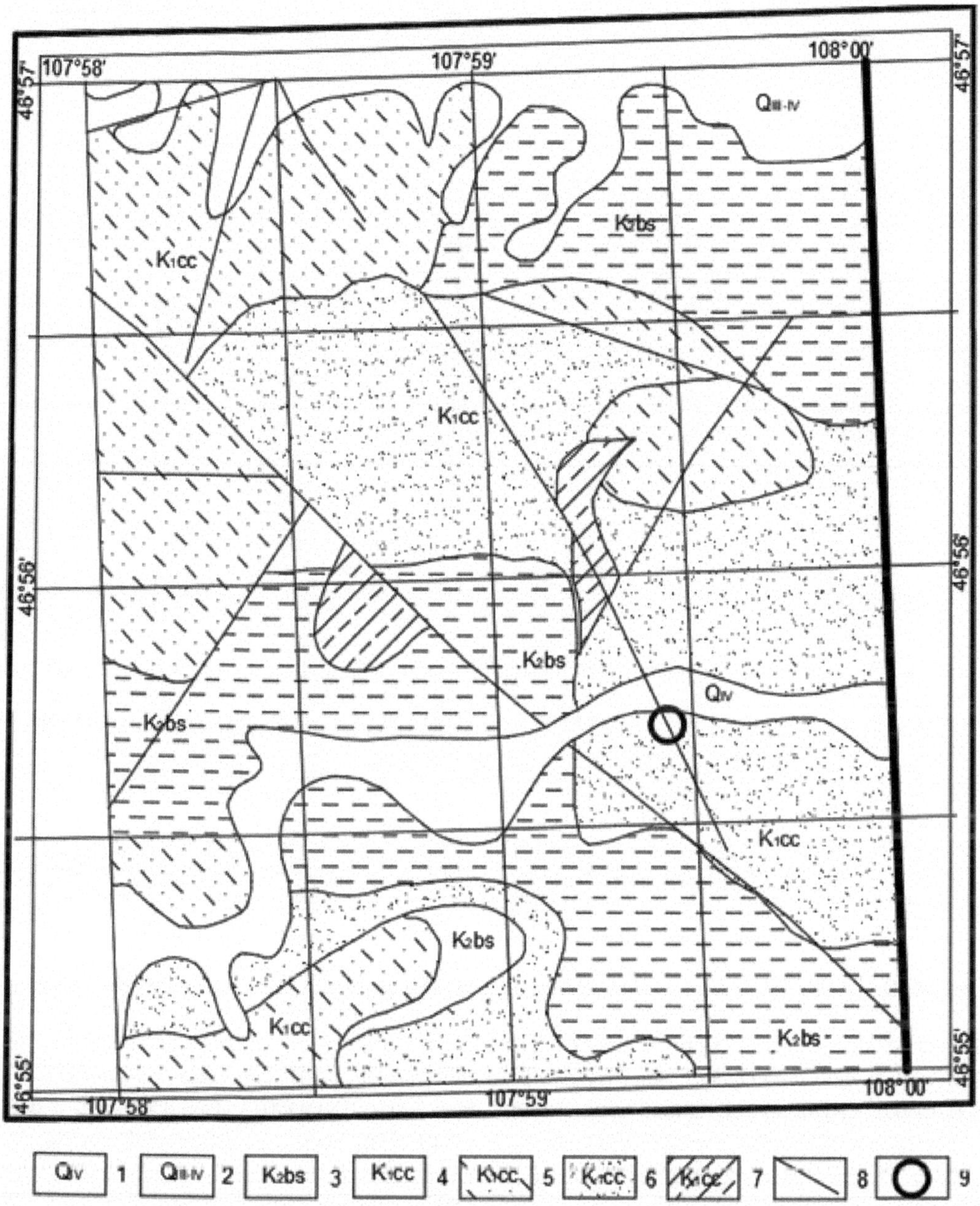

Fig. 4.64 Geology map of the Engerar deposit (Tungalag 2006): 1—modern Quaternary deposits: alluvial boulders, sands, sandy loams, loams, gravel, alluvial-proluvial and alluvial-lacustrine, clays, loams, sandy loams, crushed stones; 2—Upper Pleistocene-Holocene deposits: deluvial-proluvial boulders, rubbly sediments, siltstones, sands, loams, sandy loams; 3—Upper Cretaceous. Bayanshire formation. Red- and gray-colored clays, sandstones, mudstones; 4—Lower Cretaceous. Tsagaantsev formation. Basalts, trachybasalts, trachyandesites, trachyrhyolites, sandstones, conglomerates, siltstones, mudstones; 5—sandstones; 6—gravelites; 7—gravel; 8—faults: 1—established; 2—suspected; 9—Engerar deposit

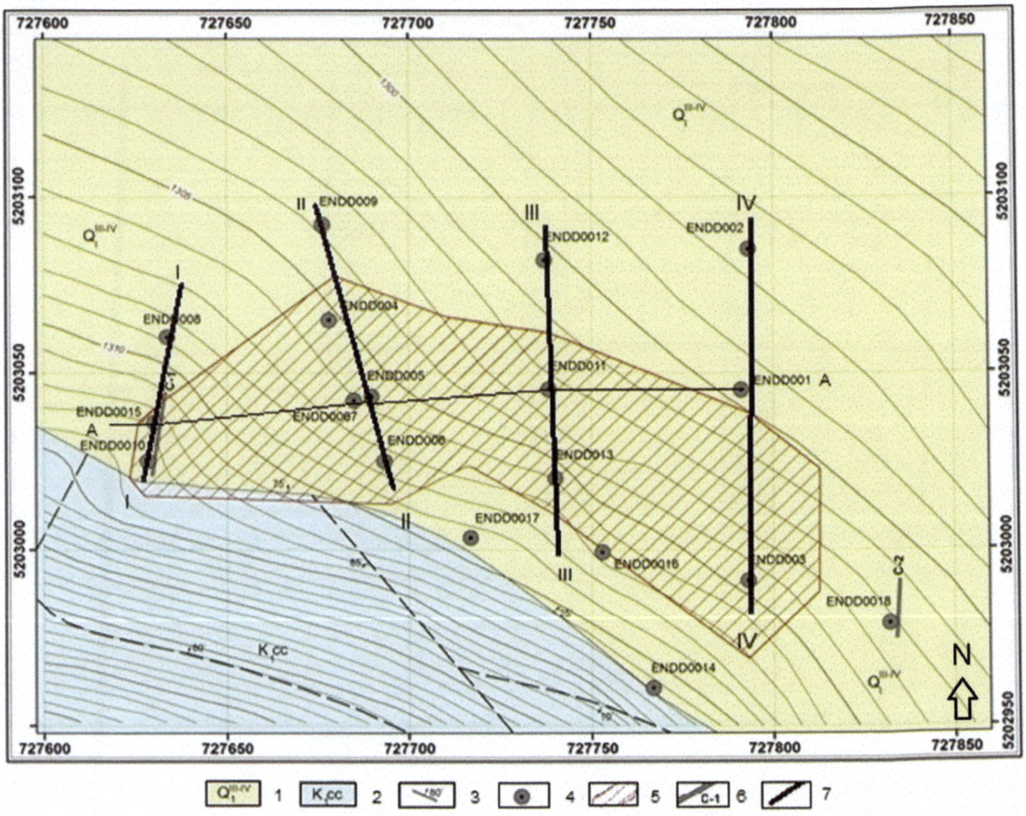

Fig. 4.65 Geological plan view of the Engerar deposit (Champan and Jamsrandorj 2018): 1—Upper Pleistocene—Holocene. Sands, clays, boulders, crushed stones, loams and sandy loams; 2—lower chalk. Tsagaantsav formation. Volcanic rocks; 3—faults and their dip angles; 4—exploration hole, their numbers; 5—trenches, their numbers; 6—horizontal projection of ore zones; 7—exploration lines and their numbers

Ore Body 1. Has a length of 100–105 m, is elongated in a latitudinal direction and has been studied to a depth of 25–80 m (Fig. 4.66). The thickness of the ore body in holes ENDD0015, ENDD007, END005, ENDD0010 is 3, 4, 15, 27 m, respectively, the average thickness is 15.3 m. The average uranium grade in the body is 0.06%.

Ore Body 2. Consists of several small subhorizontal lenses (Figs. 4.66 and 4.67). The average width of the ore body is 40 m, the average thickness is 2.5 m with an average uranium grade of 0.036%.

In the lower part of the section of exploration lines II, III and IV, of Fig. 4.67, there are numerous lenses with a thickness of 0.25–9.3 m intersected in the drill holes, which, due to the lack of a sufficient amount of drilling, cannot be classified as reserves. They are combined into ore zones I and II. Ore zone I is intersected by bore holes ENDD0013, ENDD0016 at a depth of 40–60 m from the surface and consists of 3 lenses with a thickness of 0.5 m and 3 lenses with a thickness of 2.0–4.6 m (Fig. 4.67). The uranium grade in these intervals ranges from 0.01 to 0.06%.

Ore zone II is established on lines IV in holes ENDD0003, ENDD0014 at a depth of 85–150 m. The thickness of 5 lenses varies from 0.25 to 1.0–4.5 m. The average uranium grade ranges from 0.016 to 0.238%.

Deeper than ore-bearing zone II, hole ENDD006 at a depth of 180 m intersected an

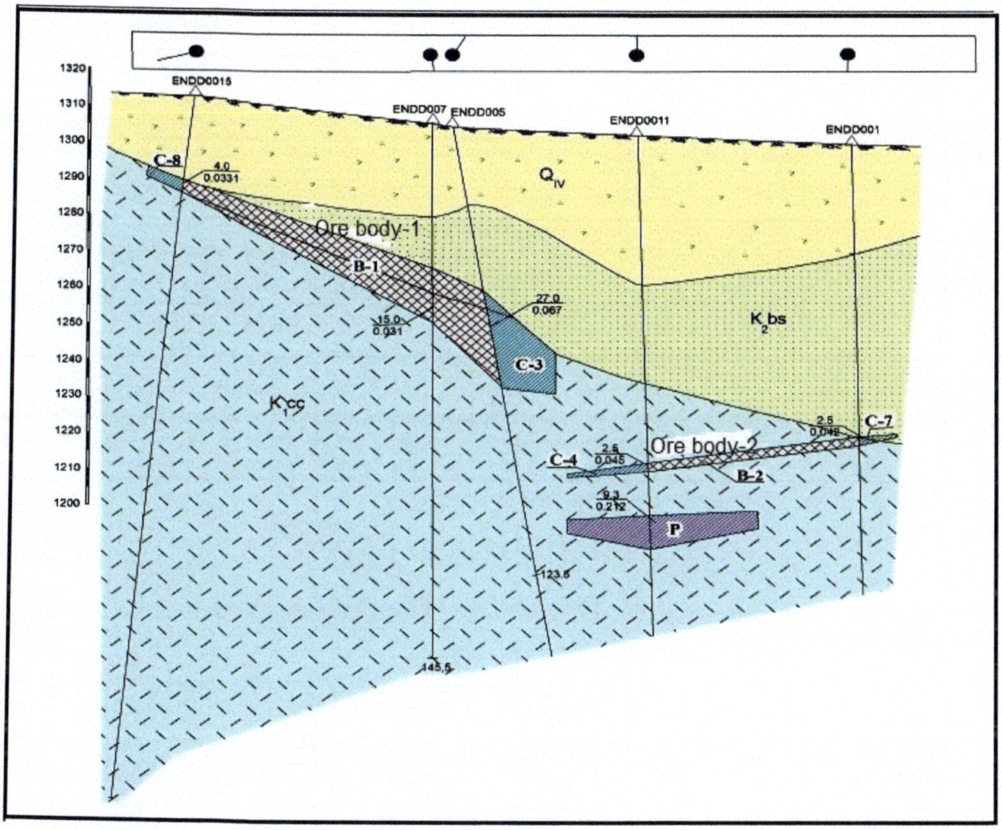

Fig. 4.66 Section through line AA (see Fig. 4.65) (Champan and Jamsrandorj 2018). For legend see Fig. 4.65

interval of 6 m, and in hole ENDD0017 at a depth of 48–72 m an interval of 2.05–4.4 m with a uranium grade of 0.116–0.303%. Mineralization is localized in the volcanic sedi mentary strata.

The primary minerals of uranium are uraninite, coffinite, uranium-containing leucoxene, uranium blacks, and in oxidation zones—uranophane and otenite. The uranium grad is directly dependent on the content of carbonized materials. Uraninite is fine-grained, its grain sizes are very small (0.004–0.02 mm) (Figs. 4.68 and 4.69).

Magnetite, goethite and martite are accessory minerals from primary rocks. Hydrogoethite forms dust-like aggregates embedded in quartz and feldspar crystals. Hydrogoethite was probably formed after biotites and is represented by small scaly grains.

The Engerar uranium deposit is hydrothermal in genesis, similar to the deposits of the Dornod uraniumdistrict. In terms of reserves, the deposit is classified as small (up to 10 thousand tons), although there are prospects for increasing reserves.

Conclusions

1. Sandstone uranium deposits are the primary deposits of Mongolia. These include deposits formed in terrigenous sediments of the Taphrogenic basins of Eastern and partly Central Mongolia, associated with the processes of formation and surface oxidation and

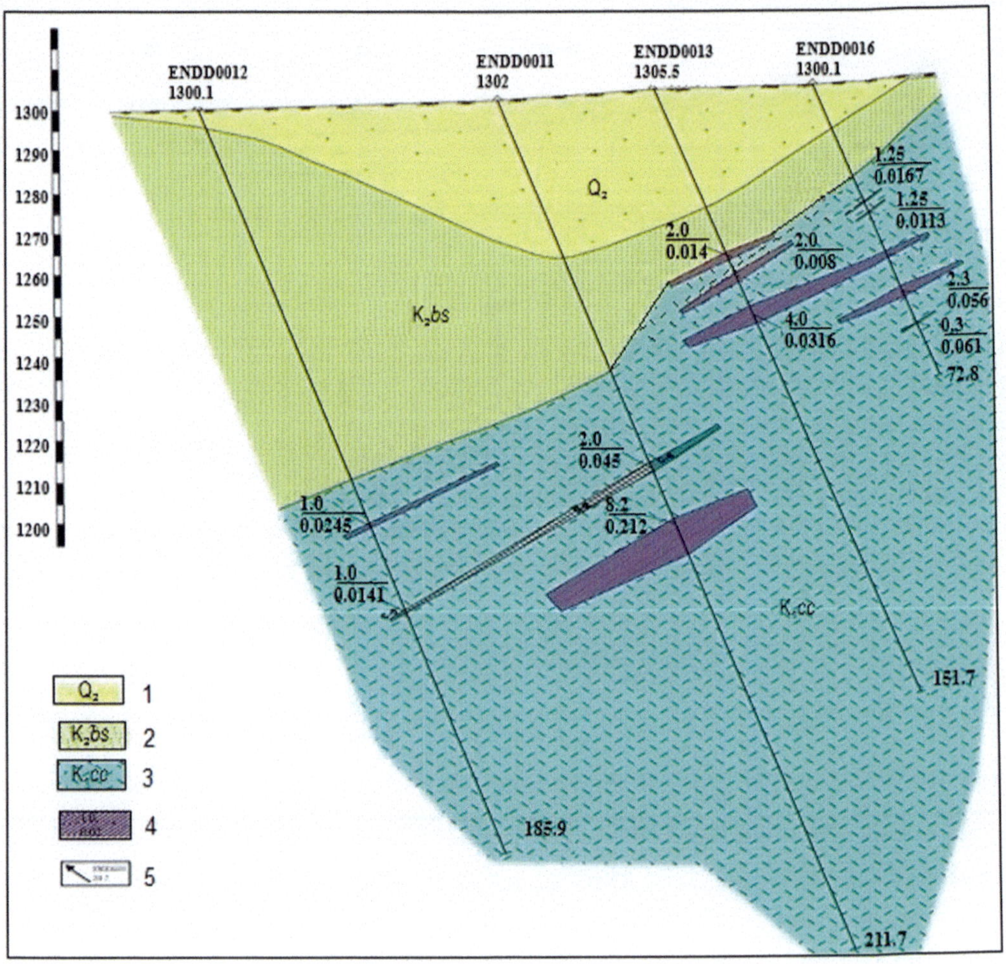

Fig. 4.67 Section through line III–III (see Fig. 4.65) (Champan and Jamsrandorj 2018). Quaternary deposits. 2—Upper Cretaceous, Bayanshire formation; 3—Lower Cretaceous, Tsagantsav formation; 4—ore bodies showing thickness and uranium content; 5—holes, their numbers and wellhead height

secondary reduction. Examples of such deposits are Kharaat, Khairkhan, Gurvansaikhan, Ulziit, Dulaan Uul, Zoovch Ovoo, etc. They contain 70.4% of Mongolia's uranium reserves. The Zoovch Ovoo deposit is being prepared for production.

2. Deposits in volcanic rocks are associated with basalt-rhyolite volcanism in the VTS of Mesozoic time and are distributed mainly in Eastern Mongolia. Examples are the deposits of the North Choibalsan uranium district. These deposits are categorized in the reserve classification and they account for about 30% of Mongolia's reserves.

Fig. 4.68 Scattered uraninite. Hole ENDD005, 58.1 m

Fig. 4.69 Uranium blacks. Hole ENDD003, 130.6 m

References

Budunov AA, Jamsrandorj G (2012) Kharaat uranium deposit. Khaiguulchin 47:76–81 (in Mongolian)

Budunov AA, Jamsrandorj G (2013) Khairkhan uranium deposit. Khaiguulchin 48:205–209 (in Mongolian)

Budunov AA, Jamsrandorj G (2015) Uranium deposit Ulziit. Khaiguulchin 53:183–186 (in Mongolian)

Budunov AA, Jamsrandorj G (2016) Uranium deposit Gurvansaichan. Khaiguulchin 55:165–170 (in Mongolian).

Byamba J (ed) (2012) Geology and Minerals of Mongolia. Volume 4. Ulaanbaatar (in Mongolian)

Champan K, Jamsrandorj G (2018) Geology structure of Engerar uranium deposit. Khaiguulchin No.59, 53–160 (in Mongolian)

Devyatkin EV (2004) Geochronological data of Cenozoic basalts of Mongolia and their connection with structures of the newest stage. Stratirgaphy. Geological correlation. vol 12, no. 2. pp 102–114. (in Russian)

Enkh-Amgalan S, Baldorj B, Saikhanjargal S, Munkhnasan Ch, Munkh-Erdene B, Gunbadral G, Batbayar Ts (2011) Report on the results of exploration work carried out in 2002–2010 at the Dulaan uul uranium deposit located in the 4817KH, 10860KH, 11921KH license areas in the teritory of Ulaanbadrakh soum of East gobi province. Uaanbaatar

Enkh-Amgalan S, Baldorj B, Saikhanjargal S, Monkhnasan Ch, Munkh-Erdene B, Dagvasuren D, Gunbadral G, Batbayar Ts, Tsogt-Orgil B (2012) Report on the results of exploration work carried out in 2010–2012 at the Zovch Ovoo uranium deposit located in license area No. 11964X in Ulaanbaatar

Gerel O, Pirajno F, Batkhishig B, Dostal J (2021) Mineral resources of Mongolia. Springer, p 461

Grechuchin MN (2008) Usloviya loкalizatsii uranovogo mestorojdeniya Ulziit v riftogennom osadochnom basseine Vostochnoi Mongolii. In: Avtoreferat dissertatsii na soisкanie uchёnoi steпeni кanditata geologo-mineralogichesкich nauк. Moscow, p 26 (in Russian)

Jamsrandorj G (2009) Uranium deposit Dornod. Mongolyn Geosudlaach No 33 (in Mongolian)

Jamsrandorj G (2013) Geological structure of Dornod uranium deposit. Geosudlaach, No 38., pp 47–56 (in Mongolian)

Jamsrandorj G (2019) Uranium deposits of Mongolia, p 244 (in Mongolian)

Jamsrandorj G, Baatartsogt B, Altankhuayg D (2016) Resent discoveries and resources of uranium Mongolia. J East China Univ Technol 39(Suppl.)

Jamsrandorj G, Dorjgotov D, Tuyatsetseg Yu (2017a) Geology structure of Nemer uranium deposit. Khaiguulchin 57:130–135 (in Mongolian)

Jamsrandorj G, Dorjgotov D, Tuyatsetseg Yu (2017b) Geology structure of Mardaingol uranium deposit. Khaiguulchin 58:83–88 (in Mongolian)

Jamsrandorj G, Ench-Amgalan S, Nolf JL (2013) Dulaan Uul roll front type uranium deposit. Khaiguulchin 1(48):269–276 (in Mongolian)

Jamsrandorj G, Khaliunaa S, Amar O, Davaademid D, Purev-Ochir M (2019) Geological formation and uranium mineralization of the Dald deposit. Khaiguulchin, Issue 60. pp 118–126 (in Mongolian)

Mironov YuB (2003) Uran Mongolii. Izd-vo Anatoliya, Spb., p 326

Mironov YuB (2005) Uranovye mestorojdeniya Mongolii. In: Geologiya i metodiкa izucheniya uranovych mestorojdenii: Mejdunarodnaya shкola-seminar. Spb., p 15 (in Russian)

Mironov YuB, Afanasьev AM, Bulosev AV, Buzovкin SV, Vysoкoostrovsкaya EB, Gerel O, Кarpunin AM, Кorshunov AM, Tamiraa A, Shor GM, Shuvalov YuM (2012) Uranovye mestrojdeniya Mongolii. Sanкt-Peterburg, pp 304 (in Russian)

Mironov YuB, Naumov SS, Jamsrandorj G, Chuluun O (1995) The uranium resource of Mongolia. In: Resent development in uranium resources and supply. TECDOC 823. IAEA, pp 177–192

Mironov YuB, Oleynikov RM, Lemeshchenko GV (1986) Principles of forecasting uranium deposits in volcano-tectonic structures (using the Far Eastern Tectonic System as an example) // Materials on the geology of uranium deposits: Inf. collection. Issue 116. M. VIMS. pp 23–29. (in Russian)

Mironov YuB, Solovьev NS, Lvov VK, Pecherкin YuN (1989) Osobennosti geologichesкogo stroeniya i rudonosnosti Dornodsкoi vulкano-teкtonichesкoi struкturi (Vostochnaya Mongoliya). Geol Geofiz 9:22–32 (in Russian)

Samovich DA (1984) Otchet o vozdushno-geofizicheskom i geologicheskom kartografirovanii na territorii Altaya v masshtabe 1: 1,000,000 [Report on airborne geophysical and geological mapping of Altai area at a scale of 1:1,000,000]. Ministry of Geology USSR internal report no 2456

Samovich DA, Markov AI et al (1972) Report on the results of airborne gamma-spectrometric searches at a scale of 1:25,000 in the territory of Eastern Mongolia for 1972. — Irkutsk: FSUE Sosnovgeologiya (in Russian)

Samovich DA, Markov AI et al (1982) Report on the results of a comprehensive airborne geophysical survey at a scale of 1:200,000 in the Central Region of the Mongolian People's Republic for 1982 (MGE Geological Assignment No. 27). Irkutsk: FSUE Sosnovgeologiya. (in Russian)

Tungalag B (2006) Geological exploration report 2006 on the uranium project, Mongolia. East Asia minerals Corp

Correction to: Uranium Deposits of Mongolia

Correction to:
J. Galsan, *Uranium Deposits of Mongolia*, Mineral Resource Reviews, https://doi.org/10.1007/978-3-031-86481-0

Author provided belated revised figures and tables has been replaced in the following chapters

Chapter 3

Figures 3.6, 3.8, 3.10, 3.11 are revised

Tables 3.8 and 3.9 are updated with revised data

Chapter 4

Figures 4.1, 4.3, 4.6, 4.7, 4.8, 4.14, 4.15, 4.22, 4.23, 4.24, 4.26 to 4.35 and 4.43, 4.44, 4.46 to 4.59 and 4.65 are revised

Figure captions 4.8 and 4.22 are updated as per revised figures

The Book and the chapters are updated with the changes

The updated versions of these chapters can be found at
https://doi.org/10.1007/978-3-031-86481-0_3
https://doi.org/10.1007/978-3-031-86481-0_4

J. Galsan, *Uranium Deposits of Mongolia*, Mineral Resource Reviews,
https://doi.org/10.1007/978-3-031-86481-0_5

The manufacturer's authorised representative in the EU is Springer Nature Customer Service Centre GmbH, Europaplatz 3, 69115 Heidelberg, Germany. If you have any concerns regarding our products, please contact ProductSafety@springernature.com

Printed and bound by CPI Group (UK) Ltd, Croydon, CR0 4YY
01/10/2026
02225958-0019